Renewable Energy Devices and Systems with Simulations in MATLAB® and ANSYS®

Renewable Energy Devices and Systems with Simulations in MATLAB® and ANSYS®

Frede Blaabjerg
Aalborg University
Aalborg, Denmark

Dan M. Ionel
University of Kentucky
Lexington, USA

CRC Press
Taylor & Francis Group
Boca Raton   London   New York

CRC Press is an imprint of the
Taylor & Francis Group, an **informa** business

CRC Press
Taylor & Francis Group
6000 Broken Sound Parkway NW, Suite 300
Boca Raton, FL 33487-2742

First issued in paperback 2020

© 2017 by Taylor & Francis Group, LLC
CRC Press is an imprint of Taylor & Francis Group, an Informa business

No claim to original U.S. Government works

ISBN-13: 978-1-4987-6582-4 (hbk)
ISBN-13: 978-0-367-65621-8 (pbk)

Library of Congress Cataloging-in-Publication Data

Names: Blaabjerg, Frede, author. | Ionel, Dan M., author.
Title: Renewable energy devices and systems with simulations in MATLAB and ANSYS / Frede Blaabjerg and Dan M. Ionel.
Description: Boca Raton : Taylor & Francis, a CRC title, part of the Taylor & Francis imprint, a member of the Taylor & Francis Group, the academic division of T&F Informa, plc, [2017] | Includes bibliographical references and index.
Identifiers: LCCN 2016036879| ISBN 9781498765824 (hardback : acid-free paper) |
ISBN 9781498765831 (ebook)
Subjects: LCSH: Electric power systems. | Renewable energy sources. | Electric power systems--Computer simulation. | Renewable energy sources--Computer simulation. | MATLAB. | ANSYS (Computer system)
Classification: LCC TK1005 .B58 2017 | DDC 621.0420285/53--dc23
LC record available at https://lccn.loc.gov/2016036879

Visit the Taylor & Francis Web site at
http://www.taylorandfrancis.com

and the CRC Press Web site at
http://www.crcpress.com

Contents

Preface vii

Editors ix

Contributors xi

Chapter 1 Renewable Energy Systems: Technology Overview and Perspectives 1

Frede Blaabjerg, Dan M. Ionel, Yongheng Yang, and Huai Wang

Chapter 2 Solar Power Sources: PV, Concentrated PV, and Concentrated Solar Power 17

Katherine A. Kim, Konstantina Mentesidi, and Yongheng Yang

Chapter 3 Overview of Single-Phase Grid-Connected Photovoltaic Systems 41

Yongheng Yang and Frede Blaabjerg

Chapter 4 Three-Phase Photovoltaic Systems: Structures, Topologies, and Control 67

Tamás Kerekes, Dezső Séra, and László Máthé

Chapter 5 Overview of Maximum Power Point Tracking Techniques for Photovoltaic
Energy Production Systems 91

Eftichios Koutroulis and Frede Blaabjerg

Chapter 6 Design of Residential Photovoltaic Systems 131

Tamás Kerekes, Dezső Séra, László Máthé, and Kenn H.B. Frederiksen

Chapter 7 Small Wind Energy Systems 151

Marcelo Godoy Simões, Felix Alberto Farret, and Frede Blaabjerg

Chapter 8 Power Electronics and Controls for Large Wind Turbines and Wind Farms 177

Ke Ma, Udai Shipurkar, Dan M. Ionel, and Frede Blaabjerg

Chapter 9 Electric Generators and their Control for Large Wind Turbines 209

*Ion G. Boldea, Lucian N. Tutelea, Vandana Rallabandi,
Dan M. Ionel, and Frede Blaabjerg*

Chapter 10 Design Considerations for Wind Turbine Systems 251

Marcelo Godoy Simões, S.M. Muyeen, and Ahmed Al-Durra

Chapter 11 Marine and Hydrokinetic Power Generation and Power Plants 267

Eduard Muljadi and Yi-Hsiang Yu

Chapter 12 Power Conversion and Control for Fuel Cell Systems in Transportation and
Stationary Power Generation 291

Kaushik Rajashekara and Akshay K. Rathore

Chapter 13 Batteries and Ultracapacitors for Electric Power Systems with Renewable
Energy Sources 319

Seyed Ahmad Hamidi, Dan M. Ionel, and Adel Nasiri

Chapter 14 Microgrid for High-Surety Power: Architectures, Controls, Protection, and
Demonstration 355

Mark Dehong Xu, Haijin Li, and Keyan Shi

Index 393

Preface

In the twenty-first century, we are facing a major transition in the way energy is generated, transmitted, distributed, and utilized. At this moment, we are in the rising phase of shifting from fossil fuels to renewable sources and experiencing a large movement toward a truly sustainable society for ourselves and for future generations. Nowadays, in terms of annual newly installed capacity, renewable generation is already higher than the conventional one, and this trend is expected to rise in the future. Changes in generation are accompanied by increased measures and new technologies to use energy as efficiently as possible. Both efforts rely on power electronics as an enabling technology and on substantial research and development efforts.

This book covers the main topics in the field of renewable energy that play an important role in modern technology, with emphasis on electric power engineering and power electronics. State-of-the-art solutions together with ongoing developments are reviewed at an introductory level in Chapter 1. Solar energy, including concentrated and PV systems, maximum power point techniques, single and three configurations, and associated design aspects of utility and residential systems are covered in Chapters 2 through 6. Wind turbine systems, including their electric generators, controls, power converters, and related design and implementation aspects, are presented in Chapters 7 through 10. Chapters 11 and 12 are devoted to the topics of marine and hydrokinetic and fuel cell technologies, respectively. The fast-developing field of energy storage with batteries and ultracapacitors is covered in Chapter 13. The last chapter, 14, discusses very timely topics of architecture, controls, protection, and application for microgrids.

The book is suitable for both undergraduate and graduate levels and can serve as a useful reference for academic researchers, engineers, managers, and other professionals in the industry. Each chapter includes descriptions of fundamental and advanced concepts, supported by many illustrations and numerical and practical examples. Also included, in order to provide better understanding for the reader, are two design examples for a small wind turbine system and PV power system, respectively, and simulation models for the MATLAB® and ANSYS® software, which are available for download from the book companion website at https://www.crcpress.com/product/isbn/9781498765824.

We acknowledge the tireless efforts and assistance of the Taylor & Francis Group/CRC Press editorial staff. We especially thank Nora Konopka and Kyra Lindholm. Special thanks to Dr. Yongheng Yang of Aalborg University for his assistance in the final stages of editing the manuscript. We also gratefully acknowledge the expert contributions of our worldwide group of collaborators and the support of Aalborg University, CORPE Consortium, and University of Kentucky, the L. Stanley Pigman Endowment.

Frede Blaabjerg
Dan M. Ionel
January 2017

Editors

Frede Blaabjerg is professor of power electronics and drives at Aalborg University in Denmark. Earlier in his career, he was with ABB-Scandia, Randers, Denmark, from 1987 to 1988, and afterward joined Aalborg University where he became an assistant professor in 1992, an associate professor in 1996, and a full professor in 1998. He has been a part time research leader in Research Center Risoe in wind turbines. From 2006 to 2010, he was the dean of the Faculty of Engineering, Science, and Medicine at Aalborg University and became a visiting professor in Zhejiang University, Hangzhou, China, in 2009. He is also guest professor at Harbin Institute of Technology, Shandong University, and Shanghai Maritime University in China. He is currently heading the Center of Reliable Power Electronics (CORPE) located at Aalborg University. He has published extensively with major emphasis on power electronics and its applications, such as in wind turbines, PV systems, reliability in power electronics, harmonics, power quality, and adjustable speed drives.

He received the 1995 Angelos Award for his contribution in modulation techniques and the Annual Professor Award from Aalborg University. In 1998, he received the Outstanding Young Power Electronics Engineer Award from the IEEE Power Electronics Society and over the years has received 17 IEEE Prize Paper Awards. He received the IEEE PELS Distinguished Service Award in 2009, the EPE-PEMC Council Award in 2010, and the IEEE William E. Newell Power Electronics Award in 2014. He has also received a number of major research awards in Denmark, such as the Villum Kann Rasmussen Research Award in 2014.

Dr. Blaabjerg is an IEEE fellow. He was the editor-in-chief of the *IEEE Transactions on Power Electronics* from 2006 to 2012 and distinguished lecturer for the IEEE Power Electronics Society from 2005 to 2007 and for the IEEE Industry Applications Society from 2010 to 2011. He was the chairman of EPE in 2007 and PEDG in 2012, both held in Aalborg. In 2002, he founded the IEEE Danish joint chapter for PELS/IAS/IES. He has also been ADCOM member of PELS twice, serving the last term from 2013 to 2015. He was elected and served as vice president for products in the Power Electronics Society for 2015–2016. He was also chairman of IEEE FEPPCON in 2015 held in Italy and is a member of a couple of technical committees in PELS as well as in the EPE organization. For many years, he has been involved in research funding and policies in Denmark and Europe.

Dan M. Ionel is professor of electrical engineering and the L. Stanley Pigman chair in power at the University of Kentucky in Lexington, Kentucky. Previously, he worked in industry for 25 years, most recently as chief engineer for Regal Beloit Corp. and, before that, as the chief scientist for Vestas Wind Turbines. Concurrently with his industrial appointments, Dr. Ionel also served as visiting and research professor at the University of Wisconsin and Marquette University in Milwaukee, Wisconsin. He is currently the director of the SPARK Laboratory and of the PEIK Institute at the University of Kentucky, teaching and researching on topics of sustainable and renewable energy technologies, electric machines and power electronic drives, electromagnetic devices, electric power systems, smart grids, and buildings.

He has contributed to technology developments with long-lasting industrial impact, including the world's most powerful wind turbine and United States' most successful range of PM motor drives. His research benefited from multimillion dollar support directly from industry as well as from the NSF, NIST, and DOE. In the industry, he received the Innovation Award from AO Smith Corp. and the Archer Inventor Award from Regal Beloit Corp. During his PhD studies, he was a Leverhulme visiting fellow at the University of Bath in England. He was a keynote and invited speaker at major international conferences. Three of his publications received IEEE Best Paper Awards. He holds more than 30 patents, including a medal winner at the Geneva Invention Fair.

Dr. Ionel is an IEEE fellow. He was the inaugural chair of the IEEE Industry Applications Society, Renewable and Sustainable Energy Conversion Systems Committee; chair of the IEEE Power and Energy Society, Electric Motor Subcommittee; and chair of the IEEE Power Electronics, Milwaukee Chapter. He also served as editor of *IEEE Transactions on Sustainable Energy* and technical program chair for IEEE ECCE 2015. He is the general chair of the 2017 anniversary edition of the IEEE IEMDC Conference and the editor-in-chief of the *Electric Power Components and Systems* journal, which is published by the Taylor & Francis Group.

Contributors

Ahmed Al-Durra
Department of Electrical Engineering
The Petroleum Institute
Abu Dhabi, United Arab Emirates

Frede Blaabjerg
Department of Energy Technology
Aalborg University
Aalborg, Denmark

Ion G. Boldea
Department of Electrical Engineering
Politehnica University of Timisoara
Timisoara, Romania

Felix Alberto Farret
Department of Energy Processing
Federal University of Santa Maria
Santa Maria, Brazil

Kenn H.B. Frederiksen
Kenergy
Horsens, Denmark

Seyed Ahmad Hamidi
Milwaukee Electric Tool
Brookfield, Wisconsin

Dan M. Ionel
Department of Electrical and Computer
    Engineering
University of Kentucky
Lexington, Kentucky

Tamás Kerekes
Department of Energy Technology
Aalborg University
Aalborg, Denmark

Katherine A. Kim
School of Electrical and Computer Engineering
Ulsan National Institute of Science and
    Technology
Ulsan, Korea

Eftichios Koutroulis
School of Electrical and Computer Engineering
Technical University of Crete
Chania, Greece

Haijin Li
Institute of Power Electronics
Zhejiang University
Hangzhou, Zhejiang, People's Republic of
    China

Ke Ma
Department of Energy Technology
Aalborg University
Aalborg, Denmark

László Máthé
Department of Energy Technology
Aalborg University
Aalborg, Denmark

Konstantina Mentesidi
Department of Electrical and Electronic
    Engineering
Public University of Navarra
Pamplona, Spain

Eduard Muljadi
Power Systems Engineering Center
National Renewable Energy Laboratory
Golden, Colorado

S.M. Muyeen
Department of Electrical and Computer
    Engineering
Curtin University
Perth, Western Australia, Australia

Adel Nasiri
Department of Electrical and Computer
    Engineering
University of Wisconsin–Milwaukee
Milwaukee, Wisconsin

Kaushik Rajashekara
Department of Electrical & Communication
    Engineering
University of Houston
Houston, Texas

Vandana Rallabandi
Department of Electrical & Computer
    Engineering
University of Kentucky
Lexington, Kentucky

Akshay K. Rathore
Department of Electrical and Computer
    Engineering
Concordia University
Montréal, Québec, Canada

Dezső Séra
Department of Energy Technology
Aalborg University
Aalborg, Denmark

Keyan Shi
Institute of Power Electronics
Zhejiang University
Hangzhou, Zhejiang, People's Republic of
    China

Udai Shipurkar
Department of Electrical Sustainable Energy
Delft University of Technology
Delft, The Netherlands

Marcelo Godoy Simões
Department of Electrical Engineering and
    Computer Science
Colorado School of Mines
Golden, Colorado

Lucian N. Tutelea
Department of Electrical Engineering
Politehnica University of Timisoara
Timisoara, Romania

Huai Wang
Department of Energy Technology
Aalborg University
Aalborg, Denmark

Mark Dehong Xu
Institute of Power Electronics
Zhejiang University
Hangzhou, Zhejiang, People's Republic of
    China

Yongheng Yang
Department of Energy Technology
Aalborg University
Aalborg, Denmark

Yi-Hsiang Yu
National Wind Technology Center
National Renewable Energy Laboratory
Golden, Colorado

1 Renewable Energy Systems: Technology Overview and Perspectives

Frede Blaabjerg, Dan M. Ionel,
Yongheng Yang, and Huai Wang

CONTENTS

Abstract 1
1.1   Introduction 1
1.2   State of the Art 2
1.3   Examples of Recent Research and Development 7
1.4   Challenges and Future Trends 12
1.5   Further Readings 12
References 14

ABSTRACT

In this chapter, essential statistics demonstrating the increasing role of renewable energy generation are first discussed. A state-of-the-art review section covers the fundamentals of wind turbine and photovoltaic (PV) systems. Schematic diagrams illustrating the main components and system topologies are included. Also, the increasing role of power electronics is explained as an enabler for renewable energy integration and for future power systems and smart grids. Recent examples of research and development, including new devices and system installations for utility power plants and for residential and commercial applications, are provided. Fuel cells, solar thermal, wave generators, and energy storage systems are also briefly presented and illustrated. Challenges and future trends for the technologies in 2025 are summarized in a table for onshore and offshore wind energy; solar power, including PV and concentrating solar power; wave energy; fuel cells; and storage with batteries and hydrogen, respectively. Recommended further readings on topics of electric power engineering for renewable energy are included in the final section.

1.1   INTRODUCTION

Sustainable energy supply remains to be a main requirement of modern society in order to respond to the increased energy demand caused by the larger consumption especially due to population growth. For a long time, the energy boom has been based on fossil fuels. Unfortunately, not only the supply of oil, coal, and natural gas is limited, but there exist also major pollution and environmental concerns associated with using the traditional energy sources and alternatives are important.

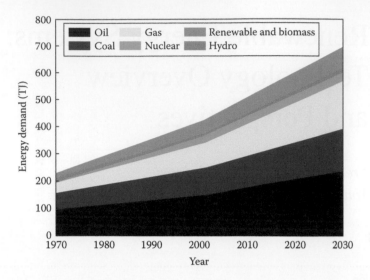

FIGURE 1.1   Worldwide energy demand since 1970 and the estimation till 2030. (Based on data from the International Energy Agency (IEA), World energy outlook 2004, OECD, Paris, France, 2004, http://www.worldenergyoutlook.org/media/weowebsite/2008-1994/weo2004.pdf.)

Renewable energy technologies are seen as some of the most important solutions for the future, and they need to be further developed in this century in order to take over most of the energy production. Many emerging technologies exist and they have different levels of maturity. The scale of implementation is not the same either. Most of the renewable technologies are dependent on weather conditions, and they are challenging in respect to the integration into the grid system. These issues need to be fully solved—how to make a power system that is able to cope with a very high penetration of renewables, which also involves the development of smart grid systems. Such systems may include microgrids and energy storage facilities, and, in many cases, they will combine the electric power system with other energy carriers like heating/cooling and gas as well as look at how to use transportation as a resource.

Figure 1.1 shows the worldwide energy demand in the past decades and also the estimated energy demand until 2030. As it can be observed, due to the continuous increase in gross domestic product, the overall energy demand is expected to increase by more than 50% by 2030 [1]. To achieve this primary goal, renewable energy will be an important part in the foreseeable future energy production (hydro, renewable and biomass, etc.).

This chapter provides an overview of the penetration of renewable energy generation and schematically illustrates the basic principles of the state-of-the-art technologies. Emphasis is placed on power electronics as a major technology enabler of the ongoing transformation of the electric power systems. The presentation is complemented with examples of recent research, developments, and significant achievements worldwide. This chapter also summarizes possible trends for the next decade and includes a final section with suggested further readings.

1.2   STATE OF THE ART

Worldwide research, development, and major implementation efforts are focused on renewable energies. Historically, hydropower has accounted for most of the installed renewable generation capacity, which is by now in excess of 1000 GW, as it is illustrated in Figure 1.2. In recent years, other sources, such as wind (onshore and offshore), solar (photovoltaics [PV] and concentrated solar power [CSP]), geothermal, bioenergy (solid biomass, biogas, and liquid biofuels), and marine (tide, wave, and ocean) energy, accounted all together for a substantial proportion of the newly installed

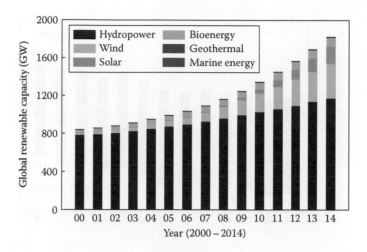

FIGURE 1.2 Annual evolution (2000–2014) of the worldwide installed renewable energy generation capacity based on the data available from IRENA statistics (hydropower also includes pumped storage and mixed plants; marine energy covers tide, wave, and ocean energy).

capacity as shown in Figure 1.3. For example, in 2014, for which recent statistics are available [2], the wind together with solar generation made the largest contribution with approximately 50 GW and 40 GW, respectively, and it is still growing.

It is interesting to note in Figure 1.3 that from 2011 to 2014, the worldwide annual installed renewable capacity exceeded 100 GW, a figure that is comparable with the generation capacity for the entire country of Germany. Remarkable is also that recently more than half of the new annual generation installations are based on renewables and that the trend is increasing as it is indicated in Figure 1.4.

Behind these achievements are years of dedicated work, which resulted in new devices and systems and lowered continuously the cost of renewable energy as calculated on a complete life cycle of generating systems. The generating capacity installed in 2013 for PV is larger than for wind, reflecting, among other things, the recent technology evolution and substantial cost reductions. However, in 2014, wind becomes again the number one in terms of annual installations, as it is shown in Figure 1.3.

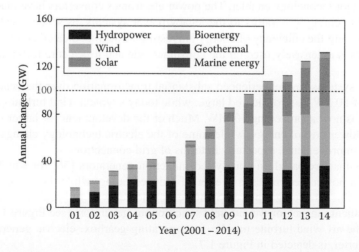

FIGURE 1.3 Annual worldwide installations of renewable energy generating capacity based on data available from IRENA statistics. (Hydropower also includes pumped storage and mixed plants; marine energy covers tide, wave, and ocean energy.)

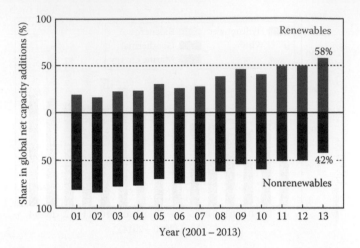

FIGURE 1.4 Annual installations of renewables and nonrenewables as a share of the net total additions based on data available from IRENA statistics.

In Europe, Denmark has now 60% of its electricity supplied from renewable resources, mostly wind and a little from solar [3]. This is the result of pioneering efforts started in the 1970s and of the systematically favorable governmental policy over the decades since then. Two other countries, Portugal and Spain, generate today approximately 30% of their electricity from renewables.

Some of the early technological developments of PV cells, wind turbines and farms, and large solar thermal power plants originated in the United States. Although here the contribution of renewables remains relatively low, at only 10% of the total power sector electricity supply in 2010, extensive studies have been done focusing on 80% of all U.S. electricity generation from renewables in 2050 [4].

Worldwide, the shift toward renewables is most significant. This is, for example, documented in two papers reporting on recent developments and state-of-the-art technology in China and India published in a special issue, 43(8–10), 2015, of the *Electric Power Components and Systems* journal [5, 6].

Among the technologies that supported the growth of renewables, power electronics has been representing a major technology enabler. The power electronics converters have made it possible to connect renewable energy generators to the legacy power systems, as schematically illustrated in Figure 1.5, improving the efficiency of energy harvesting through dedicated controls. Furthermore, power electronics is extensively used on the consumer side and it is a core technology for the new smart grid.

Wind turbine systems have experienced substantial transformation over the years. In the 1980s, a wind turbine of 50 kW was considered large, while today's typical wind turbines are rated at 2–3 MW and design is now approaching 10 MW. Much of the development for larger units was driven by the need for lower cost of energy, while some of the electric technology changes were imposed by performance improvements, especially in terms of grid connection.

Older designs that are based on doubly fed induction generators [36] for which only the rotor circuit employs a power electronics converter of reduced rating, while the stator winding is connected to the AC line, are being replaced by newer topologies with full-scale power converters and induction, permanent magnet, or wound-field synchronous generators (see Figure 1.6). An example of the state-of-the-art wind turbine nacelle, incorporating gearbox, electric generator, and power electronics converter, is depicted in Figure 1.7.

For solar energy, it can be used in different ways. The power generation is based on the PV effect, where the voltage or current level is very low for a single solar PV cell. Hence, in practical applications, a considerable number of solar PV cells have been connected in parallel and in series in order to

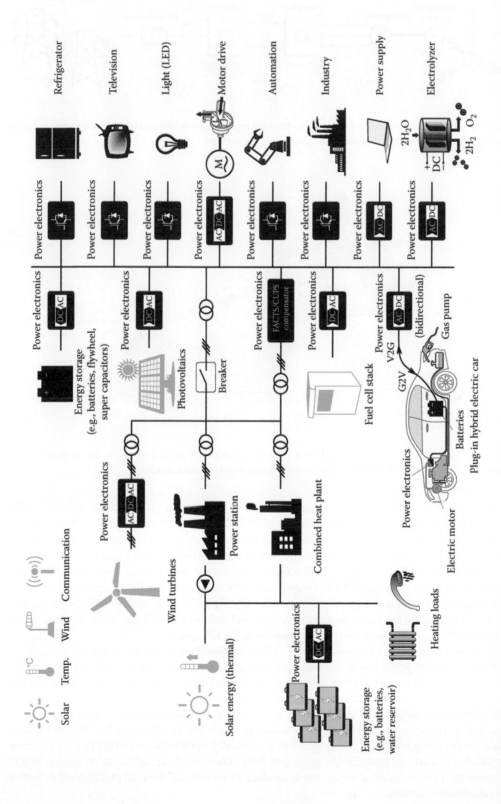

FIGURE 1.5 Schematic representation of a modern power system, which incorporates renewable energy sources, distributed generation, and smart grid functions. Integration is made possible through the extensive use of power electronics (G2V: grid to vehicle; V2G: vehicle to grid).

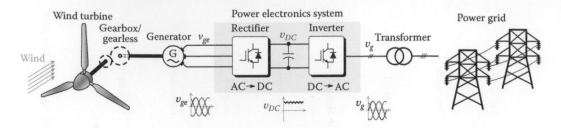

FIGURE 1.6   Power electronics–enabled wind turbine energy conversion system, where the generator can be induction, permanent magnet, or wound-field synchronous generators.

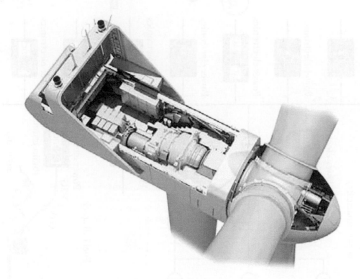

FIGURE 1.7   Bird's-eye view of the nacelle of a state of the multi-MW wind turbine, including electric generator and power electronics converters. (Courtesy of Vestas Wind Systems A/S.)

form PV panels and thereby PV arrays. Figure 1.8a shows the one using conventional solar PV cells, whose efficiency is around 20% [7]. To improve the efficiency, the concentrator PV is also observed in the market, as shown in Figure 1.8b, where highly efficient multi-junction solar cells can be used.

Grid-connected PV systems, such as the one schematically illustrated in Figure 1.9, comprise a power electronics DC/DC converter, which ensures a maximum solar energy harvesting through a maximum power point tracking (MPPT) control, and a DC/AC converter for interconnection to the grid. PV systems have gained large popularity not only for multi-MW utility-scale power plants/farms but also as rooftop installations on commercial and residential buildings with ratings as small as hundreds of Watts (see Figure 1.10), but typically in the kW range.

The dramatic reduction in the price of PV cells contributed to recent very large-scale deployments. National and regional tax incentives were another contributing factor to the deployment. At the same time, technological progress made possible the increase in reliability, such that some PV systems now have an expected life span of 25 years or even longer.

In addition, electricity can be generated in CSP (which is called concentrating solar power or concentrated solar thermal power) systems [8], as it is illustrated in Figure 1.11. CSP can concentrate sunlight by using reflective components (e.g., typically mirrors) to receivers (e.g., a tower or pines), which can carry or transfer the generated heat. Then, the heat can drive an engine that is further connected to an electrical generator to produce electricity; or the heat can be used to power thermal–chemical reactions.

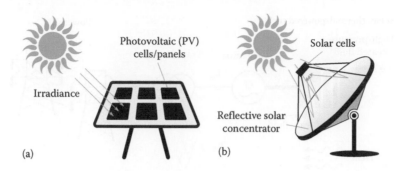

FIGURE 1.8 Two solar power generator technologies: (a) solar PV technology and (b) concentrator PV technology.

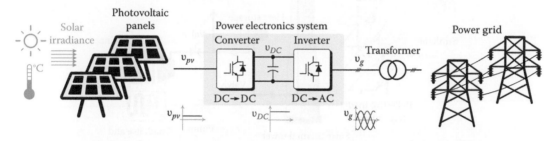

FIGURE 1.9 Power electronics–enabled PV energy conversion system with DC/DC and DC/AC converters, where the solar PV technology shown in Figure 1.8a is adopted.

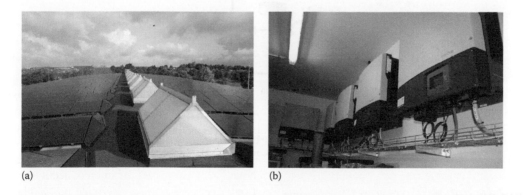

FIGURE 1.10 Rooftop-installed PV systems: (a) PV arrays with a total rating of 60 kW installed on the roof of Aalborghus High School in Denmark and (b) power electronic converters with the schematic shown in Figure 1.9 are installed within the building and are connected to the AC grid.

1.3 EXAMPLES OF RECENT RESEARCH AND DEVELOPMENT

The natural potential for renewable energy is enormous. For example, covering just a small region of Arizona with state-of-the-art PV cells would generate enough electricity for the entire U.S. consumption [6], and installing modern wind turbines in regions of the North Sea would generate enough electricity to satisfy the entire European Union (EU) requirements [9]. It comes therefore as no surprise that proposals for very high penetration of renewable energy have been made, for example, going all the way to a future 100% renewable energy society [10, 11]. In order to enable

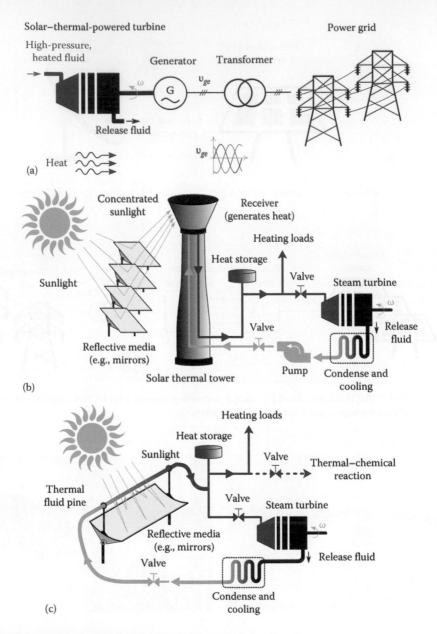

FIGURE 1.11 Concentrated solar power (CSP) systems: (a) electric power generation system, (b) central solar tower as a receiver for heat generating, and (c) sunlight concentrated by a parabolic trough line structure, where the heat can also be utilized to power thermal–chemical reactions depending on the fluid type.

such large-scale deployments, ongoing research and development efforts are very much needed with a major thrust to ensure economic feasibility. Larger devices and power plants hold the promise of the economy of scale, including lower cost of energy.

An example of technological achievement is represented by the completion and installation in early 2014 of the first Vestas V164, the world's most powerful wind turbine. With an 8 MW rated power generator and a three-bladed rotor having a 164 m diameter, this record-setting turbine is primarily dedicated to the development of large offshore wind farms. The 20 × 8 × 8 m nacelle, which weighs together with the hub approximately 390 tons, includes the gearbox, medium-speed generator, and an AC/DC rectifier. The reminder of the electric power system, including a DC/AC inverter, together with power supply cabinets, transformer, and switchgear, is fitted in a multilevel 14 m tall structure within the base of the turbine tower [12].

Most of the new installations of multi-MW turbines are part of large wind farms, which aggregate the individual contributions into a power station. Traditionally, most wind farms have been onshore, but present plans and ongoing developments exist for very large offshore deployments that would benefit of better wind conditions and will be remotely located from the countryside. Examples include the Anholt 400 MW wind farm commissioned in 2013, which is the Denmark's largest off-shore park and one of the largest in the world. This installation, which is illustrated in Figure 1.12, is operated by the Danish utility company Dong Energy and comprises 111 Siemens wind turbines, each rated at 3.6 MW.

For solar energy, there were also many recent record-breaking developments. These include, for example, the recent completion in early 2015 of the 550 MW Desert Sunlight solar PV power plant built by First Solar near Joshua Tree National Park in California, which generates enough electricity to power 160,000 homes [13]. The long-standing record of leading innovations in solar PV technologies of companies such as SunPower now includes smart grid solutions for residential and commercial installations and integrated micro-inverters [14].

Another example is from the Atacama Desert in Chile, South America, where Abengoa developed a 110 MW solar thermal plant using tower technology. Remarkable is that this includes 17.5 hours of thermal energy storage using molten salts, which aims to ensure dispatchable electricity throughout 24-hour cycles, responding to all periods of demand [15].

While wind and solar may already be cost-competitive, depending of course on regional and particular conditions, other renewable energy technologies still require substantial developments. Fuel cell systems may be included in this category, as they have been expected for many years to increase their presence in applications over a wide range of power ratings. The typically low-temperature proton exchange membrane (PEM) technology and the higher-temperature solid oxide fuel cells (SOFC) type can be applied for large power supplies, with some demonstrators being completed for uninterruptible power supply (UPS) systems. The fuel, represented by hydrogen in combination with air/oxygen, is converted into electricity and some heat using a system topology of the type schematically presented in Figure 1.13. An example of hardware implementation is further shown in Figure 1.14.

The power electronics system for fuel cell applications incorporates a DC/DC converter, which ensures power extraction maximization, and a DC/AC inverter for grid connection. Challenges being addressed by ongoing research programs include improvements in efficiency and specific

FIGURE 1.12 Denmark's largest offshore wind farm of 400 MW installed at Anholt. (Courtesy of Dong Energy Wind Power.)

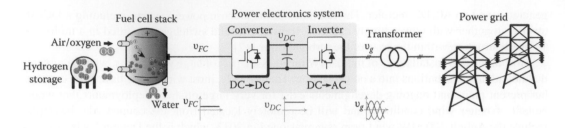

FIGURE 1.13 Fuel cell system with DC/DC and DC/AC power electronics converters for AC power grid connection.

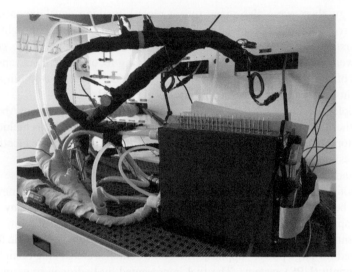

FIGURE 1.14 Low-temperature PEM fuel cell setup at Aalborg University with approximate 1.2 kW electrical capacity (28 cells each of 43 W) and hydrogen as fuel.

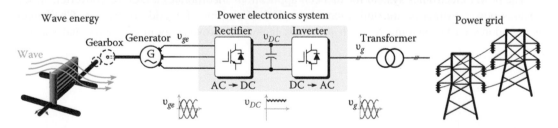

FIGURE 1.15 Wave energy system including an electric generator and AC/DC and DC/AC power electronics converters for AC power grid connection.

power output, increasing lifetime, simplification, and cost reduction. A major potential advantage is considered to be the fact that energy can be stored in the form of hydrogen that could be produced using different forms of renewable energy but which in turn would require a different infrastructure than the legacy power systems.

More than 70% of the earth surface is covered by water, making oceans and seas a potentially huge energy resource, which is yet largely untapped. In the example solution, as depicted in Figure 1.15, the movement of waves engages a mechanical transmission that is coupled to an electric generator. The very low speeds and large power variations require special solutions and may result in relatively reduced conversion efficiencies. Long-term reliability under very harsh environmental conditions and survivability during storms are major challenges that drive up the investment, operation and maintenance efforts, and ultimately the final cost of energy for such system. Although systems such as the one shown in Figure 1.16 have been in operation for years, much of the wave energy technology is still considered to be at the level of demonstration and it is costly.

Matching the inherent weather-dependent variability of renewable energy generation with the load demand in modern power systems and the smart grid remains a major challenge. This general problem benefits of great attention and sustained research programs with emphasis on both power electronics and energy storage devices and systems. Two alternative solutions are exemplified in Figures 1.17 and 1.18.

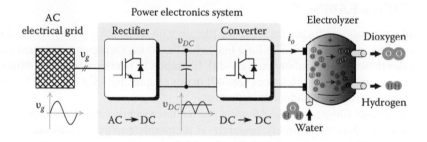

FIGURE 1.16 Wave Star wave energy generator located at the Hanstholm test site in Denmark. (Courtesy of Wave Star Company.)

FIGURE 1.17 System for producing hydrogen using an electrolyzer supplied with electricity from the AC electrical grid or renewable energy sources. The hydrogen can be stored under pressure and/or used with fuel cells.

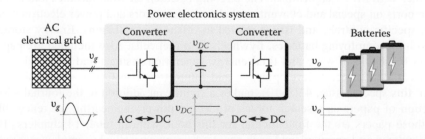

FIGURE 1.18 System using batteries for storing electric energy from the AC electrical grid or renewable energy sources. Power electronics control ensures bidirectional energy flow.

In one approach, electricity from the AC grid or renewable resources is used to produce hydrogen using an electrolyzer system. The hydrogen can then be stored at high pressure and used for fuel cell systems, such as the one schematically depicted in Figure 1.13. In principle, provided that suitable infrastructure would be in place for hydrogen, a large variety of consumers, both stationary and mobile, can be supplied.

The other approach relies on the use of batteries for which chemistries continue to be developed and perfected such that energy density and lifetime are increasing, while the cost is decreasing. With the use of bidirectional power electronics converters in the battery-based system from Figure 1.18, the electric energy can flow both from the grid to the battery and in the opposite direction, and the functionality can include active grid support and stabilization.

1.4   CHALLENGES AND FUTURE TRENDS

Table 1.1 summarizes the research challenges and opportunities and future trends over the next 10 years to 2025, alongside the present status for the main renewable energy technologies. Sources of information include references such as [16–23]. General expectations are that through technological improvements, the renewable energy devices and systems are expected to become even more cost competitive with fossil fuel–based generators.

Apart from the particular technologies, systems, and devices that have been previously described, a special note is due to the substantial ongoing efforts for integrating renewables into the power system, where a proper balance between production and consumption exists. Being able to control the entire energy system as a whole, including not only the electrical power system but also the thermal energy and the water flow, is also an ongoing challenge for research and the society. In this context, the smart grid functions and their facilities, such as communications, load control, and energy storage, are seen as solution enablers.

1.5   FURTHER READINGS

With such wide interest over the last decades, the literature on renewables is very rich and many new journals are coming up in the field. The long list of books, which focuses on electrical engineering aspects, includes those listed as references [24–28]. Examples of overview papers published in journals and magazines are listed as references [28–40] in this chapter, but they continue to be expanded as renewables are getting more and more important.

The *Electric Power Components and System* journal has recently (2015) published a triple special issue, 43(8–10), devoted to timely research on renewable energy devices and systems. This includes more than 30 papers from across the world. Power system topics comprise distributed generation and system integration with solar PV and other renewables, microgrid configurations, and optimal controls. Alternative topologies are presented for PV converters and system installations together with MPPT algorithms. The ongoing research on wind turbines and farms is illustrated by reports on special and conventional electric generators and power electronic converters, reliability, specific controls, and issues related to offshore installations. Energy storage studies examine systems employing batteries, flywheels, and fuel cells. Also included are papers on the control and life cycle of wave and tidal energy generation systems and two reviews reporting on the current status, recent developments, and future trends in China and India, respectively. Altogether, this special issue, 43(8–10), makes a significant addition to the journal's long-standing collection of papers on research topics of electric power engineering for renewable energy. Some of those papers are the fundamental for this book with, in total, 14 chapters. The papers have been updated and reviewed, and each chapter has some practical examples/exercises that are used to study more detailed technology. The examples are supported with the opportunity to do MATLAB®/Simulink® and ANSYS® simulations.

TABLE 1.1

Renewable Energy Technologies: Present Status, Research Challenges and Opportunities, and Future Trends to 2025

Renewable Energies	Present Status	Research Challenges and Opportunities	Trends to 2025 (Prediction)
Wind onshore	• 361 GW installed by 2014. • Average turbine size: 1.9 MW in 2013. • DFIG has the majority at about 50%, and PMSG is increasing. • Mainly low-voltage power converters (e.g., 690 V). • LCOE of USD 0.06–USD 0.12/kWh (typical).	• Reliability issues. • Wind farm interconnection technologies. • Grid code compliances. • Medium-voltage wind power converters (e.g., 3.3 kV). • Wind power converters based on new semiconductor devices.	• >700 GW global installation. • +50% wind power in Denmark. • 10–15 MW turbines available. • A larger number of MV turbines. • Improved component and system reliability. • LCOE of USD 0.05–USD 0.10/kWh (typical).
Wind offshore	• 8.8 GW installed by 2014; 91% are in Northern Europe. • Average size: 3.6 MW in 2013. • LCOE of USD 0.14–USD 0.25/kWh (typical).	• Reliability issues. • Offshore foundations. • Interconnection of onshore and offshore wind farms. • Harmonics and stability issues. • High maintenance costs.	• About 89 GW global installation. • Improved component and system level reliability. • LCOE of USD 0.10–USD 0.19/kWh (typical).
Photovoltaics (PV)	• 180 GW installed by 2014. • PV inverters with wide bandgap devices commercialized. • LCOE of USD 0.08–USD 0.36/kWh (typical).	• Reliability issues. • Medium-voltage PV systems. • High-efficiency thin-film PV. • Energy storage for PV systems. • Grid integration to fulfill more stringent codes.	• High-voltage PV systems. • Lifetime of 20–30 years, including the PV inverters. • LCOE of USD 0.06–USD 0.15/kWh (typical). • Micro-inverters.
Concentrated solar power (CSP)	• About 4 GW installed globally. • Many existing CSP plants use fossil fuel as backup. • LCOE of USD 0.20–USD 0.25/kWh (typical).	• Large upfront investment (USD 4000–USD 9000/kW). • Hybrid PV and CSP plants. • New lightweight low-cost reflector optics.	• Cost of CSP plants is expected to be half as technologies mature. • LCOE of USD 0.11–USD 0.16/kWh (typical).
Wave energy	• More than 100 pilot projects exist throughout the world. • Testing up to 5 km offshore and 50 m in depth. • LCOE of USD 0.37–USD 0.71/kWh (10 MW scale).	• Cost reduction. • Go further offshore at larger depth and higher waves. • Synergetic research with the offshore wind industry with shared infrastructures.	• Contribute to 1%–2% of global electricity generation. • LCOE will be higher than that of fossil fuel power plants, similar as that of offshore wind in 2015.
Fuel cell—low temperature	• FCV commercialized. • 2500 hours of durability in FCV application (typical). • PEMFC for UPS. • USD 51/kW for FCV.	• Improve durability and robustness. • Cost reduction to increase market acceptance. • Develop non-rare metal catalysts.	• >5000 hours lifetime for FCV. • FCV becomes economic viable. • Growth in portable and home applications. • USD 40/kW for FCV.

(Continued)

TABLE 1.1 (*Continued*)

Renewable Energy Technologies: Present Status, Research Challenges and Opportunities, and Future Trends to 2025

Renewable Energies	Present Status	Research Challenges and Opportunities	Trends to 2025 (Prediction)
Fuel cell—high temperature	• SOFC and MCFC have majority. • Fuel cell–based micro-Combined Heat and Power (m-CHP).	• Improve durability. • Stack sintering of SOFC. • Improve conversion efficiency.	• SOFC commercial DG plants. • Degradation: 0.2%/1000 hours. • USD 2500/kWe at 1 MWe.
Batteries (storage)	• Efficiency of 75%–95%. • Cost near competitive for many off-grid and remote applications. • USD 300–USD 3500/kW for distributed or off-grid storage.	• Improve reliability. • New materials to enhance energy density. • Battery management systems.	• Reliability and energy density to be improved. • Growth in on-grid applications. • USD 1000/kW for new on-grid battery systems.
Hydrogen (storage)	• Hydrogen FCV commercialized. • MW-level plants demonstrated. • Efficiency of 22%–50%. • Compressed hydrogen transport. • USD 500–USD 750/kW.	• Energy efficiency is a challenge. • New methods for hydrogen storage and transport. • Thermal management.	• Market introduction of hydrogen power plant. • Hydrogen from renewables. • 70% cost reduction for electric vehicle applications.

Notes: The LCOE of fossil fuel power plants is typically from USD 0.045 to USD 0.14/kWh.

Abbreviations: LCOE, levelized cost of energy; MV, medium voltage; USD, U.S. dollars; DFIG, doubly fed induction generator; PMSG, permanent magnet synchronous generator; PV, photovoltaic; FCV, fuel cell vehicles; PEMFC, proton exchange membrane fuel cells; SOFC, solid oxide fuel cells; MCFC, molten carbonate fuel cells; UPS, uninterruptible power supply; DG, distributed generation.

This book covers topics of

- Maximum power point techniques for PV systems
- Single-phase and three-phase grid-connected PV systems, respectively
- Fuel cells and their power conversion and control
- Small and large wind turbine systems, respectively, including power electronics and electric generators
- Marine hydrokinetic generators and power plants
- Batteries and ultracapacitors used for renewable energy support
- Architectures, controls, and other microgrid issues

The papers on related topics of smart grid published in the double special issue 42(3–4) are also recommended for further reading [41].

REFERENCES

1. International Energy Agency (IEA), World energy outlook 2004, OECD, Paris, France, 2004, http://www.worldenergyoutlook.org/media/weowebsite/2008-1994/weo2004.pdf, Retrieved on October 26, 2016.
2. International Renewable Energy Agency (IRENA), Renewable energy capacity statistics 2015, June 2015, http://www.irena.org/publications, last accessed June 6, 2015.

3. C. Roselund and J. Bernhardt, Lessons learned along Europe's road to renewables, *IEEE Spectrum*, May 2015, http://spectrum.ieee.org/energy/renewables/lessons-learned-along-europes-road-to-renewables, last accessed June 6, 2015.

4. M. M. Hand, S. Baldwin, E. DeMeo, J. M. Reilly, T. Mai, D. Arent, G. Porro, M. Meshek, and D. Sandor, Eds., *Renewable Electricity Futures Study*, Vol. 4, National Renewable Energy Laboratory (NREL), Golden, CO, 2012, NREL/TP-6A20-52409, http://www.nrel.gov/analysis/re_futures/, last accessed June 6, 2015.

5. Y. Jia, Y. Gao, Z. Xu, K. P. Wong, L. L. Lai, Y. Xue, Z. Y. Dong, and D. J. Hill, Powering China's sustainable development with renewable energies: Current status and future trend, *Electric Power Components and Systems*, 43(8–10), 1193–1204, 2015.

6. C. Nagamani, G. Saravana Ilango, M. J. B. Reddy, M. A. A. Rani, and Z. V. Lakaparampil, Renewable power generation Indian scenario: A review, *Electric Power Components and Systems*, 43(8–10), 1205–1213, 2015.

7. National Renewable Energy Laboratory (NREL), Best research-cell efficiencies, Technical Report August 12, 2016, http://www.nrel.gov/ncpv/images/efficiency_chart.jpg, Retrieved on October 26, 2016.

8. General Electric PV Systems, http://www.ge.com/products_services/energy.html, last accessed April 21, 2011.

9. European Wind Energy Association (EWEA), Deep water—The next step for offshore wind energy, July 2013, ISBN: 978-2-930670-04-1.

10. M. Z. Jacobson and M. A. Delucchi, A path to sustainable energy by 2030, *Scientific American*, 301(5), 58–65, November 2009.

11. D. Abbott, Keeping the energy debate clean: How do we supply the world's energy needs?, *Proceedings of the IEEE*, 98(1), 42–66, January 2010.

12. E. de Vries, Close up—Vestas V164, http://www.windpowermonthly.com/article/1211056/close—vestas-v164-80-nacelle-hub, September 9, 2013, last accessed October 21, 2014.

13. S. Roth, World's largest power plant opens in California desert, *US Today*, February 10, 2015, http://www.usatoday.com/story/tech/2015/02/10/worlds-largest-solar-plant-california-riverside-county/23159235/, last accessed June 6, 2015.

14. SunPower invests in the smart energy home, http://newsroom.sunpower.com/2014-12-18-SunPower-Invests-in-the-Smart-Energy-Home, last accessed June 6, 2015.

15. Abengoa to develop South America's largest solar-thermal plant in Chile, http://www.prnewswire.com/news-releases/abengoa-to-develop-south-americas-largest-solar-thermal-plant-in-chile-239427421.html, accessed June 6, 2015.

16. IRENA Report, Renewable power generation costs in 2014, January 2015.

17. ITRPV Working Group Report, International technology roadmap for photovoltaic (ITRPV) 2014 results, SEMI, Berlin, Germany, April 2015.

18. USDRIVE Report, Hydrogen storage technical team roadmap, June 2013.

19. IEA Report, Technology roadmap—Energy storage, 2014.

20. IEA Report, Technology roadmap—Solar thermal electricity, 2014.

21. European Ocean Energy Association Report, Oceans of energy—European ocean energy roadmap 2010–2050, 2010.

22. M. Melaina, Hydrogen infrastructure expansion: Consumer demand and cost-reduction potential, *Presentation at the Hydrogen Infrastructure Investment Forum*, April 2014, Palo Alto, CA.

23. D. Rosenwirth and K. Strubbe, Integrating variable renewables as Germany expands its grid, March 21, 2013, http://www.renewableenergyworld.com/articles/2013/03/germanys-grid-expansion.html, accessed November 12, 2014.

24. M. P. Kazmierkowski, R. Krishnan, and F. Blaabjerg, *Control in Power Electronics. Selected Topics*, Academic Press, New York, 2002.

25. R. A. Messenger and J. Ventre, *Photovoltaic Systems Engineering*, 3rd edn., CRC Press, Boca Raton, FL, 2010.

26. R. Teodorescu, M. Liserre, and P. Rodriguez, *Grid Converters for Photovoltaic and Wind Power Systems*, John Wiley & Sons, 2011.

27. F. Lin Luo and Y. Hong, *Renewable Energy Systems: Advanced Conversion Technologies and Applications*, CRC Press, Boca Raton, FL, 2012.

28. H. Abu-Rub, M. Malinowski, and K. Al-Hadad, Eds., *Power Electronics for Renewable Energy Systems, Transportation and Industrial Applications*, Wiley, Hoboken, NJ, 2014.

29. F. Blaabjerg, Z. Chen, and S. B. Kjaer, Power electronics as efficient interface in dispersed power generation systems, *IEEE Transactions on Power Electronics*, 19(4), 1184–1194, September 2004.

30. S. B. Kjaer, J. K. Pedersen, and F. Blaabjerg, A review of single-phase grid connected inverters for photovoltaic modules, *IEEE Transactions on Industry Applications*, 41(5), 1292–1306, September 2005.

31. J. M. Carrasco, L.-G. Franquelo, J. T. Bialasiewicz, E. Galvan, R. C. P. Guisado, M. A. M. Prats, J. I. Leon, and N. Moreno-Alfonso, Power-electronic systems for the grid integration of renewable energy sources: A survey, *IEEE Transactions on Industrial Electronics*, 53(4), 1002–1016, June 2006.

32. H. Li and Z. Chen, Overview of different wind generator systems and their comparisons, *IET Renewable Power Generation*, 2(2), 123–138, June 2008.

33. T. Kerekes, R. Teodorescu, M. Liserre, C. Klumpner, and M. Sumner, Evaluation of three-phase trans-formerless photovoltaic inverter topologies, *IEEE Transactions on Power Electronics*, 24(9), 2202–2211, September 2009.

34. M. Tsili and S. Papathanassiou, A review of grid code technical requirements for wind farms, *IET Renewable Power Generation*, 3(3), 308–332, September 2009.

35. F. Blaabjerg, M. Liserre, and K. Ma, Power electronics converters for wind turbine systems, *IEEE Transactions on Industry Applications*, 48(2), 708–719, March–April 2012.

36. F. Blaabjerg and K. Ma, Future on power electronics for wind turbine systems, *IEEE Journal of Emerging and Selected Topics in Power Electronics*, 1(3), 139–152, 2013.

37. O. Ellabban, H. Abu-Rub, and F. Blaabjerg, Renewable energy resources: Current status, future prospects and their enabling technology, *Renewable and Sustainable Energy Reviews*, 39, 748–764, 2014.

38. S. Kouro, J. L. Leon, D. Vinnikov, and L.-G. Franquelo, Grid-connected photovoltaic systems: An overview of recent research and emerging PV converter technology, *IEEE Industrial Electronics Magazine*, 9(1), 47–61, March 2015.

39. E. Romero-Cadaval, B. Francois, M. Malinowski, and Q. Zhong, Grid-connected photovoltaic plants: An alternative energy source, replacing conventional sources, *IEEE Industrial Electronics Magazine*, 9(1), 18–32, March 2015.

40. V. Yaramasu, B. Wu, P. C. Sen, S. Kouro, and M. Narimani, High-power wind energy conversion systems: State-of-the-art and emerging technologies, *Proceedings of the IEEE*, 103(5), 740–788, May 2015.

41. M. E. El-hawary, The smart grid—State-of-the-art and future trends, *Electric Power Components and Systems*, 42(3–4), 239–250, 2014.

2 Solar Power Sources: PV, Concentrated PV, and Concentrated Solar Power

Katherine A. Kim, Konstantina Mentesidi, and Yongheng Yang

CONTENTS

Abstract..17
2.1 Introduction...18
2.2 PV Power...20
    2.2.1 PV Cell Basics..20
    2.2.2 PV Operating Characteristics 22
    2.2.3 PV Panel Configuration 26
2.3 Concentrated Photovoltaic Power...28
2.4 Concentrated Solar Power 29
    2.4.1 Solar Thermal Power Concepts 30
        2.4.1.1 Parabolic Trough...31
        2.4.1.2 Linear Fresnel Reflector..31
        2.4.1.3 Solar Towers..31
        2.4.1.4 Parabolic Dish Reflector...32
    2.4.2 CSP and Storage 33
    2.4.3 Grid Integration and Stability Considerations 34
    2.4.4 CSP Modeling for Transient Stability Analysis................................34
        2.4.4.1 Inertial Response...35
        2.4.4.2 Electrical Power...36
        2.4.4.3 Mechanical Power...36
    2.4.5 CSP Large-Scale Integration: Technical Challenges and Impacts.....37
2.5 Summary..38
References...38

ABSTRACT

Solar power is highly abundant, relatively reliable, and not limited to a geographic region, making it one of the most important renewable energy sources. Catering for a clean and green energy system, solar energy will be an active player in the future mixed power grid that is also undergoing a significant change. Beyond this energy transition, the still declining cost of the solar technology has become an important driving force for more solar-powered systems. However, high penetration of solar-powered systems also brings technical challenges to the entire energy systems. In order to fully address those issues, the technological properties of solar power should be investigated. Thus, the basics of solar power technology will be introduced and discussed in this chapter.

2.1 INTRODUCTION

Conventional energy resources may run out in the future, along with the fast-growing economy worldwide. Energy consumption and efficiency have become two issues of high concern in today's society. Through developing renewable energies, societal sustainability may be maintained, which also prevents possible energy crisis. As an important source of renewable energies, solar energy as a consequence is broadly utilized, in either a passive or an active manner [1]. At present, for active solar power techniques, depending on how the solar energy is converted into electricity, the following three technologies are commonly used: photovoltaic (PV), concentrated photovoltaic (CPV), and concentrated solar power (CSP). The three solar techniques will thus be covered and addressed in this chapter.

First, the PV technology will be presented. Basically, according to the PV effect [2], abundant solar energy can be directly converted into direct current (DC) electricity with the help of semiconducting materials (e.g., crystalline silicon), forming a PV cell. Unfortunately, the voltage level of a single PV cell is relatively low compared to the voltage required for many loads (e.g., charging a battery, powering a lamp, the grid). Hence, the PV cells are connected in series and in parallel to form PV modules to increase the voltage level as well as the power rating, respectively. In practical applications, the PV modules are also connected in the same way, which enables grid connection and thereby powering heavy loads. Details of the grid-connection issues for PV systems will be covered in Chapters 3 and 4.

Although a lot of applications using PV technology can be seen in our daily lives, the energy conversion efficiency is still not satisfactory. Figure 2.1 shows the best research cell efficiencies of different PV technologies, where there are mainly five groups of solar cell technologies. In general, the multi-junction PV cells can achieve higher efficiency compared to the crystalline silicon cells. By contrast, in terms of cost, thin-film solar cells are the cheapest among the five, but is still at the early stage of massive utilization. Intensive research is ongoing to push forward the penetration of the thin-film solar technology into the PV market [4, 5]. Additionally, the emerging PV technologies have been developed for certain applications with attractive features like low cost and transparency. However, most of the emerging technologies are still not commercialized, and the efficiencies are the lowest as shown in Figure 2.1 [3].

Nevertheless, in order to improve the conversion efficiency, CPV (also referred to as concentrator PV) technology was developed and employed, where sunlight is concentrated using lenses and/or mirrors onto small solar cells (typically using multi-junction PV cells for high efficiency). As it can be observed in Figure 2.1, the CPV structure is typically multi-layered using a wider spectrum of irradiance. However, since the electricity generation is based on the PV effect, according to the PV characteristics, cooling of the CPV cells is an important issue for maintaining high efficiency. Nevertheless, although the number of CPV installations is quite low compared to that of the PV technology, systems using CPV have the potential to become competitive in the near future [6]. Hence, the power electronics technology for CPV systems has been also touched in Chapter 1, and the CPV technology is also detailed in this chapter.

The PV and CPV technologies mentioned earlier are based on the PV effect. Alternatively, the solar energy can be utilized and converted using the CSP technology. In contrast to the CPV system, the CSP technology electricity generation systems first concentrate a large area of sunlight to a small area and then heat up the thermal-carrying materials (e.g., water or thermochemical materials). Then, the heat is transferred to either drive a heat engine (e.g., a steam turbine) or power a thermochemical reaction, and hereafter, electricity is generated. The role of power electronics here is thus to convert and transmit the energy in a reliable and efficient way. However, the CSP systems can provide large thermal energy that requires specific technology or systems to store. As CSP systems are growing at a fast pace, this chapter will also conceptually introduce the CSP technology, where its impact on the entire grid is discussed. It is worth mentioning that synchronous generators can be used in the CSP systems, where the structure is similar to the classical power generation system.

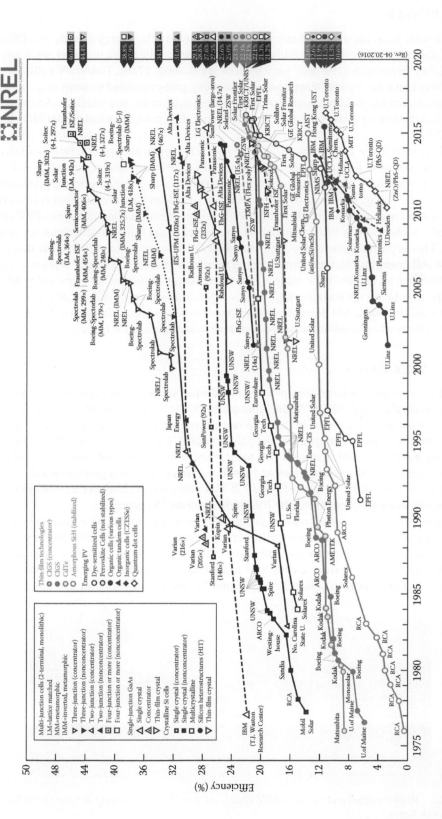

FIGURE 2.1 Best research cell efficiencies for solar PV cells from National Renewable Energy Laboratory (NREL) during the last 40 years. (From NREL National Center for Photovoltaics, Best research-cell efficiencies for solar PV cells from NREL, Available: http://www.nrel.gov/ncpv/, Retrieved on May 17, 2016.)

2.2 PV POWER

PV cells, or solar cells, are semiconductor devices that convert solar energy directly into DC electric energy. In the 1950s, PV cells were initially used for space applications to power satellites, but in the 1970s, they began also to be used for terrestrial applications [7, 8]. Today, PV cells are used to provide power in a wide variety of applications, including grid-connected systems (e.g., utility scale and residential), remote buildings, outdoor traffic-related equipment, and satellites. An example of a roof-mounted residential grid-connected PV system providing power to a campus building is shown in Figure 2.2.

PV cells can be made from many different types of materials and using a range of fabrication techniques. As shown in Figure 2.1, the major categories of PV materials are crystalline silicon (Si), thin film, multi-junction, and various emerging technologies like dye-sensitized, perovskite, and organic PV cells. Today, there is a significant amount of research that focuses on both increasing efficiency and decreasing manufacturing cost [9]. However, the most dominant type of PV cell used in large-scale applications is still crystalline silicon, which is the same basic technology as used in the 1970s. This is partially due to the high availability of low-cost silicon PV panels that has prevented new and emerging cell types from gaining significant presence in the PV market. PV materials and fabrication techniques have made significant headway in the last 15 years and a shift in the PV cell type may be on the horizon, but, for now, crystalline silicon is still the dominant cell type. This section will introduce and detail the basic characteristics and operating principles of crystalline silicon PV cells as some considerations for designing systems using PV cells.

2.2.1 PV Cell Basics

A PV cell is essentially a large-area p–n semiconductor junction that captures the energy from photons to create electrical energy. At the semiconductor level, the p–n junction creates a depletion region with an electric field in one direction. When a photon with sufficient energy hits the material in the depletion region, the energy from the photon excites a valance electron into the conduction band, leaving a hole in the valance band. Due to the electric field in the depletion region, the electron and hole will travel in opposite directions and generate a net current. This process of a photon generating an electron–hole pair is shown in Figure 2.3. This generated current over the voltage generated by the semiconductor junction allows the PV cell to generate DC power.

There are two basic types of crystalline silicon cells: mono-crystalline (m-c) and poly-crystalline (p-c). The m-c cells have one uniform lattice through the entire cell and allow electronics to flow easily through the materials, while p-c cells have multiple crystalline structures, or grains, which can impede electron flow. Thus, p-c cells tend to have lower conversion efficiency than m-c cells, but they are slightly cheaper to manufacture. Figure 2.4 shows images of an m-c and p-c PV cell

FIGURE 2.2   Roof-mounted grid-connected PV system at Ulsan National Institute of Science and Technology in Ulsan, South Korea.

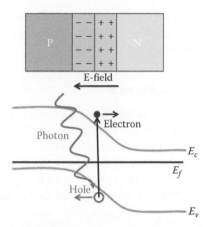

FIGURE 2.3 Process of a photon generating an electron–hole pair in a PV cell.

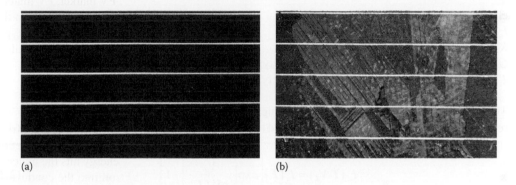

(a)

(b)

FIGURE 2.4 A PV cell with (a) a mono-crystalline (m-c) and (b) poly-crystalline (p-c) structure.

close-up, where the m-c material structure is uniform but the p-c materials have many different grain regions. Both m-c and p-c cells are widely used in PV panels and in PV systems today.

The basic structure of a PV cell can be broken down and modeled as basic electrical components [10–12]. Figure 2.5 shows the semiconductor p–n junction and the various components that make up a PV cell. The photon-to-electron flow process explained previously can be modeled as a current source, I_{ph}, where the generated current depends on the intensity of the light hitting the cell. The p–n semiconductor junction is modeled as a diode, D, with the direction as shown in Figure 2.5. The current source and diode make up the ideal model of a PV cell, but in real life there are additional parasitic components. The p–n junction will have associated parallel capacitance, C_p, and parallel resistance (also called shunt resistance), R_{sh}, while the wire leads attached to the PV cell will have

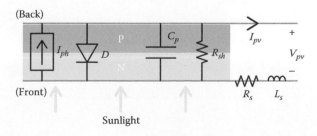

FIGURE 2.5 PV cell basic structure electrical model components with parasitic components.

associated series resistance, R_s, and series inductance, L_s. These parasitic components are often ignored when only a simple representation of a PV cell or panel is needed, but they should be taken into consideration when more accurate modeling is required.

2.2.2 PV Operating Characteristics

While there are many environmental factors that affect the operation characteristics of a PV cell and its power generation, the two main factors are solar irradiance G, measured in W/m², and temperature T, measured in degree Celsius (°C). The relation between these two factors and the PV operating characteristics can be modeled mathematically. First, we examine the ideal model that consists of just the photocurrent source I_{ph} and a diode, as it is shown in Figure 2.6.

The photocurrent I_{ph} depends on both the irradiance and temperature according to

$$I_{ph}(G,T) = \left[I_{scn} + K_i \left(T - T_n \right) \right] \frac{G}{G_n} \tag{2.1}$$

where

I_{scn} is the nominal short-circuit current
K_i is the current temperature coefficient
G_n is the nominal solar irradiance, which is typically 1000 W/m²
T_n is the nominal cell temperature, which is typically 25 °C

These values can be determined from the ratings listed for commercial PV cells or panels. Also, the current I_d and voltage V_d of the diode are expressed by an exponential relation, and they are represented as

$$I_d(T, V_d) = I_s(T) \left[\exp\left(\frac{V_d}{a V_t(T)} \right) - 1 \right] \tag{2.2}$$

where

I_s is the diode saturation current
a is the diode ideality constant
V_d is the diode voltage
V_t is the thermal voltage of the semiconductor junction

The diode saturation current depends on temperature and can be defined as

$$I_s(T) = \frac{I_{scn} + K_i \left(T - T_n \right)}{\exp\left(\dfrac{V_{ocn} + K_v \left(T - T_n \right)}{a V_t(T)} \right) - 1} \tag{2.3}$$

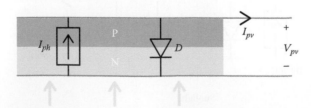

FIGURE 2.6 Ideal PV model with a current source and diode.

where

    I_{scn} is the nominal short-circuit current
    K_i is the current temperature coefficient
    T_n is the nominal cell temperature
    V_{ocn} is the nominal open-circuit voltage
    K_v is the voltage temperature coefficient
    a is the diode ideality factor
    V_t is the thermal voltage

Note that the diode voltage V_d is the same as the PV voltage V_{pv} for the ideal model. Also, the thermal voltage V_t depends on temperature T and is defined by

$$V_t(T) = \frac{kT}{q} N_s \tag{2.4}$$

where

    k is Boltzmann's constant (approximately 1.3807×10^{-23} J·K^{-1})
    q is the electron charge ($1.60217662 \times 10^{-19}$ C)
    N_s is the number of PV cells in series

Using Kirchhoff's circuit laws, the relation between the PV current I_{pv} and PV voltage V_{pv} for the ideal PV model is

$$I_{pv} = I_{ph}(G,T) - I_d(T,V_{pv}) \tag{2.5}$$

where the photocurrent I_{ph} is defined in (2.1) and the diode current I_d is defined in (2.2). Based on the PV current I_{pv} equation, given in (2.5), it is clear that the PV output current is related to the solar irradiance G and temperature T.

Given the solar irradiance and temperature, this explicit equation in (2.5) can be used to determine the PV current for a given voltage. These equations can also be rearranged using basic algebra to determine the PV voltage based on a given current. The simple PV model can be implemented in MATLAB®/Simulink®, as it is shown in Figure 2.7, where the inputs are the solar irradiance G, temperature T, and the PV voltage V_{pv} and the outputs are the PV current I_{pv} and power P_{pv}.

The I–V curve of a PV cell simulated in MATLAB®/Simulink is plotted, as shown in Figure 2.8. The star indicates the maximum power point (MPP) of the I–V curve, where the PV will produce its maximum power. At voltages below the MPP, the current is a relative constant as voltage changes such that it acts similar to a current source. At voltages above the MPP, the voltage is relatively constant as current changes such that it acts similar to a voltage source. The open-circuit voltage of a PV is the voltage when the PV current is 0 A, and it is labeled as V_{OC} in Figure 2.8. The short-circuit

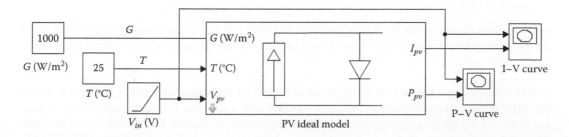

FIGURE 2.7 MATLAB®/Simulink® model of a PV cell or panel with solar irradiance G and temperature T as inputs.

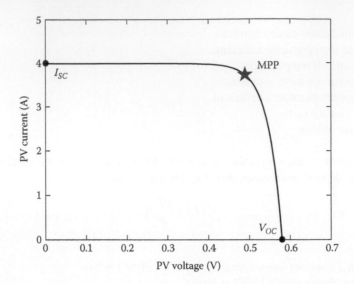

FIGURE 2.8   I–V curve for an example PV cell ($G = 1000$ W/m² and $T = 25$ °C; V_{OC}: open-circuit voltage; I_{SC}: short-circuit current).

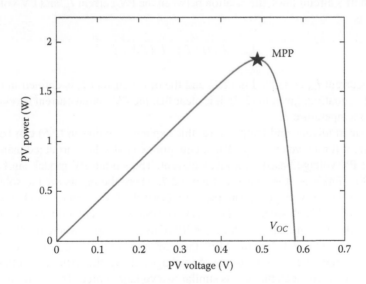

FIGURE 2.9   Power–voltage curve, for example, PV cell under a specific constant irradiance and temperature condition (i.e., $G = 1000$ W/m² and $T = 25$ °C; V_{OC}: open-circuit voltage).

current is the current when the PV voltage is 0 V, labeled as I_{SC}. These parameters are often listed on the rating labels for commercial panels and give a sense for the approximate voltage and current levels to be expected from a PV cell or panel.

Based on the I–V curve of a PV cell or panel, the power–voltage curve can be calculated. The power–voltage curve for the I–V curve shown in Figure 2.8 is obtained as given in Figure 2.9, where the MPP is the maximum point of the curve, labeled with a star. The I–V curve and power–voltage curve shown are under a specific irradiance and temperature condition. Over a day, both the irradiance and temperature will change, sometimes gradually (minutes to hours) and sometimes quickly (seconds), for example, due to passing clouds [13]. Think of the I–V curves as the characteristics for just an instant of time.

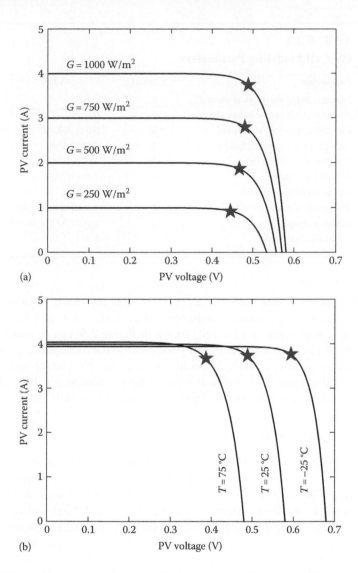

FIGURE 2.10 Effects of (a) solar irradiance and (b) temperature changes on a PV's I–V curve.

As irradiance and temperature change, the I–V curve will also change, as shown in Figure 2.10. The irradiance is directly proportional to the current characteristics. As the irradiance increases, the short-circuit current and MPP current will also increase. Conversely, the temperature is inversely proportional to the voltage characteristics. As the temperature increases, the open-circuit voltage (V_{OC}) and MPP voltage will decrease. According to these trends, a cell will produce the most power when the sunlight intensity is high but the temperature is low. However, these kinds of environmental conditions are not common, as higher-intensity light hitting an object tends to increase its temperature at the same time.

Exercise 2.1

Equations 2.1 through 2.5 and MATLAB®/Simulink are used to model the PV cell with the parameters given in Table 2.1. These are the same PV parameters used to model the graphs shown in Figures 2.8 and 2.10, such that you can compare your results with these figures. Also, an example of Simulink implementation is shown in Figure 2.11 that can be used as a reference.

TABLE 2.1
PV Cell Modeling Parameters

Parameter	Symbol	Value
Nominal string short-circuit current	I_{scn}	4 A
Nominal string open-circuit voltage	V_{ocn}	0.58 V
Current temperature coefficient	K_i	0.001 A/K
Voltage temperature coefficient	K_v	−0.002 V/K
Number of cells in series	N_s	1
Nominal irradiance	G_n	1000 W/m²
Nominal cell temperature	T_n	298.15 K
Boltzmann's constant	k	1.3807×10^{-23} J·K⁻¹
Electron charge	q	$1.60217662 \times 10^{-19}$ C
Diode ideality factor	a	1.3

The previously described ideal model is a relatively straightforward mathematical model but does not take into account any parasitic components; thus, it does not model a real PV cell very accurately. Parasitic components in a PV cell, shown in Figure 2.5, can be added to increase the model accuracy, but it also increases the model's complexity. For example, if the series and shunt resistance are added to the model, the relationship between the PV voltage and current becomes a nonlinear equation that requires a nonlinear solver. Often, a nonlinear solver, like the Newton–Raphson method, can be employed to solve the more complex PV models [10].

2.2.3 PV PANEL CONFIGURATION

In most applications using PV cells, not just a single cell but many cells are combined to produce a larger amount of power. When added together, the PV cells can be connected in series, parallel, or a combination of both. When cells are connected in series, their voltages are summed together. When cells are connected in parallel, their currents are summed together. A single silicon cell tends to produce around 0.5 V for voltage and current that is proportional to the cell's surface area. The decision to connect PV cells in series or parallel is often determined by the voltage and current needs of the application. Most PV applications today, like rooftop PV installations and satellites, require a higher voltage than a single cell, so the cells are typically connected in series. The standard commercial PV panel is typically made up of 72–96 cells connected in series so that the sum of their voltages achieves overall voltages in the range of 30–60 V.

An example of connection configuration of a PV panel is shown in Figure 2.12 that has 72 cells in series. To cover the area of a panel (one series string), cells are typically connected down one column and back up the next. Each series connection that goes down and back up is called a substring. Figure 2.12 shows three substrings with a diode that is connected across the ends of each substring. This diode is called a bypass diode, which allows for an additional current path around the substring if PV cells in that string experience a problem, such as partial shading, permanent degradation, or a connection break in the substring. Bypass diodes help to reduce the degree that these kinds of problems affect the total output power of the panel, but there are various types of faults that can occur in a panel [14].

Many PV systems are installed on rooftops and/or in remote areas that are not easily serviceable, so PV systems that include both the panels and the associated power electronics must be able to work properly with minimal maintenance requirements. PV panels are typically rated to work continuously for 25 years or even longer, with some expected level of natural degradation in output power, generally around 0.5%–1.0% per year [10]. However, in certain conditions, the PV panels

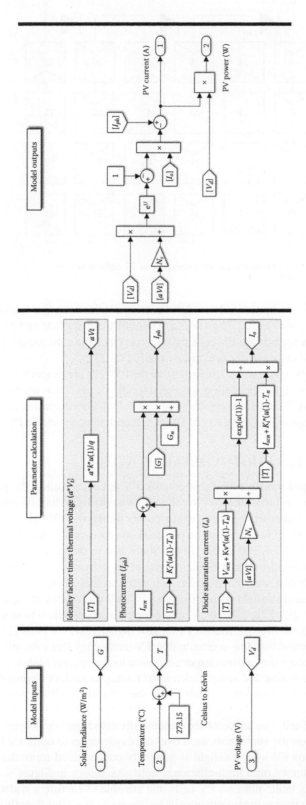

FIGURE 2.11 MATLAB®/Simulink® model of PV cell/module with the parameters shown in Table 2.1.

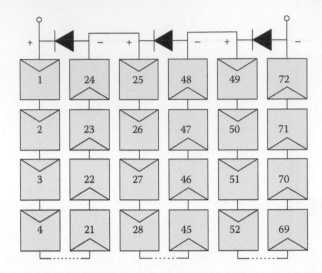

FIGURE 2.12   Diagram of a typical PV panel consisting of 72 cells in series.

tend to degrade more quickly, specifically in high-heat (e.g., in the desert) or high-humidity environments (e.g., tropical or seaside areas). High heat and humidity can degrade and corrode the materials that protect the cells and degrade the PV cells themselves. Once the cells are directly exposed to the outside environment, PV cells can degrade more rapidly [15–17].

When working with PV systems, it is important that the PV cells are properly connected to meet the voltage and current needs of the application and that the cells are protected according to the expected environmental conditions. Today, there are ongoing research and development efforts on fault detection and protection to improve the longevity of PV panels and, thus, the entire PV system [18–21].

2.3   CONCENTRATED PHOTOVOLTAIC POWER

The basic idea of CPV power, also called concentrator PV or concentrating PV power, is to capture sunlight from an area that is larger than a PV cell and focus the light directly onto a small PV cell. Using a concentrator, the intensity of the light (in W/m^2) hitting the PV cell is significantly higher without a concentrator, which leads to a more potential energy capture. In the 1970s, CPV systems began to be explored and developed in the United States due to the push for energy independence and its superior performance in terms of efficiency over all existing PV technologies. During that time period, the PV cell cost was much higher compared to that of mirrors and reflectors. Thus, concentrating a wide area of sunlight onto a high-cost PV cell was overall cheaper than covering the equivalent area with PV cells [22, 23]. Today, this concept has been taken to much higher levels of solar light concentration and higher-efficiency (but also higher-cost) PV cells [23].

Generally, CPV is divided into low-concentration PV technology that uses silicon cells and high-concentration PV technology that utilizes larger concentration setups and higher-efficiency cells. One example of a low-concentration PV setup uses a mirror setup to achieve approximately three suns, or three times the normal incident light intensity, onto the PV cell. The basic concept is illustrated in Figure 2.13, where the normal incident light either directly hits the PV cell or is reflected off the adjacent mirrors onto the PV cell. The concentrator systems often have tracker systems that move the face of the CPV array to follow the sun's movement over the day in order to optimize the output power.

For high-concentration PV systems, light is generally concentrated more than 100 times using specialized lenses or mirrors onto high-efficiency cells. As shown in Figure 2.1, these high-efficiency cells are typically multi-junction PV cells that are able to capture a wider range of sunlight to produce more energy than normal silicon cells but come with a significantly higher fabrication

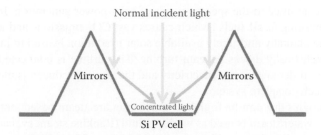

FIGURE 2.13   A low-concentration PV setup using mirrors to concentrate light onto silicon PV cells.

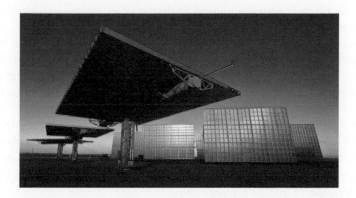

FIGURE 2.14   An example of a concentrated PV system in Aurora, Colorado. (From National Renewable Energy Laboratory, NREL, The Amonix 7700 Solar Power Generator is an example of a concentrating PV system that is well-matched to utility-scale projects, Available: http://www.nrel.gov/continuum/utility_scale/leading_solar.html, Retrieved on January 18, 2017.)

cost [24]. Nevertheless, concentrator PV technology is becoming even more competitive than some other PV technologies [24] and it has great potential to take a significant share of electricity generation in the future energy source mix. The principle of CPV power generation is still based on the PV effect. Thus, the electrical model is similar to that of the PV system, except for the expected irradiance level and temperature that are much higher. However, due to the highly concentrated sunlight hitting the small CPV cell, more thermal challenges come with the CPV technology, which have to be addressed for each specific application [25, 26]. In order to accurately model a CPV application, an electrical model linked to a thermal model of the PV setup is recommended.

One of the challenges in CPV is the high temperature that is associated with high-intensity light hitting the PV cell. Consequently, CPV setups must have a proper heat sink system to dissipate excess heat. As it can be seen in Figure 2.10, not only does high temperature decrease the voltage characteristics of the CPV cell and the resulting output power, but it can also degrade the PV materials more quickly than non-CPV cells. While there are a number of CPV installations throughout the world [22], it is still an emerging technology that has not yet gained significant ground in the worldwide PV system market. Figure 2.14 shows an example of a CPV system at the Solar Technology Acceleration Center in Aurora, Colorado, USA, where 13 solar power generators have been installed.

2.4   CONCENTRATED SOLAR POWER

Concentrated solar power (CSP), also known as concentrating solar power, solar thermal power, or solar thermal electricity, uses glass mirrors of different architectures to collect the sun's thermal energy and convert it into electrical energy. This is achieved via conventional thermodynamic power cycles that are adjusted to the specific needs of solar power generation. Instead of providing thermal energy via burning fossil fuels (which causes vast CO_2 emissions and air pollution) or via nuclear reactions, the naturally and freely available solar irradiation is used to increase the working fluid's enthalpy, which finally drives a steam turbine. The turbine is connected to the shaft of an electrical generator in order to generate electricity and the power produced is then fed into the grid. Figure 1.11 shows such complete systems.

The main technologies that can be found on the market are steam based, engine based, and PV based. In most cases, water/steam is used as working fluid (Rankine steam cycles), but there do exist some concepts that are still under consideration in which air or CO_2 is used as working fluid, which drives a gas turbine (Brayton gas turbine cycle). This last design enables the possibility to run an attached bottoming steam cycle and by using heat recovery boilers results to the so-called solar thermal combined cycle plant. In addition, typical Stirling engines are non-steam-based technologies, and they are used to convert solar thermal energy into mechanical work.

CSP plants have the inherent energy storage capability in the form of heat, and with further support from additional thermal storage systems or a hybrid system, CSP plants may continue to generate electricity even when clouds block the sun rays. Although CSP has better performance for grid integration than other renewable energy options, the relatively immature technology and the high cost restrict its large-scale deployment. In addition, one disadvantage of CSP is that it requires strong direct sunlight. Consequently, the highest share of CSP resources is mostly limited to semi-arid, hot regions or deserts. However, similar to other thermal power generation plants, CSP requires water for cooling, which constitutes a big challenge in exploiting CSP resources in arid regions.

2.4.1 SOLAR THERMAL POWER CONCEPTS

Solar thermal power plant concepts can be divided into two basic categories, namely, line and point focusing technologies. On one hand, line focusing systems concentrate the incident solar direct irradiation onto a focal line, in particular onto solar receiver tubes that are placed all around the focal lines of the solar collectors. On the other hand, point focusing systems concentrate the incoming solar irradiation onto a single focal point or better, onto a focal area that is very small compared to the total size of the reflecting mirror surface. One single solar receiver is therefore placed at this focal point or area [27–29]. Generally speaking, the concentration of the incoming solar irradiation is the fundamental principle in order to reach high temperatures, thus providing high-quality thermal power that enables the operation of conventional thermodynamic power cycles. In this sense, the CSP is not based on the PV effect anymore.

Nowadays, the most used design is the parabolic trough collector technology (line focusing). There, the power cycle's working fluid is heated indirectly, using a heat transfer fluid that is heated within the focal lines of the parabolic troughs [27]. Temperatures reached a range between 400 °C and 550 °C, depending on the used heat transfer fluid (thermal oil or molten salts). Another concept that has been tested and operated successfully is the solar power tower or central receiver concept (point focusing) [28, 29]. In that case, the solar radiation is concentrated onto a central receiver area, which is placed at the top of a tower, using many slightly curved mirror elements. At the receiver, the concentrated solar energy is transferred to the working fluid or heat transfer fluid. Depending on the concept, temperature levels can exceed 1000 °C [28]. Power towers promise high efficiency if used with Brayton cycles [30], as they can be extended to solar-combined cycle plants. Additional concepts are the linear Fresnel collector concept (line focusing) and the parabolic dish collector concept (point focusing). The linear Fresnel concept is similar to the parabolic trough systems, also line focusing. In this architecture, many flat mirror elements concentrate solar radiation onto a horizontal receiver tube [29]. Parabolic dish collectors concentrate the sunlight onto the focal point, where Stirling engines transform the thermal energy into electrical energy. They provide good efficiency rates in small power classes, ideal for isolated and stand-alone applications [31]. The most characteristic CSP designs are briefly described in the following:

2.4.1.1 Parabolic Trough

The focal point in solar parabolic linear reflectors can be found along their length (see Figure 2.15). A receiver pipe with a fluid (oil or molten salt) is running along the inner part of the curved surface. The concentrated solar energy heats the fluid flowing through the pipe, and the heat energy is then used to generate electricity in a conventional steam generator. Trough designs can incorporate thermal storage allowing for electricity generation during the evening [27, 32], and the thermal storage will also be briefly discussed later in this chapter.

2.4.1.2 Linear Fresnel Reflector

This kind of technology approaches the parabolic trough collectors but uses an array of flat or slightly curved mirrors to collect the sun rays onto a fixed receiver mounted on a linear tower (see Figure 2.16) [28, 33]. The linear Fresnel reflector (LFR) field can be pictured as a broken-up parabolic trough reflector, but it needs less land to generate a specific output. Its major asset is that it uses flat or elastically curved reflectors, which are cheaper compared to parabolic glass reflectors. However, LFRs are less efficient than troughs in converting solar energy to electricity and it is more difficult to incorporate storage capacity into their design [33].

2.4.1.3 Solar Towers

Solar towers (see Figure 2.17), also known as central receiver systems, utilize hundreds or thousands of small reflectors (called heliostats) to collect the sunlight on a central receiver mounted on the top of a fixed tower. The working fluid (mostly sea water) is heated and later drives a turbine to produce electrical power [29]. In this design, very high temperatures are reached; therefore, an increase in efficiency is marked resulting in less cost of thermal storage. Temperatures reached in a solar power tower can be greater than 1000 °C [32].

FIGURE 2.15 Parabolic trough line concentrator for CSP. (From Argonne National Laboratory, Solar Energy Development Programmatic EIS (Information Center), Solar Energy and Electric Transmission Photos, Available: http://www.solareis.anl.gov/guide/photos/index.cfm, Retrieved on January 18, 2017.)

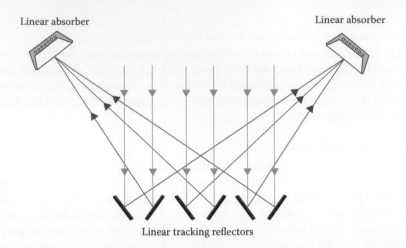

FIGURE 2.16 Scheme of a linear Fresnel solar collector. (Adapted from Mertins, M., Technische und wirtschaftliche Analyse von horizontalen Fresnel-Kollektoren, PhD thesis, Universitat Karlsruhe, Karlsruhe, Germany, October 2013, Available: http://digbib.ubka.uni-karlsruhe.de/volltexte/documents/1067166, Retrieved on April 13, 2016.)

FIGURE 2.17 Central solar tower receiver. (From Argonne National Laboratory, Solar Energy Development Programmatic EIS (Information Center), Solar Energy and Electric Transmission Photos, Available: http://www.solareis.anl.gov/guide/photos/index.cfm, Retrieved on January 18, 2017.)

2.4.1.4   Parabolic Dish Reflector

This topology consists of a stand-alone parabolic dish [31], as shown in Figure 2.18. The parabolic dish collects the sun rays and converts solar radiation into thermal energy in a running fluid. The thermal energy can then either be converted into electricity using an engine genset coupled directly to the receiver or be transported through pipes to a central power-conversion system. The Stirling machine is the most frequent type of heat engine used in parabolic dish systems that can reach temperatures more than 1500 °C [34].

The pressure of the steam fluid drives a generator to produce electrical energy and for a large Stirling power plant, a lot of these dishes are connected to attain megawatts of electrical power. The greatest advantage of dishes is their high efficiency at thermal-energy absorption and thus the

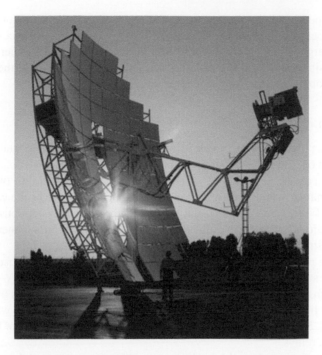

FIGURE 2.18 Schematic view of a parabolic dish solar system. (From Sandia National Laboratories, Image from Argonne National Laboratory, Solar Energy Development Programmatic EIS (Information Center), Solar Energy and Electric Transmission Photos, Available: http://www.solareis.anl.gov/guide/photos/index.cfm, Retrieved on January 18, 2017.)

highest solar-to-electric conversion performance of any CSP system [32]. However, the need for a heat transfer fluid throughout the collector field raises design issues, such as piping layout, pumping requirements, safety, and thermal losses.

2.4.2 CSP AND STORAGE

The intermittency and unreliability of solar energy may impact the performance of a solar thermal plant, which consists of the solar field, the heat transfer fluid transport system, and the power block. Consequently, there will be a mismatch in energy supply and demand, especially at night or due to clouds' shading effect during a day. In operated solar thermal plants, this imbalance is evened by using either auxiliary fossil fuel burners or heat storage systems. Therefore, the requirement for thermal-energy storage systems strongly depends on the daily and yearly variation of solar radiation and the electrical consumption profile.

Heat storage systems store the excess of thermal energy when solar radiation is above the base load output of the CSP and deliver it on the power block's steam generator when it is needed. In that way, the power output of the plant during cloudy days is increased, in addition to its operating time beyond sunset. Hence, not only is energy supply secured, but also solar thermal power plants become more competitive for grid connection [35].

The selection of the appropriate storage capacity is site and system dependent. Feasibility studies need to be performed considering statistical analysis of the electrical demand, as well as the site-specific weather conditions. Moreover, economic trade-off analysis is necessary to be carried out in order to select the optimum storage capacity and the optimum system.

Heat storage systems can be found into two basic classifications: active and passive systems. In active systems, the heat storage medium is a moving, circulating liquid, which stores and delivers the sensible heat, whereas passive heat storage systems can store either sensible or latent heat

[36, 37]. As the name already implies, in passive storage systems the heat storage medium is in a passive state and does not circulate. An additional heat storage concept successfully tested is the thermochemical energy storage. This system relies on endothermic and exothermic chemical reactions that are reversible. However, this type of thermal-energy storage is at the moment at a very early stage of development and still subject of research.

2.4.3 Grid Integration and Stability Considerations

In order to analyze a power grid from an electrical point of view, examine its stability and define the potential integration of distributed generation, that is, CSP into the grid, static, and dynamic analysis needs to be carried out [38]. Stability considerations can be divided into two case studies: the steady-state and dynamic stability analysis. Steady-state analysis involves the calculation of power flows on the network lines, transformers, and the voltage profiles of system bus bars. This study is important for the planning and design of the connection of any distribution generation system to the transmission and distribution grid.

Contingency and particularly $N - 1$ analysis is also essential in order to ensure the security and reliability of power supply, where N is the number of generation units. The $N - 1$ criterion is that the system must withstand on its own a loss of any single element. That means that all the line flows must be below their limits not only for a given normal operating state but also when any of the lines is disconnected [39]. Short-circuit calculations introduce the proper selection of high-voltage equipment and protection relays.

A transient event occurs undesirably and instantly in a power system and can either be an oscillatory or impulsive disturbance. For instance, branches and bus faults, lines, loads, and generators tripping may induce frequency, change rotor angle, and cause voltage oscillations and can lead to the failure of the distribution system where the CSP is planned to be connected to or even can cause damage to the power electronics equipment. Therefore, a transient stability study is necessary to investigate the power system response to disturbances and check over its fault ride through capability to see if the network elements have adequate stability margins [40, 41].

2.4.4 CSP Modeling for Transient Stability Analysis

Thermal and solar plants work on identical principles as fossil fuel–based power plants, where the only difference is the primary source like shown in Figure 1.11. Moreover, the dynamic modeling of conventional generators requires the definition of the standard generator model, the automatic voltage regulator modeling for voltage regulation, and the governor model (e.g., speed control). The steady-state voltage, power, and speed outputs need to be regulated and retain their equilibrium during normal conditions or obtain a new one even after the occurrence of critical disturbances in the system [42, 43].

CSP plants are considered as conventional thermal power plants using standard models of synchronous generators, governors, exciters, etc. In CSPs, sunlight produces high-pressure steam, which is used to generate electrical power by a turbine and an electrical generator that is a conventional synchronous machine. Table 2.2 concentrates the parameters utilized for the dynamic modeling of synchronous machines in order to study and analyze the transient stability of power systems.

The single-machine infinite-bus model illustrated in Figure 2.19 is the basis for transient stability analysis for systems driven by synchronous machines, where P_m is the generator's electrical power output. The grid is represented by an infinite bus like a constant voltage source. The key assumptions considered here are the following [43]:

- A three-phase balanced system is considered.
- The network is assumed to be in a steady-state equilibrium prior the fault incident.
- The resistive part of the transmission line is neglected.
- The synchronous machine is modeled only with the positive sequence components.

TABLE 2.2

Synchronous Machine Parameters for Dynamic Modeling (Calculated Considering a Diesel Generator Set of 16 kVA)

Parameters	Symbols	Values
Resistances		
Stator winding	R_a	0.0645 pu
Reactances		
Stator leakage reactance	X_a	0.06 pu
Synchronous: direct axis	X_d	1.734 pu
Synchronous: quadrature axis	X_q	0.861 pu
Transient: direct axis	X'_d	0.177 pu
Transient: quadrature axis	X'_q	0.228 pu
Subtransient: direct axis	X''_d	0.111 pu
Subtransient: quadrature axis	X''_q	0.199 pu
Time constants		
d: Transient	T'_d	0.018 s
d: Subtransient	T''_d	0.0045 s
q: Transient	T''_q	0.0045 s

FIGURE 2.19 Single-machine infinite-bus model. (From Pihl, E., Concentrated Solar Power prepared for the Energy Committee of the Royal Swedish Academy of Sciences, Technical Report, Chalmers University, Gothenburg, Sweden, 2009.)

In power systems, the frequency should remain within statutory limits, and, therefore, the generation has to meet the demand in order to ensure the safe operation of the system. Grid codes are country specific and define a certain band of over and under frequency limits so that frequency values remain close to the rated value [42].

Frequency variations result from disturbances such as short circuits, loss of generation units, or unbalances between production and demand. For instance, when the system experiences an unexpected generation loss or a connection of a large load, the frequency drops rapidly [43]. The dynamic characteristic of the power systems has impact on the rate of frequency decline, the depth of frequency drop, and the time for recovery to the prefault value when the formers are subject to a frequency event [44].

Normally, synchronous generators have inertia response that inherently and subsequently reacts to these frequency events by activating frequency control and by changing the power of their prime mover in governors. In more details, during an unexpected disturbance, synchronous machines and turbines will inject or absorb kinetic energy into/from the grid according to the incurred frequency deviation.

2.4.4.1 Inertial Response

The total kinetic energy E of a system's rotating masses, including spinning loads, is given by

$$E = \frac{1}{2} I \omega^2 \tag{2.6}$$

where

I is the moment of the system's inertia in kg·m²
ω is the rotational speed in rad/s [44]

The rate of change of rotational speed is dependent on torque balance of any spinning mass and yields from

$$T_m - T_e = I \frac{d\omega}{dt} \tag{2.7}$$

where T_m and T_e are the mechanical and electrical torques, respectively. The stored kinetic energy is often indicated as a proportional to its power rating (S_{rated}) and is called the inertia constant $H(s)$ as described by

$$H = \frac{\frac{1}{2} I \omega^2}{S_{rated}} \tag{2.8}$$

Consequently, the rate of speed/frequency change is given by

$$\frac{d\omega}{dt} = \frac{\omega^2 (T_m - T_e)}{2HS_{rated}} \tag{2.9}$$

and is expressed in per-unit as

$$\frac{d\omega}{dt} = \frac{(P_m - P_e)}{2HS_{rated}} \tag{2.10}$$

in which P_m and P_e are the mechanical and electrical torques (in per-unit), respectively.

The equilibrium of a synchronous machine can therefore be perturbed if the mechanical or electrical power is changed. For instance, in CSPs and due to the intermittent nature of solar energy higher than the inertia constant, the mechanical and electrical outputs may fluctuate and consequently affect the synchronous machine's equilibrium. The higher the inertia constant, the higher the system's resistance is to abrupt changes [44].

2.4.4.2 Electrical Power

The electrical output of the single-machine infinite-bus model is given by (2.11), where the maximum electrical power (P_{max}) depends on the reactance (X_{eq}) of the transmission line interfacing the generator to the grid, the machine's voltage (E), and the grid voltage (V). Hence, any change in one of these parameters affects the system's transient stability since the accelerative power described in (2.10) becomes nonzero [42, 43]:

$$P_e = \frac{|E||V|}{X_{eq}} \sin \delta = P_{max} \sin \delta \tag{2.11}$$

where
P_e is the electrical torque in per-unit
X_{eq} is the equivalent line reactance in per-unit
E and V are the machine's and grid voltage, respectively, in per-unit
P_{max} and δ are the maximum electrical power and rotor angle, respectively, in per-unit

2.4.4.3 Mechanical Power

The mechanical power in a steam-based CSP plant is based on the pressure that drives the turbine. The governor controller regulates the mechanical torque applied to the shaft of the generator by monitoring the amount of steam and pressure that passes and turns the turbine. Any pressure imbalance is mainly induced by abrupt changes in solar radiation; thus, the system's transient stability is affected. In some cases, a failure in governor controller may also lead to the increase of the mechanical torque of the synchronous machine. A solution to mitigate abrupt changes in mechanical torque is to equip the CSP with storage, since a storage system helps to maintain the steam-based pressure and hence the rotor speed within nominal values.

For instance, when a large disturbance occurs, such as a three-phase fault, the rotor speed increases giving a negative slip and starts to accelerate. Actually, there is a difference among the electrical and mechanical power, where the former tends to be zero (decrease).

Concluding distributed generation units such as CSP do not contribute to frequency control since they are generally equipped with power electronics interfaces that decouple electrically these devices from the grid. Consequently, they cannot deliver inertial response for short-term frequency stabilization.

2.4.5 CSP LARGE-SCALE INTEGRATION: TECHNICAL CHALLENGES AND IMPACTS

As it has been already mentioned that CSP plants are considered as conventional thermal power plants using standard models of synchronous generators, governors, exciters, etc. Generally, requirements for renewable energy system generators (including the provision of ancillary services) have to be adapted in order to facilitate their grid integration, in line with the provisions of the respective European network codes [44].

Moreover, in the medium to long term, additional measures have to be taken in order to increase the flexibility of the electricity systems, including measures to further develop the flexibility of the conventional generators, demand-side response, interconnections, and also storage options.

In brief, important challenges related to grid integration of large-scale CSP energy are the following [44]:

- Solar PV and CSP systems are variable in nature with no control over the energy source. This variability needs to be balanced by controllable generation units or storage systems or taken into account via demand management schemes.
- The optimal generation, reserve, and power exchange capacities must be redefined, taking into account that variable renewable generation may be zero for an extended time or at maximum value when the power demand is minimal.
- It necessitates a more dynamic operation of the power system, as a situation with variable/predictable demand and controllable generation is replaced by a situation where also the generation is variable and uncontrollable (but predictable). In other words, there is uncontrollable variability on both sides of the *generation=demand* equation.
- This kind of generation needs to be installed where the energy resources are good. In these places, the grid infrastructure is not necessarily strong, putting strain on the power system. A challenge is therefore to determine the limits and the best strategies for increasing these limits.
- Grid infrastructure planning depends on forecasted generation capacities, whereas development of a renewable energy plant depends on the grid infrastructure. This circular dependence represents an uncertainty for both grid and plant development.
- Grid code compliance is increasingly important for renewable energy plants, and therefore, the standardization and international harmonization of grid codes are important for simplified planning process.

PV (for modern installations) power plants have an interface to the electricity grid via power electronic converters. Thus, they interact very differently with the grid compared with conventional power plants based on directly connected synchronous generators and necessitating new control strategies.

Nevertheless, along with the technological development and the advancement in control, more solar energy systems will be seen in the future with mixed power grid, where the solar-powered systems should also perform various functionalities beyond basic power generation.

2.5 SUMMARY

In this chapter, the basic technological issues for solar power systems have been reviewed in order to cater for more installations of solar power generations. The solar power technologies include photovoltaic (PV), concentrated PV (CPV), and concentrated solar power (CSP). The first two technologies are based on the photovoltaic effect, and thus the modeling of the generating unit is identical, which has also been provided in this chapter. In contrast, the CSP using the solar energy is used to heat up thermal fluids that can drive an electrical generator or thermal–chemical reaction units. Then, the solar thermal energy is converted to electrical energy. Since the solar thermal plants work on identical principles as fossil fuel–based conventional power plants, the modeling of CSP systems is briefly discussed in this chapter. More important, through this technological review, it is known that the power electronics technology is vital for solar energy systems.

REFERENCES

1. Wikipedia, Solar energy, Online available: https://en.wikipedia.org/wiki/Solar_energy, Retrieved on May 17, 2016.
2. J. Nelson, *The Physics of Solar Cells*, London, U.K.: Imperial College Press, 2003.
3. NREL National Center for Photovoltaics, Best research-cell efficiencies for solar PV cells from NREL, Online available: http://www.nrel.gov/ncpv/, Retrieved on May 17, 2016.
4. J. Y. Ye, T. Reindl, A. G. Aberle, and T. M. Walsh, Effect of solar spectrum on the performance of various thin-film PV module technologies in tropical Singapore, *IEEE Journal of Photovoltaics*, 4(5), 1268–1274, September 2014.
5. J. Y. Ye, T. Reindl, A. G. Aberle, and T. M. Walsh, Performance degradation of various PV module technologies in tropical Singapore, *IEEE Journal of Photovoltaics*, 4(5), 1288–1294, September 2014.
6. Wikipedia, Concentrator photovoltaics, Online available: https://en.wikipedia.org/wiki/Concentrator_photovoltaics, Retrieved on May 17, 2016.
7. M. A. Green, Silicon photovoltaic modules: A brief history of the first 50 years, *Progress in Photovoltaics: Research and Applications*, 13(5), 447–455, 2005.
8. W. T. Jewell and R. Ramakumar, The history of utility-interactive photovoltaic generation, *IEEE Transactions on Energy Conversion*, 3(3), 583–588, September 1988.
9. M. A. Green, K. Emery, Y. Hishikawa, W. Warta, and E. D. Dunlop, Solar cell efficiency tables (version 47), *Progress in Photovoltaics*, 24(1), 3–11, November 2015.
10. M. G. Villalva, J. R. Gazoli, and E. Filho, Comprehensive approach to modeling and simulation of photovoltaic arrays, *IEEE Transactions on Power Electronics*, 24(5), 1198–1208, May 2009.
11. M. C. D. Piazza and G. Vitale, *Photovoltaic Sources: Modeling and Emulation*. London, U.K.: Springer-Verlag, 2013.
12. K. A. Kim, C. Xu, L. Jin, and P. T. Krein, Dynamic photovoltaic model incorporating capacitive and reverse-bias characteristics, *IEEE Journal of Photovoltaics*, 3(4), 1334–1341, October 2013.
13. R. J. Serna, B. J. Pierquet, J. Santiago, and R. C. N. Pilawa-Podgurski, Field measurements of transient effects in photovoltaic panels and its importance in the design of maximum power point trackers, in *Proceedings of IEEE APEC*, pp. 3005–3010, Charlotte, NC, March 2015.
14. K. A. Kim and P. T. Krein, Reexamination of photovoltaic hot spotting to show inadequacy of the bypass diode, *IEEE Journal of Photovoltaics*, 5(5), 1435–1441, September 2015.
15. E. L. Meyer and E. E. van Dyk, Assessing the reliability and degradation of photovoltaic module performance parameters, *IEEE Transactions on Reliability*, 53(1), 83–92, March 2004.
16. M. Vazquez and I. Rey-Stolle, Photovoltaic module reliability model based on field degradation studies, *Progress in Photovoltaics: Research and Applications*, 16(5), 419–433, 2008.
17. D. C. Jordan and S. R. Kurtz, Photovoltaic degradation rates: An analytical review, *Progress in Photovoltaics: Research and Applications*, 21(1), 12–29, January 2013.
18. J. Johnson and K. Armijo, Parametric study of PV arc-fault generation methods and analysis of conducted DC spectrum, in *Proceedings of IEEE PVSC*, pp. 3543–3548, Denver, CO, June 2014.

19. C. B. Jones, J. S. Stein, S. Gonzalez, and B. H. King, Photovoltaic system fault detection and diagnostics using Laterally Primed Adaptive Resonance Theory neural network, in *Proceedings of IEEE PVSC*, pp. 1–6, New Orleans, LA, June 2015.
20. M. K. Alam, F. Khan, J. Johnson, and J. Flicker, A comprehensive review of catastrophic faults in PV arrays: Types, detection, and mitigation techniques, *IEEE Journal of Photovoltaics*, 5(3), 982–997, May 2015.
21. K. A. Kim, G.-S. Seo, B.-H. Cho, and P. T. Krein, Photovoltaic hot-spot detection for solar panel substrings using AC parameter characterization, *IEEE Transactions on Power Electronics*, 31(2), 1121–1130, February 2016.
22. L. M. Fraas and L. D. Partain, *Solar Cells and Their Applications*, 2nd edn., Hoboken, NJ: Wiley, 2010.
23. S. Kurtz, Opportunities and challenges for development of a mature concentrating photovoltaic power industry, Technical Report, NREL/TP-5200-43208, National Renewable Energy Laboratory, 2012.
24. L. J. Young, Concentrator photovoltaics: The next step towards better solar power, *IEEE Spectrum*, Online available: http://spectrum.ieee.org/energywise/green-tech/solar/concentrator-photovoltaics-the-next-step-towards-better-solar-power, August 31, 2015.
25. N. Hayashi, A. Matsushita, D. Inoue, M. Matsumoto, T. Nagata, H. Higuchi, Y. Aya, and T. Nakagawa, Nonuniformity sunlight-irradiation effect on photovoltaic performance of concentrating photovoltaic using microsolar cells without secondary optics, *IEEE Journal of Photovoltaics*, 6(1), 350–357, January 2016.
26. R. J. Linderman, Z. S. Judkins, M. Shoecraft, and M. J. Dawson, Thermal performance of the SunPower Alpha-2 PV concentrator, *IEEE Journal of Photovoltaics*, 2(2), 196–201, April 2012.
27. H. Müller-Steinhagen and F. Trieb, Concentrating solar power. A review of the technology, Institute of Technical Thermodynamics, German Aerospace Centre, Stuttgart, Germany, February/March 2004. Online available, Retrieved on April 13, 2016 at http://www.trec-uk.org.uk/resources/ingenia_18_Feb_March_2004.pdf.
28. M. Mertins, 2009, Technische und wirtschaftliche Analyse von horizontalen Fresnel-Kollektoren, PhD thesis, Universitat Karlsruhe, Karlsruhe, Germany, October 2013, Online, Retrieved on April 13, 2016 at http://digbib.ubka.uni-karlsruhe.de/volltexte/documents/1067166.
29. M. Bolinger and S. Weaver, 2013, Utility-scale solar 2012: An empirical analysis of project cost, performance, and pricing trends in the United States, Technical Report, Environmental Energy Technologies Division, Lawrence Berkeley National Laboratory, Berkeley, CA.
30. M. Atif and F. A. Al-Sulaiman, Performance analysis of supercritical CO_2 Brayton cycles integrated with solar central receiver system, in *Proceedings of IEEE IREC*, pp. 1–6, Hammamet, Tunisia, 2014.
31. C. K. Ho and B. D. Iverson, Review of high-temperature central receiver designs for concentrating solar power, *Renewable and Sustainable Energy Reviews*, 29, 835–846, January 2014.
32. Y. Chu, *Review and Comparison of Different Solar Energy Technologies*, San Diego, CA: Global Energy Network Institute, August 2011.
33. S. A. Kalogirou, Solar thermal collectors and applications, *Progress in Energy and Combustion Science*, 30(3), 231–295, 2004.
34. P. Breeze et al., *Renewable Energy Focus Handbook*, Oxford, U.K.: Academic Press, Elsevier, pp. 333–401, 2009.
35. D. Laing, W. D. Steinmann, R. Tamme, and C. Richter, Solid media thermal storage for parabolic trough power plants, *Solar Energy*, 80(10), 1283–1289, 2006.
36. S. Relloso and Y. Gutierrez, Real application of molten salt thermal storage to obtain high capacity factors in parabolic trough plants, in *Proceedings of SolarPACES*, Las Vegas, NV, 2008.
37. M. J. Hale, Subcontractor report: Survey of thermal storage for parabolic trough power plants, Technical Report, NREL, National Renewable Energy Laboratory, Golden, CO, 2000.
38. A. M. Mohamad, N. Hashim, N. Hamzah, N. F. Nik Ismail, and M. F. Abdul Latip, Transient stability analysis on Sarawak's grid using power system simulator for engineering (PSS/E), in *Proceedings of IEEE ISIEA*, pp. 521–526, Langkawi, Malaysia, September 2011.
39. I. D. Margaris et al., Methods for evaluating penetration levels of wind generation in autonomous systems, in *Proceedings of IEEE Bucharest PowerTech Conference*, Bucharest, Romania, June 28–July 2, 2009.
40. N. Hemdan and M. Kurrat, Influence of distributed generation on different loadability aspects of electric distributions systems, in *Proceedings of 20th International Conference on Electricity Distribution (CIRED)*, Prague, Czech Republic, June 8–11, 2009.
41. M. Emranjeet and I. Syed, Stability considerations of distributed power systems with low inertia generators, in *Proceedings of International Conference on Communication, Conference and Power (ICCCP'07)*, Muscat, Oman, February 19–21, 2007.

42. E. Pihl, Concentrated Solar Power prepared for the Energy Committee of the Royal Swedish Academy of Sciences, Technical Report, Chalmers University, Gothenburg, Sweden, 2009.

43. R. Xezile, N. Mbuli, J. H. C. Pretorius and D. Matshidza, Network studies for the interconnection of a concentrating solar power plant in a weak distribution network in the upington area, in *Proceedings of IEEE Africon*, Livingstone, Zambia, September 13–15, 2011.

44. J. Licari, J. Ekanayake, and I. Moore, Inertia response from full-power converter-based permanent magnet wind generators, *Journal of Modern Power Systems and Clean Energy*, 1(1), 26–33, June 2013.

3 Overview of Single-Phase Grid-Connected Photovoltaic Systems

Yongheng Yang and Frede Blaabjerg

CONTENTS

Abstract...41
3.1 Introduction...41
3.2 Demands for Grid-Connected PV Systems 43
3.3 Power Converter Technology for Single-Phase PV Systems....................44
    3.3.1 Transformerless AC-Module Inverters (Module-Integrated PV Converters)..............44
    3.3.2 Transformerless Single-Stage String Inverters 47
    3.3.3 DC-Module Converters in Transformerless Double-Stage PV Systems 50
3.4 Control of Single-Phase Grid-Connected PV Systems..............................52
    3.4.1 General Control Objectives and Structures....................................52
    3.4.2 Grid Synchronization..53
    3.4.3 Operational Example 58
3.5 Summary..60
References..62

ABSTRACT

A continuous booming installation of solar photovoltaic (PV) systems has been witnessed worldwide. It is mainly driven by the imperative demand of "clean" power generation from renewables. Grid-connected PV systems will thus become an even more active player in the future mixed power systems, which are linked together by a vast of power electronics converters and the power grid. In order to achieve a reliable and efficient power generation from PV systems, more stringent demands have been imposed on the entire PV system. It, in return, advances the development of the power converter technology in PV systems. This chapter thus gives an overview of the advancement of power electronics converters in single-phase grid-connected PV systems, being commonly used in residential applications. Demands to single-phase grid-connected PV systems and the general control strategies are also addressed in this chapter.

3.1 INTRODUCTION

Traditional power generations based on fossil fuel resources are considered to be unsustainable in long-term national strategies. This has been one of the main driving forces for an extensive installation of renewable energies like wind power, solar photovoltaic (PV) power, hydropower, biomass power, geothermal power, and ocean power into public grids in the last decade [1, 2]. Among the major renewables, solar PV power generation has continued to be expanded at a rapid rate over the past several years, and it already plays a substantial role in electricity generation in some countries [3]. As an example, approximately 7.9% of the annual electricity demand was covered by solar PV systems throughout 2014 in Italy [3]. In Germany, the total installed capacity had reached 38.2 GW

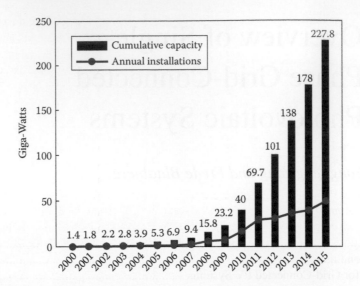

FIGURE 3.1 Evolution of global cumulative PV capacity from 2000 to 2015 based on the data available online from SolarPower Europe and PV-Insider. (Data from SolarPower Europe, Global market outlook 2015–2019, 2015, Available: http://www.solarpowereurope.tv/insights/global-market-outlook/, Retrieved on March 1, 2016; PV-Insider, US solar jobs to hike 15% in 2016, Global new capacity forecast at 65 GW, January 19, 2016, Available: http://analysis.pv-insider.com/us-solar-jobs-hike-15-2016-global-new-capacity-forecast-65-gw, Retrieved on March 1, 2016.)

by the end of 2014, where most of the PV systems are residential applications [3–5]. Figure 3.1 illustrates the worldwide solar PV capacity evolution in the past 15 years [5], which shows increasing worldwide expectations from energy production by means of solar PV power systems. Therefore, as the typical configuration for residential PV applications, the single-phase grid-connected PV systems have been in focus in this chapter in order to describe the technology catering for a desirable PV integration into the future mixed power grid.

The power electronics technology has been acknowledged to be an enabling technology for more renewable energies into the grid, including solar PV systems [7]. Associated by the advancements of power semiconductor devices [8], the power electronics part of entire PV systems (i.e., power converters) holds the responsibility for a reliable and efficient energy conversion out of the clean, pollution-free, and inexhaustible solar PV energy. As a consequence, a vast array of grid-connected PV power converters have been developed and commercialized widely [9–15].

However, the grid-connected PV systems vary significantly in size and power—from small-scale DC modules (a few hundred watts) to large-scale PV power plants (up to hundreds of megawatts). In general, the PV power converters can simply be categorized into module-level (AC-module inverter and DC-module converter), string, multistring, and central converters [9, 10]. The multistring and central converters are intensively used for solar PV power plants/farms as three-phase systems [16–18]. In contrast, the module and string converters are widely adopted in residential applications as single-phase systems [19, 20]. Although the PV power converters are different in configuration, the major functions of the power converters are the same, including PV power maximization, DC to AC power conversion and power transfer, synchronization, grid code compliance, reactive power control, and islanding detection and protection [7, 21]. It also requires advanced and intelligent controls to perform these PV features and also to meet customized demands, where the monitoring, forecasting, and communication technology can enhance the PV integration [18, 21].

As mentioned previously, PV systems are still dominant in the residential applications and will even be diversely spread out in the future mixed grid. Therefore, state-of-the-art developments of single-phase grid-connected PV systems are selectively reviewed in this chapter. The focus has been on power converter advancements in single-phase grid-connected PV systems for residential applications, which will be detailed in Section 3.3. First, demands from grid operators and consumers for single-phase PV systems are introduced in Section 3.2. In order to meet the increasing demand, the general control structures of single-phase grid-connected PV systems are discussed in Section 3.4 before the conclusion.

3.2 DEMANDS FOR GRID-CONNECTED PV SYSTEMS

The grid-connected PV systems are being developed at a very fast rate and will soon play a major role in power electricity generation in some areas [22, 23]. At the same time, demands (requirements) for PV systems as shown Figure 3.2 are increasing more than ever before. Although the power capacity of a PV system currently is still not comparable to that of an individual wind turbine system, similar demands for wind turbine systems are being transitioned to PV systems [18, 21] since the number of large-scale PV systems (power plants) is being continuously increased [24].

Nevertheless, the demands for PV systems can be specified at different levels. At the PV side, the output power of the PV panels/strings should be maximized, where a DC–DC converter is commonly used, being a double-stage PV system. This is also known as maximum power point tracking (MPPT). In this case, the DC voltage (DC-link voltage) should be maintained as a desirable value for the inverter. Moreover, for safety (e.g., fire), panel monitoring and diagnosis have to be enhanced at the PV side [25]. At the grid side, normally a desirable total harmonic distortion (THD) of the output current should be attained (e.g., lower than 5% [26]) for a good power quality. In the case of large-scale PV systems with higher power ratings, PV systems should not violate the grid voltage and the grid frequency by means of providing ancillary services (e.g., frequency regulation). Additionally, PV systems have to ride through grid faults (e.g., voltage sags and frequency variations), when a higher PV penetration level becomes a reality [18, 21, 27–33].

Since the power capacity per generating unit is relatively low but the cost of energy is relatively high, there is always a strong demand for high efficiency in order to reduce the cost of PV energy and also to optimize the energy yield. With respect to efficiency, the power electronics system (including passive components) accounts for most of the power losses in the entire PV system. Thus, possibilities to meet the efficiency demand include using advanced semiconductor devices, intelligent control, and power-lossless PV topologies. Transformerless PV technology is an example, and transformerless PV inverters can achieve a relatively high conversion efficiency when the isolation transformers are removed [11, 26]. However, minimizing the ground current in these transformerless

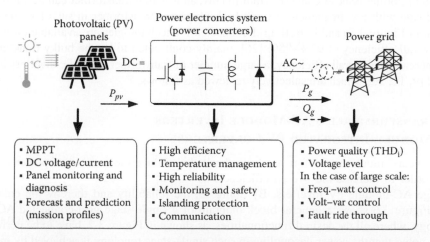

FIGURE 3.2 Demands (challenges) for a grid-connected PV system based on power electronics converters (DC, direct current; AC, alternating current; P_{pv}, PV output power; P_g, active power; Q_g, reactive power).

PV converters is mandatory for safety and reliability of the panels [11]. Moreover, reliability is of much importance in the power electronics–based PV systems, as also shown in Figure 3.2. This is motivated by extending the total energy production (service time), thereby further reducing the cost of energy [7, 34–36]. Finally, because of exposure or a smaller housing chamber, the PV converter system (power electronics system) must be more temperature insensitive (i.e., with temperature management), which will also be beneficial for the reliability performance. As has been illustrated in Figure 3.2, improving the monitoring, forecasting, and communication technologies will also be crucial to implementing future grid-friendly PV systems into a mixed power grid.

3.3 POWER CONVERTER TECHNOLOGY FOR SINGLE-PHASE PV SYSTEMS

According to the state-of-the-art technologies, there are mainly five configuration concepts [2, 9, 37, 38] to organize and transfer the PV power to the grid as is shown in Figure 3.3. Each grid-connected concept consists of a series of paralleled PV panels or strings, and they are configured by a couple of power electronics converters (DC–DC converters and DC–AC inverters) in accordance with the output voltage of the PV panels as well as the power rating.

A central inverter is normally used in a three-phase grid-connected PV plant with the power greater than tens of kWp, as it is shown in Figure 3.4. This technology can achieve a relatively high efficiency with a lower cost, but it requires high-voltage DC cables [9]. Besides, due to its low immunity to hot spots and partial shading on the panels, the power mismatch issue is significant in this concept (i.e., low PV utilization). In contrast, the MPPT control is achieved separately in each string of the string/multistring PV inverters, leading to a better total energy yield. However, there are still mismatches in the PV panels of each string, and the multistring technology requires more power electronics converters, resulting in further investments. Considering the issues mentioned earlier, the module converters (DC-module converters and/or AC-module inverters) are developed, there being a flexible solution for the PV systems of low power ratings and also for module-level monitoring and diagnostics. This module-integrated concept can minimize the effects of partial shadowing, module mismatch, and different module orientations, etc., since the module converter acts on a single PV panel with an individual MPPT control. However, a low overall efficiency is the main disadvantage in this concept due to the low power.

As it can be seen in Figure 3.3, the module concept, string inverter, and multistring inverters are the most common solutions used in single-phase PV applications, where the galvanic isolation for safety is an important issue of concern. Traditionally, an isolation transformer can be adopted either at the grid side with low frequencies or as a high-frequency transformer in such PV converters as it is shown in Figure 3.5a and b. Both grid-connected PV technologies are available on the market with an overall efficiency of 93%–95% [26], mainly contributed to by the bulky transformers. In order to increase the overall efficiency, a large number of transformerless PV converters have been developed [9, 11, 26], which are selectively reviewed as follows.

3.3.1 TRANSFORMERLESS AC-MODULE INVERTERS (MODULE-INTEGRATED PV CONVERTERS)

In the last years, much more effort has been devoted to reduce the number of power conversion stages in order to increase the overall efficiency, as well as to increase the power density of the single-stage AC-module PV inverters. By doing so, the reliability and thereby the cost may be reduced. Figure 3.6 shows a general block diagram of a single-stage grid-connected AC-module PV topology, where all the desired functionalities, as shown in Figure 3.2, have to be performed. It should be noted that the power decoupling in such single-stage topology is achieved by means of a DC-link capacitor, C_{DC}, in parallel with the PV module [9, 11].

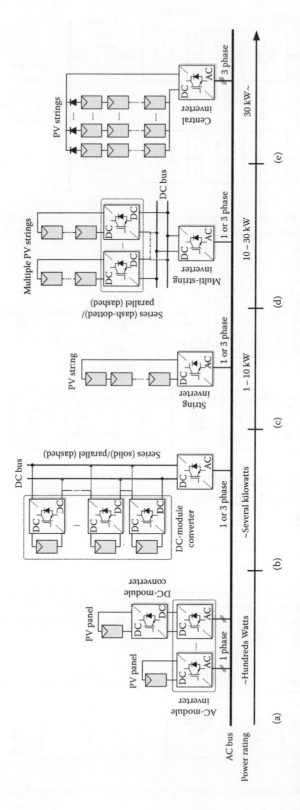

FIGURE 3.3 Grid-connected PV system concepts for: (a) small systems/residential, (b) small systems/residential, (c) residential, (d) commercial/residential, and (e) commercial/utility-scale PV applications.

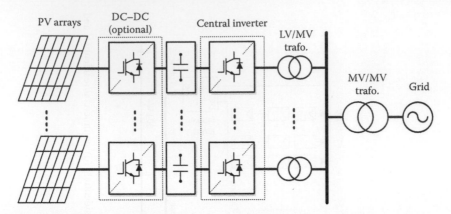

FIGURE 3.4 Large-scale PV power plant/station based on central inverters for utility applications, where multilevel converters can be adopted as central inverters to manage even higher power of up to tens of megawatts (MW).

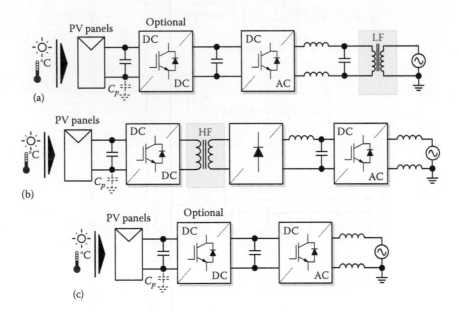

FIGURE 3.5 Single-phase grid-connected PV systems, where the AC-module inverters, the string inverters, and the multistring inverters are commonly used: (a) with a low-frequency (LF) transformer, (b) with a high-frequency (HF) transformer, and (c) without transformers. Note that C_p represents the parasitic capacitor between the PV panel and the ground.

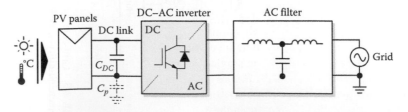

FIGURE 3.6 General block diagram of a single-stage single-phase PV topology (AC-module/string inverter system) with an AC filter.

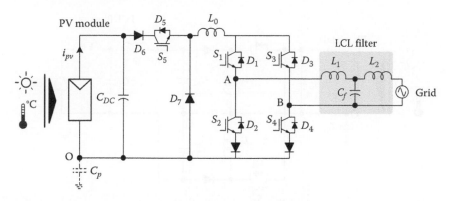

FIGURE 3.7 A universal single-stage grid-connected AC-module inverter with an LCL filter. (Based on the concept proposed by Prasad, B.S. et al., *IEEE Trans. Energy Conver.*, 23(1), 128, 2008.)

Since the power of a single PV module is relatively low and is strongly dependent on the ambient conditions (i.e., solar irradiance and ambient temperature), the trend for AC-module inverters is to integrate either a boost or a buck–boost converter into a full-bridge (FB) or half-bridge (HB) inverter in order to achieve an acceptable DC-link voltage [39–45]. As it is presented in [39], a single-stage module-integrated PV converter can operate in a buck, boost, or buck–boost mode with a wide range of PV panel output voltages. This AC-module inverter is shown in Figure 3.7, where an LCL filter is used to achieve a satisfactory THD of the injected current to the grid. A variant of the AC-module inverter has been introduced in [40], which is actually a mix of a boost converter and an FB inverter. The main drawback of the integrated boost AC-module inverter is that it may introduce a zero-crossing current distortion. In order to solve this issue, the buck–boost AC-module inverters are preferable [41–44].

Figure 3.8 shows two examples of the buck–boost AC-module inverter topologies for single-phase grid-connected PV applications. In the AC-module inverter, as it is shown in Figure 3.8a, each of the buck–boost converters generates a DC-biased unipolar sinusoidal voltage, which is 180° out of phase to the other in such a manner as to alleviate the zero-cross current distortions. Similar principles are applied to the buck–boost-integrated FB inverter, which operates for each half-cycle of the grid voltage. However, as it is shown in Figure 3.8b, this AC-module inverter is using a common source.

In addition to the topologies mentioned earlier, which are mainly based on two relatively independent DC–DC converters integrated in an inverter, alternative AC-module inverters are also proposed in the literature. Most of these solutions are developed in accordance with the impedance–admittance conversion theory and an impedance network [46–52]. The Z-source inverter is one example, which is able to boost up the voltage for an FB inverter by adding an *LC* impedance network, as it is exemplified in Figure 3.9. Notably, the Z-source inverter was mostly used in three-phase applications in the past.

3.3.2 Transformerless Single-Stage String Inverters

The AC-module inverters discussed earlier with an integration of a DC–DC boosting converter are suitable for use in low power applications. When it comes to higher power ratings (e.g., 1–5 kWp), the compactness of AC-module inverters is challenged. In such applications, the most commonly used inverter topology is the single-phase FB string inverter due to its simplicity in terms of less power switching devices. Figure 3.10 depicts the hardware schematics of a single-phase FB string inverter with an LCL filter for better power quality. It is also shown in Figure 3.10 that a leakage current will circulate in the transformerless topology, requiring a specifically designed modulation scheme to minimize it. Conventional modulation methods for the single-stage FB string inverter topology include a bipolar modulation, a unipolar modulation, and a

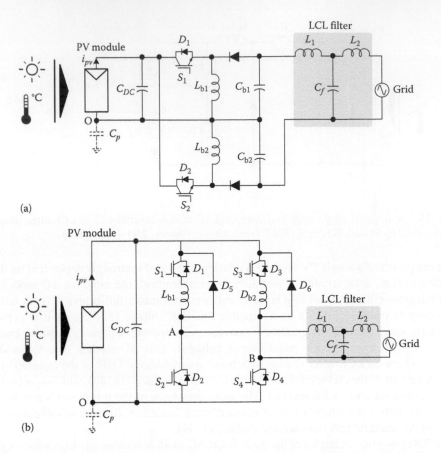

(a)

(b)

FIGURE 3.8 Buck–boost-integrated AC-module inverters with an LCL filter: (a) differential buck–boost inverter and (b) buck–boost-integrated FB inverter. ([a]: Based on the concept proposed by Vazquez, N. et al., *Proceedings of Annual IEEE Power Electronics Specialists Conference*, 1999, pp. 801–806; [b]: Based on the concept proposed by Wang, C.-M., *IEEE Trans. Power Electron.*, 19(1), 150, 2004.)

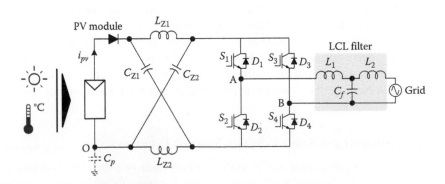

FIGURE 3.9 Single-phase Z-source-based AC-module inverter with an LCL filter.

hybrid modulation [26]. Considering the leakage current injection, the bipolar modulation scheme is preferable [26, 53]. Notably, optimizing the modulation patterns is another alternative to eliminate the ground (leakage) currents [54].

Transformerless structures are mostly derived from the FB topology by providing an AC path or a DC path by using additional power switching devices. This will result in isolation between the

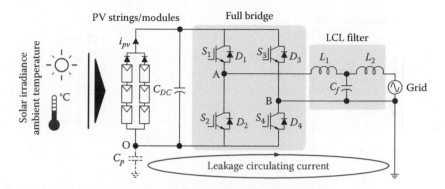

FIGURE 3.10 Single-phase transformerless FB string inverter with an LCL filter, which also indicates the ground current circulating in the path through the parasitic capacitor C_p from the panels.

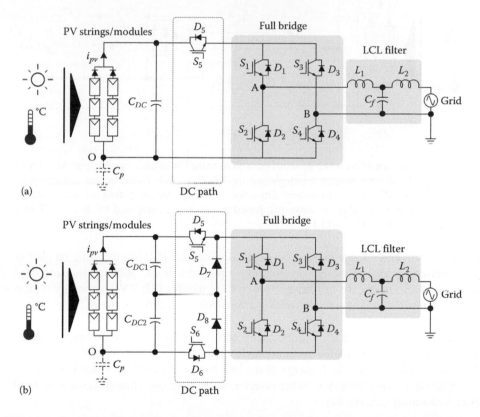

FIGURE 3.11 Transformerless string inverters derived from the FB inverter by adding a DC path: (a) H5 inverter topology and (b) H6 inverter topology. ([a]: Based on the concept proposed by Victor, M. et al., Method of converting a direct current voltage from a source of direct current voltage, more specifically from a photovoltaic source of direct current voltage, into an alternating current voltage, U.S. Patent 20050286281 A1, December 29, 2005; [b]: Based on the concept proposed by Gonzalez, R. et al., *IEEE Trans. Power Electron.*, 22(2), 693, 2007.)

PV modules and the grid during the zero-voltage states, thus leading to a low leakage current injection. Figure 3.11 shows two examples of single-stage transformerless PV inverters derived from the single-phase FB topology by providing a DC path [55, 56]. Thanks to the extra DC bypass, the PV strings/panels are isolated from the grid at zero-voltage states. Alternatively, the isolation can be achieved at the grid side by means of adding an AC path. As it is exemplified in Figure 3.12a,

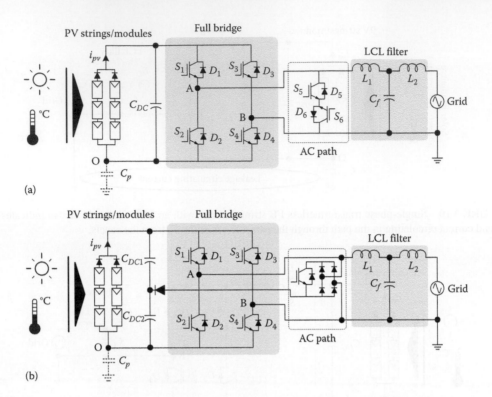

FIGURE 3.12 Transformerless string inverters derived from the FB inverter by adding an AC path: (a) highly efficient and reliable inverter concept topology and (b) FB-ZVR. ([a]: Based on the concept proposed by Schmidt, H. et al., DC/AC converter to convert direct electric voltage into alternating voltage or into alternating current, U.S. Patent 7046534, May 16, 2006; [b]: Based on the concept proposed by Kerekes, T. et al., *IEEE Trans. Ind. Electron.*, 58(1), 184, 2011.)

the highly efficient and reliable inverter concept string inverter [57] has the same number of power switching devices as that of the H6 inverter, but it provides an AC path to eliminate the leakage current injection. Similarly, the full-bridge zero-voltage rectifier (FB-ZVR) topology is proposed in [58], where the isolation is attained by adding a zero-voltage rectifier at the AC side, as it is shown in Figure 3.12b.

It should be pointed out that there are many other transformerless topologies reported in the literature in addition to the solutions mentioned previously by means of adding an extra current path [26, 54, 59–62]. Taking the Conergy string inverter as an example, which is shown in Figure 3.13, a single-phase transformerless string inverter can be developed based on the multilevel power converter technology [26, 60, 61].

3.3.3 DC-MODULE CONVERTERS IN TRANSFORMERLESS DOUBLE-STAGE PV SYSTEMS

A major drawback of the single-stage PV topologies is that the output voltage range of the PV panels/strings is limited especially in the low power applications (e.g., AC-module inverters), which thus will affect the overall efficiency. The double-stage PV technology can solve this issue since it consists of a DC–DC converter that is responsible for amplifying the voltage of the PV module to a desirable level for the inverter stage. Figure 3.14 shows the general block diagram of a double-stage single-phase PV topology. The DC–DC converter also performs the MPPT control of the PV panels, and thus, extended operating hours can be achieved in a double-stage PV system. The DC-link capacitor C_{DC} shown in Figure 3.14 is used for power decoupling, while the PV capacitor C_{pv} is used for filtering.

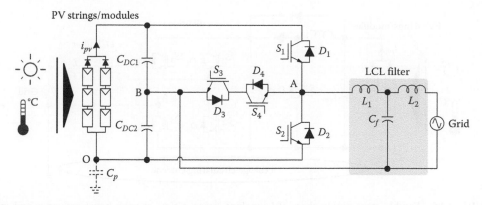

FIGURE 3.13 A transformerless string inverter derived from a neutral point–clamped (NPC) topology—The Conergy NPC inverter with an LCL filter. (Based on the concept proposed by Knaup, P., Inverter, International Patent Application, Publication Number: WO 2007/048420 A1, Issued May 3, 2007.)

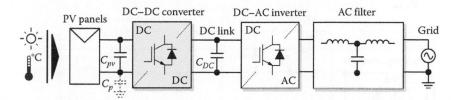

FIGURE 3.14 General block diagram of a double-stage single-phase PV topology (with a DC–DC converter).

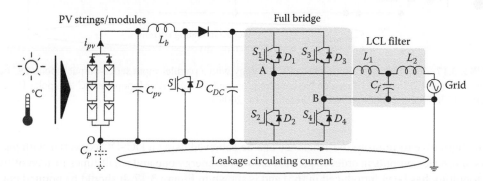

FIGURE 3.15 Conventional double-stage single-phase PV topology consisting of a boost converter and a full-bridge inverter with an LCL filter.

In general, the DC–DC converter can be included between the PV panels and the DC–AC inverters. The inverters can be the string inverters as discussed earlier or a simple half-bridge inverter. The following illustrates the double-stage PV technology consisting of a DC–DC converter and an FB string inverter. Figure 3.15 shows a conventional double-stage single-phase PV system, where the leakage current needs to be minimized as well. However, incorporating a boost converter will decrease the overall conversion efficiency. Thus, variations of the double-stage configuration have been introduced by means of a time-sharing boost converter or a soft-switched boost converter [63, 64]. The time-sharing boost converter shown in Figure 3.16 is a dual-mode converter, where the switching and conduction losses are reduced, leading to a satisfactory efficiency.

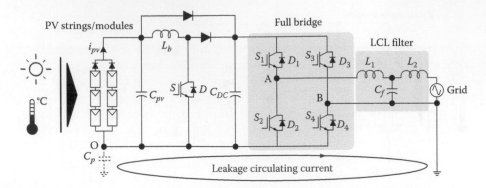

FIGURE 3.16   Double-stage single-phase PV topology using a time-sharing boost converter and an FB inverter with an LCL filter. (Based on the concept proposed by Ogura, K. et al., *Proceedings of IEEE PESC*, 2010, pp. 4763–4767.)

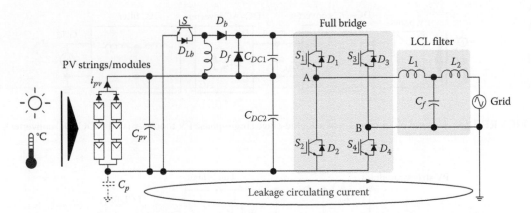

FIGURE 3.17   Double-stage single-phase PV topology with a parallel-input series-output bipolar DC output converter and an FB inverter. (Based on the concept proposed by West, R., Monopolar DC to bipolar DC to AC converter, U.S. Patent 2008/0037305 A1, February 14, 2008.)

An alternative to improve the efficiency can be achieved using a DC–DC converter with parallel inputs and series outputs in order to process the source energy one and a half times instead of twice. This topology has been introduced in [65] and is shown in Figure 3.17. It should be pointed out that the voltage step-up gain of the DC–DC converter is also improved at the same time. In addition, the impedance network–based DC–DC converters (e.g., the Z-source and Y-source networks) might be the other promising solutions for single-phase double-stage PV systems, due to the high step-up voltage gain [66–69], which might be beneficial in some applications.

3.4   CONTROL OF SINGLE-PHASE GRID-CONNECTED PV SYSTEMS

3.4.1   GENERAL CONTROL OBJECTIVES AND STRUCTURES

The control objectives of a single-phase grid-connected PV system [70] can be divided into two major parts: (1) PV-side control with the purpose to maximize the power from PV panels and (2) grid-side control performed on the PV inverters with the purpose of fulfilling the demands to the power grid as shown in Figure 3.2. A conventional control structure for such a grid-connected PV system thus consists of a two-cascaded loop in order to fulfill the demands/requirements [32, 70]—the outer power/voltage control loop generates the current references and the inner control loop is responsible for shaping the current, so the power quality is maintained, and also it might perform various functionalities, as shown in Figure 3.18.

Figure 3.19 shows the general control structure of a single-phase single-stage grid-connected PV system, where the PV inverter has to handle the fluctuating power (i.e., MPPT control) and also to control the injected current according to the specifications shown in Figure 3.18. As it can be observed in Figure 3.19, the control can be implemented in both stationary and rotating reference frames in order to control the reactive power exchange with the grid, where the Park transformation ($dq \rightarrow \alpha\beta$) or inverse Park transformation ($\alpha\beta \rightarrow dq$) is inevitable [70]. In terms of simplicity, the control in the stationary reference frame ($\alpha\beta$-reference frame) is preferable, but it requires an orthogonal system to generate a "virtual" system, which is in quadrature to the real grid. In the dq-reference frame, the MPPT control gives the active power reference for the power control loop based on proportional integral (PI) controllers, which then generate the current references as shown in Figure 3.19b. The current controller (CC) in the dq-reference frame can be PI controllers, but current decoupling is required in order to alleviate the interactive impact of the d-axis and q-axis currents in the synchronous rotating reference frame (i.e., the dq-reference frame). In contrast, enabled by the single-phase PQ theory [32, 71], the reference grid current i_g^* can be calculated using the power references and the in-quadrature voltage system. In that case, the PI controller will give an error in the controlled grid current. The controller (CC) should be designed in the $\alpha\beta$-reference frame. For example, a proportional resonant (PR) controller, a repetitive controller, or a deadbeat controller [70, 72–75] can directly be adopted as the CC as shown in Figure 3.19c.

Notably, since the CC is responsible for the current quality, it should be taken into account in the controller design and the filter design (e.g., using high-order passive filter, LCL filter). By introducing harmonic compensators [26, 32, 70, 72] and adding appropriate damping for the high-order filter [76, 77], an enhancement of the CC tracking performance can be achieved.

Similar control strategies can be applied to the double-stage system, as shown in Figure 3.20. The difference lies in that the MPPT control is implemented on the DC–DC converter, while the other functionalities are performed on the control of the PV inverter. There are other control solutions available for single-phase grid-connected PV systems [78–80]. For example, the instantaneous power is controlled in [79], where the synthesis of the power reference is a challenge; in [80], a one-cycle control method has been applied to single-stage single-phase grid-connected PV inverters for low power applications.

3.4.2 GRID SYNCHRONIZATION

It should be noted that the injected grid current is demanded to be synchronized with the grid voltage as required by the standards in this field [70]. As a result, grid synchronization is an essential grid monitoring task that will strongly contribute to the dynamic performance and the stability of the entire control system. The grid synchronization is even challenged in single-phase systems, as there is only one variable (i.e., the grid voltage) that can be used for synchronization. Nevertheless, different methods to extract the grid voltage information have been developed in recent studies [26, 70, 81–84] like the zero-crossing method, the filtering of grid voltage method, and the phase-locked loop (PLL) techniques, which are important solutions.

Figure 3.21 shows the structure of the PLL-based synchronization system. It can be observed that the PLL system contains a phase detector (PD), namely, to detect the phase difference, a PI-based loop filter (PI-LF) to smooth the frequency output, and finally a voltage-controlled oscillator (VCO). Accordingly, the transfer function of the PLL system [26, 84, 85] can be obtained as

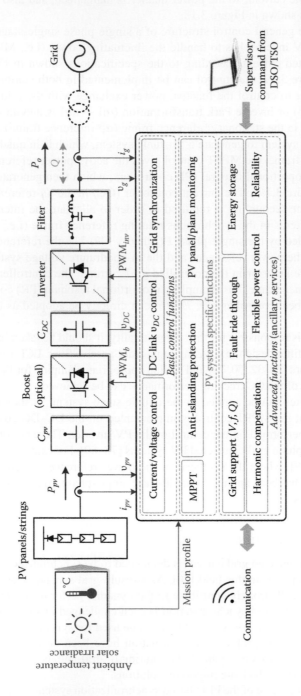

FIGURE 3.18 General control blocks (control objectives) of a grid-connected single-phase PV system (PWM, pulse width modulation; MPPT, maximum power point tracking).

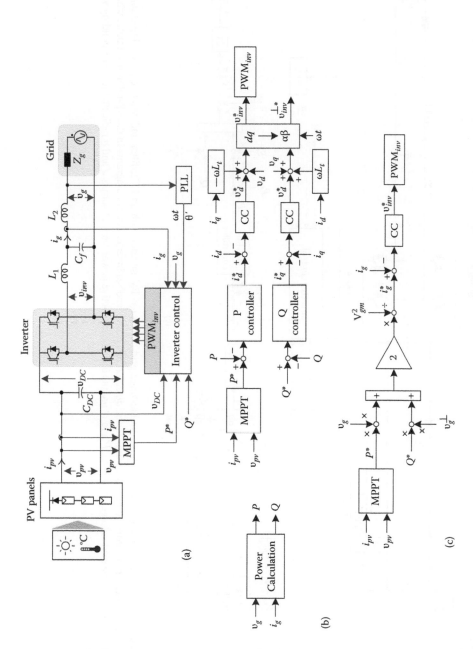

FIGURE 3.19 General control structure of a single-phase single-stage FB PV system with an LCL filter and reactive power injection capability (CC, PLL): (a) hardware schematics, (b) control block diagrams in the dq-reference frame, and (c) control block diagrams in the $\alpha\beta$-reference frame, where L_t is the total inductance ($L_t = L_1 + L_2$), V_{gm} is the grid voltage amplitude, Z_g is the grid impedance, P and Q are the active and reactive power, and subscripts d and q indicate the corresponding voltage or current in the dq-reference frame.

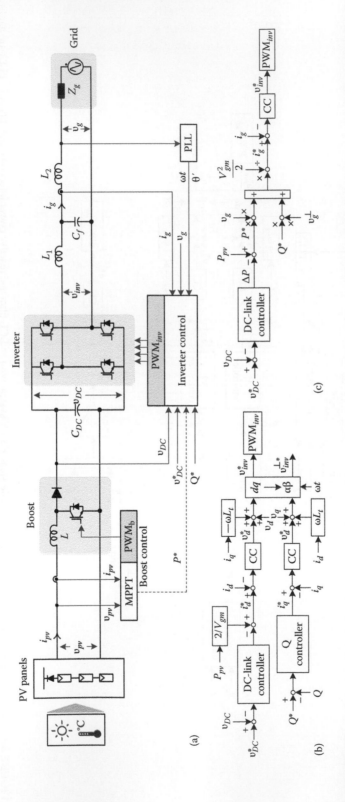

FIGURE 3.20 General control structure of a single-phase double-stage grid-connected PV inverter with an LCL filter and reactive power injection capability (CC, PLL): (a) hardware schematics, (b) inverter control diagrams in the dq-reference frame, and (c) inverter control diagrams in the $\alpha\beta$-reference frame, where L_t is the total inductance ($L_t = L_1 + L_2$), V_{gm} is the grid voltage amplitude, P and Q are the active and reactive power, and subscripts d and q indicate the corresponding voltage or current in the dq-reference frame.

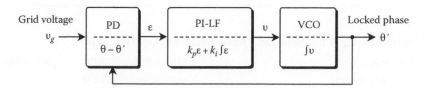

FIGURE 3.21 Basic structure of a PLL system, where $v_g = V_{gm}\cos\theta$ with θ being the phase of the grid voltage and θ' is the output (locked) phase.

$$G_{pll}(s) = \frac{\theta'(s)}{\theta(s)} = \frac{k_p s + k_i}{s^2 + k_p s + k_i} \tag{3.1}$$

which is a typical second-order system with k_p and k_i being the proportional and integral gain of the PI-LF, respectively. Subsequently, the corresponding damping ratio ζ and undamped natural frequency ω_n can be obtained, respectively, as

$$\zeta = \frac{k_p}{2\sqrt{k_i}} \quad \text{and} \quad \omega_n = \sqrt{k_i} \tag{3.2}$$

which can be used to tune the PI-LF parameters according to the desired settling time and resultant overshoot. More details about single-phase PLL synchronization techniques are directed to [26].

In literature, a vast array of PLL-based synchronization schemes has been reported [26, 70, 79, 81–88], while the major difference among the various PLL systems lies in the configuration of the PD unit. The most straightforward way is to use a sinusoidal multiplier [85], where the output contains a double-line-frequency term that requires more efforts to design a low-pass filter to filter it out. Thus, in prior-art PLL synchronization systems, more advanced PD techniques are adopted. Figure 3.22 exemplifies three possibilities for phase detection, namely,

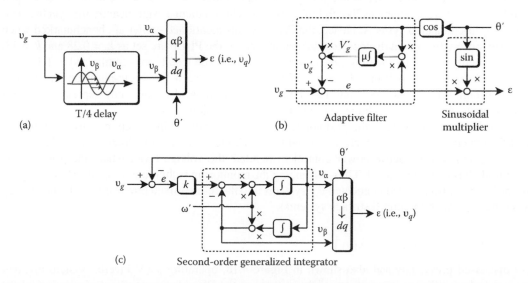

FIGURE 3.22 Phase detection structure of the three PLL systems for single-phase systems: (a) the T/4 delay PLL, (b) the EPLL, and (c) the SOGI-PLL, where $v_g = V_g\cos(\theta) = V_g\cos(\omega t)$ is the input grid voltage, V_g is the input voltage amplitude, ω is the grid angular frequency, μ and k are the control parameters, ε is the phase error, $v_{\alpha\beta}$ is the voltage component in the $\alpha\beta$-reference frame, v_q is the q-axis voltage in the dq-reference frame, and "′" indicates the corresponding estimated component of the input voltage signal v_g.

TABLE 3.1

Benchmarking of the Selected Phase-Locked Loop Techniques

PLL Technique		T/4 Delay PLL	EPLL	SOGI-PLL
Main design parameter		N_d unity delays	$\mu = 250$ s^{-1}	$k = 1.41$
Voltage sag (0.45 p.u.)	Settling time (ms)	4.7	7.8	8
	Frequency error (Hz)	0.26	0.91	0.62
Phase jump (+90°)	Settling time (ms)	75	120	72
	Frequency error (Hz)	16.1	16	19.1
Frequency jump (+1 Hz)	Settling time (ms)	Oscillate	186	111
	Frequency error (Hz)	(−1.2, 1.2)	8.4	10.4
Harmonic immunity		×	✓	✓
Implementation complexity		✓	✓✓	✓✓✓

Notes:

1. ×, no such capability; ✓, the more the better or the more complicated.
2. $N_d = f_s/f_g/4$ with f_s being the sampling frequency and f_g the grid fundamental frequency. The T/4 Delay PLL is normally implemented by cascading a number (N_d) of unit delay (z^{-1}) in digital control systems, as z^{-N_d}.

forming the T/4 Delay PLL [26, 84], the enhanced PLL (EPLL) [87], and the second-order generalized integrator (SOGI)-based PLL system [88]. It can be seen in Figure 3.22 that the phase detection of the T/4 Delay PLL and the SOGI-PLL systems is enabled by the Park transformation ($\alpha\beta \rightarrow dq$), where a virtual voltage component v_β in quadrature with the input grid voltage v_g (v_α) is also generated. In contrast, the EPLL adopts an adaptive filter and a sinusoidal multiplier to detect the phase error ε.

Practically, the grid voltage is not purely sinusoidal, and it may be distorted or it may be sagged due to various severe situations like lightning strikes. This challenges significantly the synchronization of grid-connected systems. Hence, the PLL systems mentioned earlier are benchmarked when the grid suffers from disturbances, and the parameters of the PI-LF are set as $k_p = 0.28$ and $k_i = 13$, which roughly results in a settling time of 100 ms. Comparison results are summarized in Table 3.1. The benchmarking reveals that the SOGI-PLL is a good solution for single-phase applications in terms of high tracking accuracy and fast dynamic, where the grid voltage may experience various disturbances, for example, voltage sags and frequency variations.

In addition to the phase of the grid voltage, other grid condition information is also very important for the control system to perform special functionalities, for example, low-voltage ride through [27, 32], and flexible active power control to regulate the voltage level, where the grid voltage amplitude has to be monitored. Thus, advancing the monitoring technology is another key to a grid-friendly integration of grid-connected PV systems into the future mixed power grid and other energy systems (e.g., with integrated storage systems).

3.4.3 OPERATIONAL EXAMPLE

As discussed previously and also shown in Figure 3.18, operating a PV inverter system involves the control of different components of the system: at the PV side for power maximization, at the inverter side for proper injection of a high-quality current (power), and at the grid side for ancillary services [7, 26, 70]. In all cases, the PV panels are exposed to varying environmental conditions,

TABLE 3.2

Parameters of the 3 kW Two-Stage Photovoltaic Inverter System

Parameter	Symbol	Value
Grid voltage (root mean square [RMS])	V_{gRMS}	230 V
Grid frequency	ω_g	314 rad/s
Grid impedance	Z_g (L_g, R_g)	0.5 mH, 0.2 Ω
Boost inductor	L	2 mH
DC-link capacitor	C_{DC}	2200 μF
DC-link voltage reference	v^*_{DC}	400 V
LCL filter	L_1	1.8 mH
	C_f	2.35 μF
	L_2	1.8 mH
Boost converter switching frequency	f_b	20 kHz
Inverter switching frequency	f_{inv}	10 kHz

which are also referred to as mission profiles (i.e., solar irradiance level and ambient temperature). The mission profile should be considered in the design and planning phases of a PV system, since it affects the availability of PV energy. Hereafter, referring to Figure 3.20, the MPPT operation is demonstrated on a 3-kW double-stage single-phase PV system grid, where the grid-side control has been implemented in the αβ-reference frame (i.e., Figure 3.20c). A PI controller is adopted as the DC-link controller, and a PR controller is used as the CC with a resonant harmonic compensator (RHC). A SOGI-PLL is employed to generate the in-quadrature voltage system (i.e., for synchronization). The system and control parameters are given in Table 3.2.

For the MPPT control, a perturb and observe (P&O) method [13, 89–91] has been adopted for simplicity, as it is given in Figure 3.23a (see also Chapter 5 for more details). The entire CC is shown in Figure 3.23b. As it can be seen, the P&O MPPT algorithm gives the reference voltage for the PV panels (i.e., the voltage at the maximum power point), which is controlled by a proportional controller k_m. In this chapter, k_m is designed as 0.00126. Since there might be background distortions in the grid voltage, the current quality should be enhanced by incorporating a harmonic compensator, which is an RHC as mentioned earlier. Actually, the RHC is a summation of multiple resonant controllers, whose central frequencies are placed at the targeted harmonics (e.g., the third, fifth, and seventh harmonics), as shown in Figure 3.23b. The CC and the RHC are designed as $k_{pr} = 22$, $k_{pi} = 2000$, $k_{pi}^3 = 1200$, $k_{pi}^5 = 800$, and $k_{pi}^7 = 200$. Simulation models have been built up in MATLAB®/Simulink®/Simscape, as shown in Figure 3.24. The simulation results are shown in Figure 3.25, where the solar irradiance has experienced step changes under a constant ambient temperature of 25 °C, and the system was operated at unity power factor according to demands.

The step changes in the solar irradiance are actually to reflect the PV intermittency, which leads to a continuous injection of fluctuating PV power, as it is shown in Figure 3.25. The fluctuation could be even severer during daily operations, for example, cloudy days with passing clouds. Nevertheless, the P&O MPPT algorithm shown in Figure 3.23a can effectively maximize the power production from PV panels. Moreover, the use of a boost converter can extend the operating hours for PV systems, when the solar irradiance is very weak.

Additionally, in order to test the performance of the CC, a grid with background distortions has been considered. This results in a THD of 3.4% for the grid voltage. Hence, the PR controller with

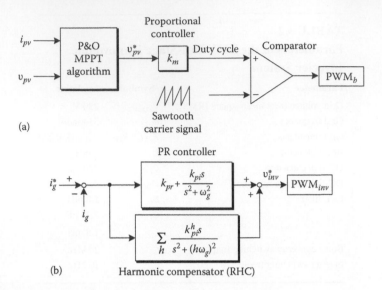

FIGURE 3.23 Details of the MPPT and CCs for a double-stage single-phase PV system (see Figure 3.20b): (a) MPPT control structure for the boost converter and (b) CC with a harmonic compensator, where k_m, k_{pr}, k_{pi}, and k_{pi}^h are the controller parameters.

the RHC is enabled, which can maintain a high-quality grid current. Specifically, when the RHC is absent, the resultant THD of the grid current is around 3.6% at the nominal operating condition. In contrast, when the harmonic compensator is added in parallel with the PR controller, the harmonics in the grid current have been suppressed, thus leading to a THD of 2.1%. It is suggested using the provided model to identify the difference as an exercise.

3.5 SUMMARY

In this chapter, a review of the recent technology of single-phase PV systems has been conducted. Demands for the single-phase PV systems, including the grid-connected standards, the solar PV panel requirements, the ground current requirements, the efficiency, and the reliability of the single-phase PV converters have been emphasized. Since achieving higher conversion efficiency is always of intense interest, both the transformerless single-stage (with an integrated DC–DC converter) and the transformerless double-stage (with a separate DC–DC converter) PV topologies have been in focus. The review reveals that the AC-module single-stage PV topologies are not very suitable for use in the European grid, since it is difficult to achieve a desirable voltage, and thus, the daily operating time is limited. The AC-module inverter concept is very flexible for small PV units with lower power ratings. In contrast, the string inverters are gaining greater popularity in Europe due to the higher efficiency that they are able to achieve. Especially, when a DC–DC boost stage is included, the MPPT control becomes more convenient, and the operating time of the PV systems is then extended. Finally, the general control structures for both single-stage and double-stage transformerless PV systems are presented, as well as a brief discussion on the synchronization and monitoring technologies. Operational examples are also provided at the end of this chapter.

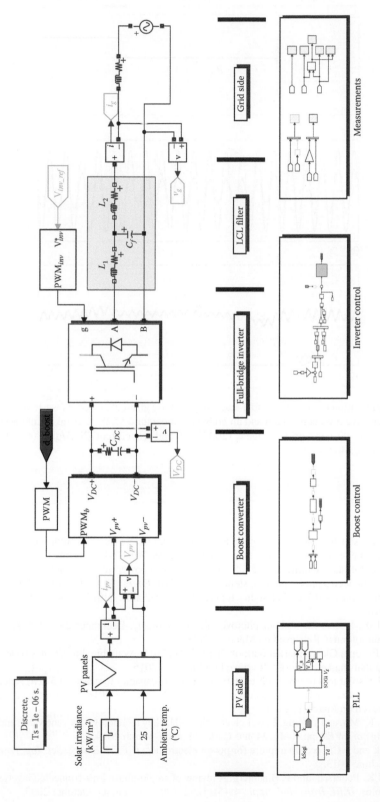

FIGURE 3.24 MATLAB®/Simulink® model of a double-stage single-phase grid-connected PV system.

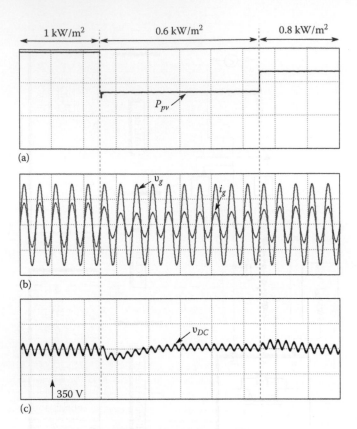

FIGURE 3.25 Simulation results of a double-stage single-phase grid-connected PV system, where solar irradiance step changes have been considered and an ambient temperature of 25 °C is taken into account: (a), PV output power P_{pv} (1 kW/div); (b), grid voltage v_g (200 V/div) and grid current (20 A/div); (c), DC-link voltage v_{DC} (25 V/div) and time (40 ms/div).

REFERENCES

1. O. Ellabban, H. Abu-Rub, and F. Blaabjerg, Renewable energy resources: Current status, future prospects and their enabling technology, *Renew. Sust. Energy Rev.*, 39, 748–764, November 2014.
2. L. Hassaine, E. OLias, J. Quintero, and V. Salas, Overview of power inverter topologies and control structures for grid connected photovoltaic systems, *Renew. Sust. Energy Rev.*, 30, 796–807, February 2014.
3. REN21, Renewables 2015 global status report, 2015. Available: http://www.ren21.net. Retrieved on March 1, 2016.
4. Fraunhofer ISE, Recent facts about photovoltaics in Germany, December 25, 2015. Available: http://www.ise.fraunhofer.de/. Retrieved on March 1, 2016.
5. SolarPower Europe, Global market outlook 2015–2019, 2015. Available: http://www.solarpowereurope.tv/insights/global-market-outlook/. Retrieved on March 1, 2016.
6. PV-Insider, US solar jobs to hike 15% in 2016; Global new capacity forecast at 65 GW, January 19, 2016. Available: http://analysis.pv-insider.com/us-solar-jobs-hike-15-2016-global-new-capacity-forecast-65-gw. Retrieved on March 1, 2016.
7. F. Blaabjerg, K. Ma, and Y. Yang, Power electronics—The key technology for renewable energy systems, in *Proceedings of EVER*, pp. 1–11, Monte-Carlo, Monaco, March 2014.
8. J.D. van Wyk and F.C. Lee, On a future for power electronics, *IEEE J. Emerg. Sel. Top. Power Electron.*, 1(2), 59–72, June 2013.
9. S.B. Kjaer, J.K. Pedersen, and F. Blaabjerg, A review of single-phase grid-connected inverters for photovoltaic modules, *IEEE Trans. Ind. Appl.*, 41(5), 1292–1306, September–October 2005.

10. H.J. Bergveld, D. Buthker, C. Castello, T. Doorn, A. de Jong, R. Van Otten, and K. de Waal, Module-level DC/DC conversion for photovoltaic systems: The delta-conversion concept, *IEEE Trans. Power Electron.*, 28(4), 2005–2013, April 2013.

11. D. Meneses, F. Blaabjerg, O. García, and J.A. Cobos, Review and comparison of step-up transformerless topologies for photovoltaic AC-module application, *IEEE Trans. Power Electron.*, 28(6), 2649–2663, June 2013.

12. B. Liu, S. Duan, and T. Cai, Photovoltaic DC-building-module-based BIPV system—Concept and design considerations, *IEEE Trans. Power Electron.*, 26(5), 1418–1429, May 2011.

13. D. Sera, L. Mathe, and F. Blaabjerg, Distributed control of PV strings with module integrated converters in presence of a central MPPT, in *Proceedings of IEEE ECCE*, pp. 1–8, Pittsburgh, PA, September 14–18, 2014.

14. M.A. Eltawil and Z. Zhao, Grid-connected photovoltaic power systems: Technical and potential problems—A review, *Renew. Sust. Energy Rev.*, 14(1), 112–129, January 2010.

15. F. Spertino and G. Graditi, Power conditioning units in grid-connected photovoltaic systems: A comparison with different technologies and wide range of power ratings, *Sol. Energy*, 108, 219–229, October 2014.

16. M. Meinhardt and G. Cramer, Multi-string-converter: The next step in evolution of string-converter technology, in *Proceedings of EPE*, pp. P.1–P.9, Graz, Austria, 2001.

17. SMA, SUNNY CENTRAL—High tech solution for solar power stations, Products category. Available: http://www.sma-america.com/. Retrieved on March 1, 2016.

18. M. Morjaria, D. Anichkov, V. Chadliev, and S. Soni, A grid-friendly plant: The role of utility-scale photovoltaic plants in grid stability and reliability, *IEEE Power Energy Mag.*, 12(3), 87–95, May–June 2014.

19. Y. Xue, K.C. Divya, G. Griepentrog, M. Liviu, S. Suresh, and M. Manjrekar, Towards next generation photovoltaic inverters, in *Proceedings of IEEE ECCE*, pp. 2467–2474, Phoenix, AZ, September 17–22, 2011.

20. Q. Li and P. Wolfs, A review of the single phase photovoltaic module integrated converter topologies with three different DC link configurations, *IEEE Trans. Power Electron.*, 23(3), 1320–1333, May 2008.

21. Y. Yang, P. Enjeti, F. Blaabjerg, and H. Wang, Wide-scale adoption of photovoltaic energy: Grid code modifications are explored in the distribution grid, *IEEE Ind. Appl. Mag.*, 21(5), 21–31, September–October 2015.

22. C. Winneker, World's solar photovoltaic capacity passes 100-gigawatt landmark after strong year, [Online], February 2013. Available: http://pr.euractiv.com/pr/worlds-solar-photovoltaic-capacity-passes-100-gigawatt-landmark-after-strong-year-93110. Retrieved on October 26, 2016.

23. M. Braun, T. Stetz, R. Brundlinger, C. Mayr, K. Ogimoto, H. Hatta, H. Kobayashi et al., Is the distribution grid ready to accept large-scale photovoltaic deployment? State of the art, progress, and future prospects, *Prog. Photovolt. Res. Appl.*, 20(6), 681–697, 2012.

24. *Renew Economy*, Global utility-scale solar capacity climbs through 21 GW in 2013, January 23, 2014. Available: http://reneweconomy.com.au/.

25. S. Spataru, D. Sera, F. Blaabjerg, L. Mathe, and T. Kerekes, Firefighter safety for PV systems: Overview of future requirements and protection systems, in *Proceedings of IEEE ECCE*, pp. 4468–4475, Denver, CO, September 15–19, 2013.

26. R. Teodorescu, M. Liserre, and P. Rodriguez, *Grid Converters for Photovoltaic and Wind Power Systems*. Hoboken, NJ, Wiley, 2011.

27. C.H. Benz, W.-T. Franke, and F.W. Fuchs, Low voltage ride through capability of a 5 kW grid-tied solar inverter, in *Proceedings of EPE/PEMC*, pp. T12-13–T12-20, Ohrid, Macedonia, September 6–8, 2010.

28. X. Bao, P. Tan, F. Zhuo, and X. Yue, Low voltage ride through control strategy for high-power grid-connected photovoltaic inverter, in *Proceedings of IEEE APEC*, pp. 97–100, Long Beach, CA, 2013.

29. H.-C. Chen, C.-T. Lee, P.T. Cheng, R. Teodorescu, F. Blaabjerg, and S. Bhattacharya, A flexible low-voltage ride-through operation for the distributed generation converters, in *Proceedings of PEDS'13*, pp. 1354–1359, Kitakyushu, Japan, April 22–25, 2013.

30. N.P. Papanikolaou, Low-voltage ride-through concept in flyback inverter based alternating current photovoltaic modules, *IET Power Electron.*, 6(7), 1436–1448, August 2013.

31. Y. Bae, T.-K. Vu, and R.-Y. Kim, Implemental control strategy for grid stabilization of grid-connected PV system based on German grid code in symmetrical low-to-medium voltage network, *IEEE Trans. Energy Convers.*, 28(3), 619–631, September 2013.

32. Y. Yang, F. Blaabjerg, and H. Wang, Low voltage ride-through of single-phase transformerless photovoltaic inverters, *IEEE Trans. Ind. Appl.*, 50(3), 1942–1952, May/June 2014.

33. G. Dotter, F. Ackermann, N. Bihler, R. Grab, S. Rogalla, and R. Singer, Stable operation of PV plants to achieve fault ride through capability—Evaluation in field and laboratory tests, in *Proceedings of IEEE PEDG*, pp. 1–8, Galway, Ireland, June 24–27, 2014.

34. H. Wang, M. Liserre, and F. Blaabjerg, Toward reliable power electronics: Challenges, design tools, and opportunities, *IEEE Ind. Electron. Mag.*, 7(2), 17–26, June 2013.
35. Fraunhofer ISE, Levelized cost of electricity–Renewable energy technologies, November 2013. Available: http://www.ise.fraunhofer.de/. Retrieved on March 1, 2016.
36. U.S. Energy Information Administration, Annual energy outlook 2014—With projections to 2040, April 2014. Available: http://www.eia.gov/forecasts/aeo/. Retrieved on March 1, 2016.
37. Solarpraxis and Sunbeam Communications, Inverter, storage and PV system technology 2013, July 2013. Available: http://www.pv-system-tech.com/. Retrieved on March 1, 2016.
38. E. Romero-Cadaval, G. Spagnuolo, I. Garcia Franquelo, C.A. Ramos-Paja, T. Suntio, and W.M. Xiao, Grid-connected photovoltaic generation plants: Components and operation, *IEEE Ind. Electron. Mag.*, 7(3), 6–20, September 2013.
39. B.S. Prasad, S. Jain, and V. Agarwal, Universal single-stage grid-connected inverter, *IEEE Trans. Energy Convers.*, 23(1), 128–137, March 2008.
40. L.G. Junior, M.A.G. de Brito, L.P. Sampaio, and C.A. Canesin, Single stage converters for low power stand-alone and grid-connected PV systems, in *Proceedings of IEEE ISIE*, pp. 1112–1117, Gdansk, Poland, 2011.
41. R.O. Caceres and I. Barbi, A boost DC–AC converter: Analysis, design, and experimentation, *IEEE Trans. Power Electron.*, 14(1), 134–141, January 1999.
42. S. Funabiki, T. Tanaka, and T. Nishi, A new buck-boost-operation-based sinusoidal inverter circuit, in *Proceedings of IEEE PESC*, pp. 1624–1629, Cairns, Queensland, Australia, 2002.
43. N. Vazquez, J. Almazan, J. Alvarez , C. Aguilar, and J. Arau, Analysis and experimental study of the buck, boost and buck-boost inverters, in *Proceedings of IEEE PESC*, pp. 801–806, Charleston, SC, 1999.
44. C. Wang, A novel single-stage full-bridge buck-boost inverter, *IEEE Trans. Power Electron.*, 19(1), 150–159, 2004.
45. S. Jain and V. Agarwal, A single-stage grid connected inverter topology for solar PV systems with maximum power point tracking, *IEEE Trans. Power Electron.*, 22(5), 1928–1940, 2007.
46. S. Yatsuki, K. Wada, T. Shimizu , H. Takagi, and M. Ito, A novel AC photovoltaic module system based on the impedance-admittance conversion theory, in *Proceedings of IEEE PESC*, pp. 2191–2196, Vancouver, British Columbia, Canada, 2001.
47. Y. Tang, S. Xie, and C. Zhang, Single-phase Z-source inverter, *IEEE Trans. Power Electron.*, 26(12), 3869–3873, 2011.
48. D. Li, P.C. Loh, M. Zhu, F. Gao, and F. Blaabjerg, Generalized multicell switched-inductor and switched-capacitor Z-source inverters, *IEEE Trans. Power Electron.*, 28(2), 837–848, 2013.
49. A.H. Rajaei, M. Mohamadian, S.M. Dehghan, and A. Yazdian, Single-phase induction motor drive system using z-source inverter, *IET Electr. Power Appl.*, 4(1), 17–25, January 2010.
50. L. Huang, M. Zhang, L. Hang, W. Yao, and Z. Lu, A family of three-switch three-state single-phase Z-source inverters, *IEEE Trans. Power Electron.*, 28(5), 2317–2329, May 2013.
51. H. Liu, G. Liu, Y. Ran, G. Wang, W. Wang, and D. Xu, A modified single-phase transformerless Z-source photovoltaic grid-connected inverter, *J. Power Electron.*, 5(5), September 2015.
52. Y.P. Siwakoti, F.Z. Peng, F. Blaabjerg, P.C. Loh, and G.E. Town, Impedance-source networks for electric power conversion Part I: A topological review, *IEEE Trans. Power Electron.*, 30(2), 699–716, February 2015.
53. Y. Yang, H. Wang, F. Blaabjerg, and K. Ma, Mission profile based multi-disciplinary analysis of power modules in single-phase transformerless photovoltaic inverters, in *Proceedings of EPE*, pp. 1–10, Lille, France, September 2013.
54. N. Achilladelis, E. Koutroulis, and F. Blaabjerg, Optimized pulse width modulation for transformerless active-NPC inverters, in *Proceedings of EPE*, pp. 1–10, Lappeenranta, Finland, August 26–28, 2014.
55. M. Victor, F. Greizer, S. Bremicker, and U. Hubler, Method of converting a direct current voltage from a source of direct current voltage, more specifically from a photovoltaic source of direct current voltage, into a alternating current voltage, U.S. Patent 20050286281 A1, December 29, 2005.
56. R. Gonzalez, J. Lopez, P. Sanchis, and L. Marroyo, Transformerless inverter for single-phase photovoltaic systems, *IEEE Trans. Power Electron.*, 22(2), 693–697, March 2007.
57. H. Schmidt, C. Siedle, and J. Ketterer, DC/AC converter to convert direct electric voltage into alternating voltage or into alternating current, U.S. Patent 7046534, May 16, 2006.
58. T. Kerekes, R. Teodorescu, P. Rodriguez, G. Vazquez, and E. Aldabas, A new high-efficiency single-phase transformerless PV inverter topology, *IEEE Trans. Ind. Electron.*, 58(1), 184–191, January 2011.
59. I. Patrao, E. Figueres, F. González-Espín, and G. Garcerá, Transformerless topologies for grid-connected single-phase photovoltaic inverters, *Renew. Sustain. Energy Rev.*, 15(7), 3423–3431, September 2011.

60. R. Gonzalez, E. Gubia, J. Lopez, and L. Marroyo, Transformerless single-phase multilevel-based photovoltaic inverter, *IEEE Trans. Ind. Electron.*, 55(7), 2694–2702, July 2008.

61. P. Knaup, Inverter, International Patent Application, Publication Number: WO 2007/048420 A1, Issued May 3, 2007.

62. Y. Zhou, W. Huang, P. Zhao, and J. Zhao, A transformerless grid-connected photovoltaic system based on the coupled inductor single-stage boost three-phase inverter, *IEEE Trans. Power Electron.*, 29(3), 1041–1046, March 2014.

63. K. Ogura, T. Nishida, E. Hiraki, M. Nakaoka, and S. Nagai, Time-sharing boost chopper cascaded dual mode single-phase sinewave inverter for solar photovoltaic power generation system, in *Proceedings of IEEE PESC*, pp. 4763–4767, Aachen, Germany, 2004.

64. Y. Kim, J. Kim, Y. Ji, C. Won, and Y. Jung, Photovoltaic parallel resonant dc-link soft switching inverter using hysteresis current control, in *Proceedings of APEC*, pp. 2275–2280, Palm Springs, CA, 2010.

65. R. West, Monopolar DC to bipolar DC to AC converter, U.S. Patent 2008/0037305 A1, February 14, 2008.

66. Y.P. Siwakoti, P.C. Loh, F. Blaabjerg, S.G. Andreasen, and G.E. Town, Y-source boost DC/DC converter for distributed generation, *IEEE Trans. Ind. Electron.*, 62(2), 1059–1069, February 2015.

67. Y.P. Siwakoti, F. Blaabjerg, P.C. Loh, and G.E. Town, High-voltage boost quasi-Z-source isolated DC/DC converter, *IET Power Electron.*, 7(9), 2387–2395, September 2014.

68. R.-J. Wai, C.-Y. Lin, R.-Y. Duan, and Y.-R. Chang, High-efficiency DC–DC converter with high voltage gain and reduced switch stress, *IEEE Trans. Ind. Electron.*, 54(1), 354–364, February 2007.

69. W.-Y. Choi, J.-S. Yoo, and J.-Y. Choi, High efficiency dc-dc converter with high step-up gain for low PV voltage sources, in *Proceedings of ICPE-IEEE ECCE Asia*, pp. 1161–1163, Jeju, Korea, May 30, 2011–June 3, 2011.

70. F. Blaabjerg, R. Teodorescu, M. Liserre, and A.V. Timbus, Overview of control and grid synchronization for distributed power generation systems, *IEEE Trans. Ind. Electron.*, 53(5), 1398–1409, October 2006.

71. M. Saitou and T. Shimizu, Generalized theory of instantaneous active and reactive powers in single-phase circuits based on Hilbert transform, in *Proceedings of IEEE PESC*, Vol. 3, pp. 1419–1424, Cairns, Queensland, Australia, 2002.

72. M. Ciobotaru, R. Teodorescu, and F. Blaabjerg, Control of single-stage single-phase PV inverter, in *Proceedings of EPE*, pp. P.1–P.10, Dresden, Germany, 2005.

73. Y. Yang, K. Zhou, H. Wang, F. Blaabjerg, D. Wang, and B. Zhang, Frequency adaptive selective harmonic control for grid-connected inverters, *IEEE Trans. Power Electron.*, 30(7), 3912–3924, July 2015.

74. T. Jakub, P. Zdenek, and B. Vojtech, Central difference model predictive current control of single-phase H-bridge inverter with LCL filter, in *Proceedings of EPE*, pp. 1–8, Lille, France, September 2–6, 2013.

75. J.-F. Stumper, V. Hagenmeyer, S. Kuehl, and R. Kennel, Deadbeat control for electrical drives: A robust and performant design based on differential flatness, *IEEE Trans. Power Electron.*, 30(8), 4585–4596, August 2015.

76. R. Pena-Alzola, M. Liserre, F. Blaabjerg, M. Ordonez, and Y. Yang, LCL-filter design for robust active damping in grid-connected converters, *IEEE Trans. Ind. Inf.*, 10(4), 2192–2203, November 2014.

77. W. Wu, Y. Sun, M. Huang, X. Wang, H. Wang, F. Blaabjerg, M. Liserre, and H.S.-H. Chung, A robust passive damping method for LLCL-filter-based grid-tied inverters to minimize the effect of grid harmonic voltages, *IEEE Trans. Power Electron.*, 29(7), 3279–3289, July 2014.

78. C.-H. Chang, Y.-H. Lin, Y.-M. Chen, and Y.-R. Chang, Simplified reactive power control for single-phase grid-connected photovoltaic inverters, *IEEE Trans. Ind. Electron.*, 61(5), 2286–2296, 2014.

79. S.A. Khajehoddin, M. Karimi-Ghartemani, A. Bakhshai, and P. Jain, A power control method with simple structure and fast dynamic response for single-phase grid-connected DG systems, *IEEE Trans. Power Electron.*, 28(1), 221–233, January 2013.

80. E.S. Sreeraj, K. Chatterjee, and S. Bandyopadhyay, One-cycle-controlled single-stage single-phase voltage-sensorless grid-connected PV system, *IEEE Trans. Ind. Electron.*, 60(3), 1216–1224, March 2013.

81. B.P. McGrath, D.G. Holmes, and J.J.H. Galloway, Power converter line synchronization using a discrete Fourier transform (DFT) based on a variable sample rate, *IEEE Trans. Power Electron.*, 20(4), 877–884, July 2005.

82. W. Li, X. Ruan, C. Bao, D. Pan, and X. Wang, Grid synchronization systems of three-phase grid-connected power converters: A complex-vector-filter perspective, *IEEE Trans. Ind. Electron.*, 61(4), 1855–1870, April 2014.

83. B. Liu, F. Zhuo, Y. Zhu, H. Yi, and F. Wang, A three-phase PLL algorithm based on signal reforming under distorted grid conditions, *IEEE Trans. Power Electron.*, 30(9), 5272–5283, September 2015.

84. Y. Yang, L. Hadjidemetriou, F. Blaabjerg, and E. Kyriakides, Benchmarking of phase locked loop based synchronization techniques for grid-connected inverter systems, in *Proceedings of ICPE-IEEE ECCE Asia*, pp. 2167–2174, Seoul, Korea, June 2015.

85. Y. Yang, Advanced control strategies to enable a more wide-scale adoption of single-phase photovoltaic systems. PhD thesis, Department of Energy Technology, Aalborg University, Aalborg, Denmark. August 2014.

86. Y. Han, M. Luo, X. Zhao, J.M. Guerrero, and L. Xu, Comparative performance evaluation of orthogonal-signal-generators-based single-phase PLL algorithms—A survey, *IEEE Trans. Power Electron.*, 31(5), 3932–3944, May 2016.

87. M. Karimi-Ghartemani and M.R. Iravani, Robust and frequency-adaptive measurement of peak value, *IEEE Trans. Power Deliv.*, 19(2), 481–489, April 2004.

88. M. Ciobotaru, R. Teodorescu, and F. Blaabjerg, A new single-phase PLL structure based on second order generalized integrator, in *Proceedings of IEEE PESC*, pp. 1–6, Jeju, South Korea, June 2006.

89. Plexim GmbH, PLECS user manual, Ver. 3.7. Available: http://www.plexim.com/download, January 7, 2016.

90. E. Koutroulis, K. Kalaitzakis, and N.C. Voulgaris, Development of a microcontroller-based, photovoltaic maximum power point tracking control system, *IEEE Trans. Power Electron.*, 16(1), 46–54, January 2001.

91. N. Femia, G. Petrone, G. Spagnuolo, and M. Vitelli, Optimization of perturb and observe maximum power point tracking method, *IEEE Trans. Power Electron.*, 20(4), 963–973, July 2005.

4 Three-Phase Photovoltaic Systems: Structures, Topologies, and Control

Tamás Kerekes, Dezső Séra, and László Máthé

CONTENTS

Abstract 67
4.1   Introduction 67
4.2   PV Inverter Structures 69
4.3   Three-Phase PV Inverter Topologies 71
4.4   Control Building Blocks for PV Inverters 72
    4.4.1   Modulation Strategies for Three-Phase PV Inverters 72
    4.4.2   Implementation of the Modulation Strategies 74
    4.4.3   Grid Synchronization 76
    4.4.4   Implementation of the PLLs for Grid Synchronization 78
    4.4.5   Current Control 81
    4.4.6   Implementation of the Current Controllers 82
    4.4.7   Maximum Power Point Tracking 82
    4.4.8   Grid Integration Functions for PV Systems 84
    4.4.9   Internal Active Power Reserve Management 85
    4.4.10 Active Power Ramp Limitation 86
4.5   Summary 87
References 87

ABSTRACT

Photovoltaic (PV) technology has experienced an unprecedented growth in the last two decades, transforming from mainly an off-grid niche generation to a major renewable energy technology, reaching approximately 227 GW of capacity worldwide at the end of 2015 with a predicted extra 50 GW of new installations for 2016 [1]. Large PV power plants interfacing the grid through a three-phase power electronic converter are now well on the way to become a major player in the power system in many countries. Therefore, this chapter gives an overview of PV systems with a focus on three-phase applications from a hardware point of view, detailing the different PV inverter structures and topologies and discussing the different control layers within a grid-connected PV plant. Modulation schemes for various PV inverter topologies, grid synchronization, current control, active and reactive power control, maximum power point tracking, and grid integration requirements and support functions are also reviewed.

4.1   INTRODUCTION

Renewable energy has become very important both worldwide and on the European market, mainly due to the decrease in the photovoltaic (PV) system cost (up to 75%) during the last decade. PV has been the top energy resource in Europe in terms of new installations in 2012, with above 16 GW of new added capacity, overtaking wind energy installation that was below 12 GW. During 2014, the PV technology lost some momentum, with only 7 GW of added capacity being the second in the line, as reported by SolarPower Europe (formerly European Photovoltaic Industry Association [EPIA]) [2].

PV has been one of the fastest growing energy technologies during the last years, thanks to the booming of the European market until its saturation in 2012. During 2014, markets outside Europe increased, especially in China and the Asia-Pacific region, as reported by EPIA. As a consequence, at the end of 2015, there were over 222 GW of PV systems installed worldwide [2].

For 2016, it is reported that the global installed capacity has increased by 50 GW, and by 2019, the worldwide capacity might reach 396 GW in the case of low installation scenario, while the optimistic forecast is 540 GW [2].

Ground-mounted PV plants have reached power levels of several hundreds of MW, and during the last years, these systems no longer worked in an "install and forget" mode, meaning that grid-connected PV systems injected all the available energy into the electrical network. Presently, extra services for supporting the grid with reactive power is a requirement even for systems in the kW range, making PV inverters a key component in the energy transformation chain of solar power.

PV inverters convert the DC power supplied by the PV panels into grid-synchronized AC power that can be injected into the electrical network. In the past, PV inverters were only passive elements in the network, working as negative loads. The injected power was proportional with the available sunshine, supplying only active power to the electrical grid. This passive behavior led to grid stability issues, mostly in low-voltage (LV) residential or rural areas, where PV penetration was very high, thereby overwhelming the LV distribution lines and transformers and resulting in an increase of the voltage in the distribution system level. This led to several outcomes and corresponding key requirements for the PV inverters [3–5]:

- The PV inverter was required to disconnect from the grid due to the fact that the grid voltage (V_g) has risen above 110% of the nominal value (V_{gn}). Standards and grid codes refer to the voltage range for normal operation of converters connected to the grid, stating that converters can only stay connected to the grid in case the grid voltage is within $0.85 * V_{gn} < V_g < 1.1 * V_{gn}$.
- The PV inverter was required to reduce the injection of active power with a rate of 40% of the actual power per 1 Hz difference from the nominal grid frequency and the output power can only be increased again in case the grid frequency reaches 50.05 Hz in the case of a nominal frequency of 50 Hz.
- The PV inverter was required to support the grid with reactive power. The amount of reactive power can be calculated from the actual active power or based on the actual grid voltage measurement.

Future PV systems should take advantage of the intelligence that can be built into the power electronic converters, thereby making renewable energy systems influential players on the energy market.

A general control scheme of the three-phase grid-connected PV inverter is shown in Figure 4.1. In case the nominal PV string voltage does not reach the peak line-to-line grid voltage, a DC–DC boost converter is added, forming a dual-stage PV inverter. For larger systems, (with sufficiently high string voltages) a single-stage DC–AC inverter is controlling the power flow from the PV panels to the grid by sensing the most important voltage and current signals.

In the next section, different converter structures and topologies are presented. This is then followed by explaining the pulse width modulation (PWM) techniques applied for the listed topologies. Synchronization is indispensable in case of grid-connected applications, and this is discussed in Section 4.4.2. The reference voltage for the PWM is supplied from a current control block, which is discussed in Section 4.4.5, followed by a presentation of different maximum power point tracking (MPPT) techniques.

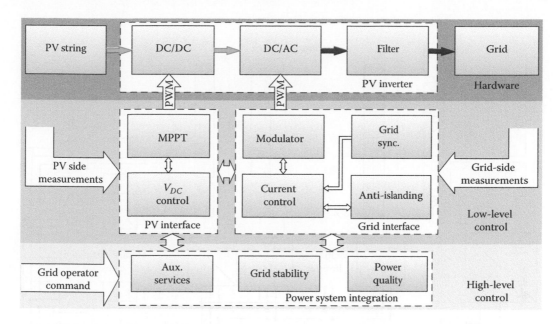

FIGURE 4.1 Three-phase grid-connected PV system, showing the hardware layer from the PV generator (PV string) to the grid; in the bottom the two control loops—Low-level control mainly related to the hardware, high-level control related mainly to the power system.

Finally, grid integration aspects with focus on grid code requirements and PV-specific functions are presented, followed by conclusions.

4.2 PV INVERTER STRUCTURES

Depending on the size of the PV plant, one can talk about the three different structures, as presented in Figure 4.2. Central inverters (CI) are usually connected to several parallel-connected PV strings, thereby having a PV array made up of several thousands of PV panels connected to the DC link of the inverter. This solution would offer the cheapest €/kW price for the inverter, but the MPPT will not be optimum for all the PV strings and the annual energy production (AEP) will suffer. CI can be found for power ranges from 100 kW to 2 MW size in a modular design, where 4–10 inverters are connected in parallel to the same DC input and the same AC output. In this way, based on the actual input power, only the required number of inverters will operate, thereby improving the efficiency of the entire PV system.

The tracking efficiency of PV systems can be improved by connecting each PV string to a separate DC input, either by having string inverters or by having a multistring inverter with several DC/DC converters and a common DC/AC stage. These inverter structures are also called minicentral inverters (MCI). With the MCI structure, each PV string has an individual MPPT that changes the working point of the PV string based on the individual atmospheric conditions, leading to an increased AEP and energy yield. The string topology structure is used within a power range of 1–10 kW. The multistring topology structure is also used in case of MCIs up to power ranges of 60 kW. This way MW-sized PV plants can be designed with a modular approach, making PV plant design flexible and simple to build. In case of plant extension or maintenance, only a single PV section needs to be disconnected, while the rest of the PV plant is still actively connected to the network, thereby reducing the unnecessary energy loss to a minimum.

To further improve the AEP of PV systems, one could use module-integrated converters (MICs). In this configuration, each PV panel has its own converter, thereby optimizing the tracking of

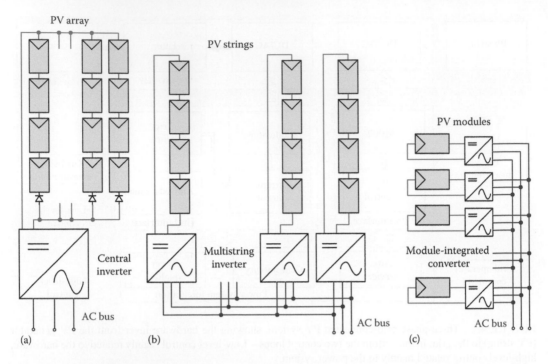

FIGURE 4.2 Different inverter structures for grid-connected PV applications: (a) central inverter structure (CI), (b) string and multistring inverter structure (MCI), and (c) module-integrated structure (MIC, microinverter).

each individual PV panel according to the atmospheric conditions, partial shading conditions, etc. Typically, these MICs are in a power range of 100 W up to 300 W and it can consist of a DC–DC or DC–AC conversion stage, depending on the application. Nevertheless, from a financial point of view, the MIC structure has the highest cost in the €/kW scale.

Although most of the microinverter systems today are single-phase ones, due to the lower power level and the possibility of having lower DC-link voltage, three-phase microinverters are also emerging [6–9]. An important disadvantage of the single-phase systems is that the power pulsates at the double the rate of the grid frequency. In order to smooth this power pulsation, a large (typically electrolytic) decoupling capacitor is necessary in the DC bus, which negatively affects the reliability of the system [10, 11]. Three-phase microinverters avoid this drawback at the expense of a higher required DC bus voltage [12, 13] and hence a higher DC–DC boost ratio and, of course, a larger number/size of components, which also increases the costs.

There are an increasing number of publications [14–19] regarding MICs, and there are a number of products offered by manufacturers (e.g., SolarEdge, Enphase) who claim that higher yield and lower levelized cost of energy (LCOE) can be achieved with their systems compared to traditional PV systems. These technologies mainly focus on rooftop installations, where the advantages of high granularity are more obvious.

In ground-mounted PV plants where shadowing is usually less prominent, this technology has usually been deemed inferior to traditional PV plants in terms of LCOE [20], due to the extra cost of the converters. On the other hand, considering the development of small-scale electronics, it is very likely that MICs will become competitive with centralized or string inverter–based systems also in large-scale ground-mounted PV plants. Today, there are commercial DC modules with peak efficiency of 99.5% [21] and prototypes of submodule-integrated converters (DC–DC) integrated in a chip with an area of less than 1 mm² as it has been reported [22], indicating high potential for cost reduction in the future.

4.3   THREE-PHASE PV INVERTER TOPOLOGIES

Three-phase PV inverters are the key elements in the transformation of the DC power from PV arrays into a grid-synchronized AC power. In the case of single-phase systems, besides the classical half-bridge and H-bridge topologies, there are many different topologies proposed in different publications during the last 10 years; some of these topologies are also used within the industry today [23–27].

Nevertheless, in the case of three-phase systems, the variety of topologies is not so big. One can find the standard three-phase full-bridge topology, presented in Figure 4.3, with six power switches, a matured topology widely used also in the case of adjustable speed drives. This topology is not suitable for transformerless PV systems, due to its common-mode behavior, as reported in [28]. It can only be used if there is a galvanic isolation between the PV panels and the grounded electrical grid. This is the case for certain CI, where the grid connection is done toward the medium-voltage (MV) network; thereby, the installation of a transformer is required. However, the common-mode behavior of the three-phase full-bridge inverter can be improved by modifying this topology and by having a split DC link, where the middle point of the capacitor bank is connected to the neutral of the grid (Figure 4.3). In this case, the potential of the PV array is fixed to the neutral, having a constant potential toward ground at the terminals of the PV array, leading to very low leakage ground currents as in the case of transformerless PV systems [28].

Nowadays, wide bandgap devices, based on silicon carbide (SiC) or gallium nitride (GaN), are commercially available. This means that the performance of classical two-level inverters can be improved [29, 30]. These new devices can operate at higher switching frequencies than traditional Si-based semiconductors, which means better controllability and smaller size grid-side filter can be

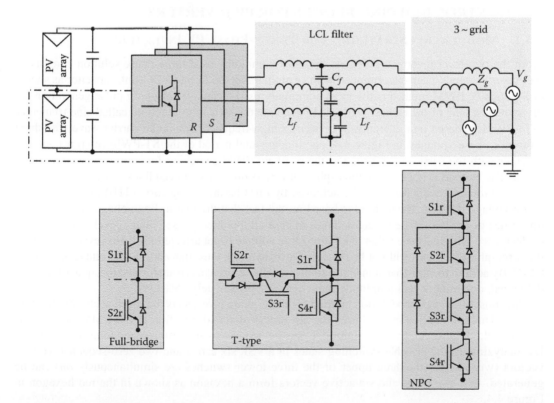

FIGURE 4.3 Three-phase PV inverter topology overview, showing the classical voltage source full-bridge, T-type, and NPC configuration for one leg of the converter.

achieved, while high efficiency is still maintained. Besides the higher costs of the SiC and GaN, new challenges need to be dealt with, like radiated and conducted EMI [31].

The drive for higher efficiencies has led to the use of multilevel topologies like neutral point clamped (NPC) [32] or T type [33], also sketched in Figure 4.3, resulting in a decrease in filter requirements and increase in the overall efficiency of the system, thanks to the multilevel output voltage. Some of these multilevel topologies are also preferred in high power applications, due to the reduced voltage stress of the power devices, compared to a traditional two-level converter. Several companies use these multilevel topologies in their MCIs within a range of 10–60 kW and these MCIs are being used in MW-sized PV plants, where the modularity given by this topology is a huge advantage during installation, operation, and maintenance.

In order to maintain high efficiency over a wide range of power levels and to ensure fault-tolerant operation for the converter, a new trend is to apply modular topologies, such as modular multilevel converters (MMC) or paralleled converters. The MMC consist of series-connected cells, usually with half- or full-bridge topology [34]. The number of cells used in a phase of the MMC will define the discrete voltage levels at the output of the converter; thus, the size of the passive filters between the MMC and grid can be reduced considerably. Another advantage of such topology is the possibility of using LV power switches. By adding redundant cells on each phase of the converter, fault-tolerant operation can be ensured [35]. By paralleling inverters, a flat efficiency profile can be achieved over a wide power range, where the number of parallel-connected inverters depends on the available power from the solar panels [36]. The parallel configuration for the central inverter ensures also fault-tolerant operation. By using interleaving between the modulation carrier wave, the current THD can be reduced considerably [37].

4.4 CONTROL BUILDING BLOCKS FOR PV INVERTERS

4.4.1 MODULATION STRATEGIES FOR THREE-PHASE PV INVERTERS

From the two-level inverter topologies, the most commonly used three-phase solution consists of six power switches, each one equipped with an antiparallel diode, and a DC-link capacitor as shown in Figure 4.3. PWM method is the most commonly used technique for *on/off* pulse generation for the switches. One of the earliest methods to generate the switching function, called sine triangular (ST)-PWM, is based on a comparison between a high-frequency triangular carrier wave with three reference phase voltages. In Figure 4.4, one fundamental period of the ST-PWM reference signals together with the carrier waveform and the resultant PWM switching function waveform is illustrated. The method is not used for three-phase inverters due to its reduced linear operating range. An improved modulation scheme can be achieved by third harmonic injection (THI) in the common-mode voltage (CMV). The method is based on the fact that the line-to-line voltage is equal to the difference of the two-phase voltages. Thus, in case all the three-phase voltage levels are increased or decreased with the same value, the line-to-line voltage is not affected [38]. However, the sum of the three-phase voltages will not be equal to zero anymore; thus, this additional voltage appears as CMV. By adding the third harmonic, the sinusoidal phase voltage is transformed to a quasi-trapezoidal signal, obtaining a peak amplitude reduction of approximately 15% [39].

The most often used modulation method for three-phase inverters is the space vector modulation (SVM). This modulation method uses the advantage of signal injection in the CMV as the THI-PWM method; however, in the beginning, it was treated as a totally different modulation technique. By analyzing all the possible switching states in a VSI, six active and two zero-sequence voltage vectors (when either the three upper or the three lower switches are simultaneously *on*) can be generated. In a *d–q* plane, the six active vectors form a hexagon as shown in the red hexagon in Figure 4.4.

With the SVM method, during one modulation period, two adjacent vectors are generated consecutively; thus, the position of the resultant voltage vector (V_{ref}) in the sector is defined by the ratio

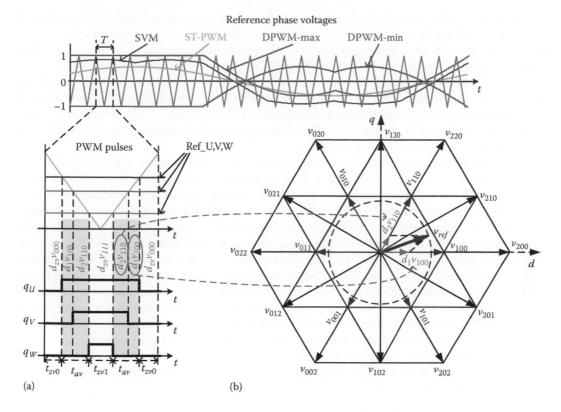

FIGURE 4.4 Pulse generation mechanism for three-phase inverters: In top ST-PWM, SVM, DPWM-MAX, and DPWM-MIN reference signals for one phase during a fundamental period, (a) pulse generation for the three phases and (b) with red the vector diagram of two-level inverters, while with black the vector diagram of the NPC is presented.

between the time spent for the generation of the two vectors. The amplitude of the voltage vector is defined by the ratio between the time spent for active and zero-sequence vector generation. The resultant voltage is not dependent on which of the two possible zero-sequence vectors are applied. However, to maintain one single switching at each transition during a modulation period, the two zero-sequence vectors have to alter. The SVM technique is based on splitting this time equally between the zero vector, and in this way, the current ripple is minimized. The SVM waveform during a fundamental period for one phase can be also tracked in Figure 4.4. In order to reduce the switching losses, the modulation can be done by using one single zero vector, a modulation method named 120° discontinuous PWM (DPWM) MAX or MIN depending on which zero vector is used (Figure 4.4). Nevertheless, by using this modulation method, the switching losses can be reduced, but the current ripple is increased and the loss distribution between the switches is not equal anymore. An improvement on the loss distribution can be achieved by alternating between the two zero vectors after each 60° [39].

Using the SVM, the CMV varies between $\pm V_{DC}$, while with DPWM, it is reduced to $+V_{DC}$ and $-1/3V_{DC}$ or $1/3V_{DC}$ and $-V_{DC}$. The CMV can be reduced even more, if the modulation is made without zero-sequence vector generation [40]. One such method is the active zero state PWM (AZSPWM), which is based on generating two equal-length opposite active vectors instead of zero vectors [41]. Near-State PWM (NSPWM) introduces only one extra active vector instead of zero vectors in such a way that the same resultant vector is achieved as with SVM [41]. With both methods, the CMV varies between $\pm 1/3V_{DC}$, at the expense of the increased current ripple.

The spectrum of the output voltage of each modulation technique can be determined by analytical equations, which can be used for efficiency calculations [38, 42]. By using different random modulation techniques, the discrete components from the output voltage spectrum can be spread, which can improve the electromagnetic compatibility (EMC) of the converter [43].

From the modulation techniques mentioned earlier, SVM is the most commonly used method due to its lowest current ripple. However, in high power applications, where the switching loss reduction is important due to the complex thermal management, DPWM or a modified version of it is preferred. The reduction of the peak-to-peak CMV is important in PV applications; thus, the AZSPWM or NSPWM methods can be a good alternative.

For high-power inverters, the NPC topology is often used. The topology is based on using two capacitors placed in series in the DC link, with the midpoint connected to the inverter neutral point, as shown in Figure 4.3. The voltage balancing of the two capacitors is an essential task for NPC topologies; the modulation has to have the ability to ensure this balance. From an implementation point of view, there are two approaches: carrier based (CB) [44] and SVM [45]. The CB methods are the phase disposition (PD-PWM), phase opposition disposition (POD-PWM), and alternate phase opposition (APOD-PWM). The difference between the CB methods is the phase shift between the carrier waves. In the PD method, both the positive and negative carriers are in phase, while in POD the carriers for the positive references are 180° shifted from the negative reference carriers. For the APOD method, the two carrier waves are for both positive and negative part of the reference voltage and they are 180° phase shifted from each other. From the three methods, PD-PWM has the lowest harmonic distortion, while APOD and POD PWM strategies are similar [44]. The SVM technique for NPC is implemented similarly as for two-level topologies shown in Figure 4.4 and uses the freedom given by the zero-sequence vector redistribution; thus, it can be easily optimized for low switching loss [46].

4.4.2 Implementation of the Modulation Strategies

The implementation of the modulation strategies can be achieved by using analog or digital components. For the analog implementation, a carrier wave generator and an analog comparator are needed between the carrier waveform and the reference signal, as it is sketched in Figure 4.4. Today, such an implementation method is rarely used because it is not flexible. In most applications, the overall system control implemented in microcontrollers is straightforward by using digitally implemented modulators. Moreover, the microcontrollers are usually equipped with a dedicated PWM unit. The PWM unit consists of a counter, which generates the carrier wave and a compare register (CR), where the reference signal is loaded at each sampling period. The counter is reset or changed from up-count mode to down-count mode (depending on whether sawtooth or triangular carrier needs to be generated) when it reaches the value loaded in the period register (PR). Thus, through the PR, the frequency of the carrier wave is established. The new values for the CR are usually loaded when the counter takes the value of zero and/or the value of PR as shown in Figure 4.5. At the time instance, when the counter reaches the CR value, the input/output port (which has to be set as output port) of the microcontroller performs the switching.

In simulation environments, such as MATLAB®/Simulink®, both analog and digital methods are easy to implement. In Figure 4.6, an example of a three-phase modulator is given by using the elementary blocks from MATLAB®/Simulink®.

The carrier wave can be generated by using the repetitive sequence block (it can be found under Simulink/Sources). In order to generate a triangular signal, three time values have to be set [1 1/ $(2F_{sw})$ 1/F_{sw}] where F_{sw} is the switching frequency, while the output of the block should also take three values [0 1 0] in case the amplitude of the carrier varies between 0 and 1 or [−1 1−1] to obtain the amplitude between −1 and 1. F_{sw} represents the value of the switching frequency in this case.

The zero-order hold block is inserted in order to take into consideration the effect of the counter from the PWM unit (in the case of analog circuit simulation, it should not be added), where the

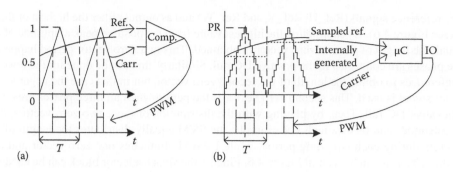

FIGURE 4.5 PWM pulse train generation: (a) implementation with analog comparator, (b) implementation using a PWM unit in the microcontroller (μC).

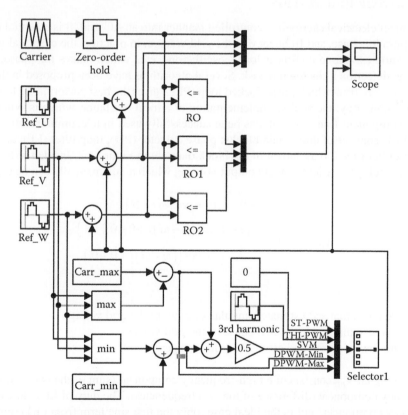

FIGURE 4.6 Simulink® model of different modulation techniques.

quantization should be equal to the counter time base. The reference signals are generated as three sinusoidal waveforms where the peak-to-peak amplitude can vary between 0 and 1 or −1 and 1, depending on the settings of the carrier wave. In closed-loop application, the control algorithm generates these reference signals. The three comparator blocks are generating the PWM output signal of the modulator. In order to simulate the behavior of the improved modulation techniques, the injection of zero-sequence voltage has to be performed. For the THI-PWM method, a sinusoidal with triple frequency has to be added to the three reference waveforms, as illustrated in Figure 4.6. In case of closed-loop application, this zero-sequence waveform should be generated based on the phase/frequency of the grid voltage, which is supplied by the grid synchronization block. For simulation of the SVM and DPWM methods, a "min" and "max" block can be used. These blocks have as input the three reference signals (Ref_U, Ref_V, and Ref_W) and as output either the highest or the lowest values (see Figure 4.6). By subtracting the highest value from the positive peak amplitude of the carrier signal, the length of the first zero vector is obtained. This first zero vector period happens when all three phases are connected to the positive DC rail. Similarly, this can be done for the lowest value of the references to obtain the length of the second zero vector, but in this case the negative peak of the carrier signal is used. This second zero vector is the period when all three phases are connected to the negative DC rail. Then, by playing with the distribution of these two zero vectors, different PWM techniques can be obtained. For example, the SVM equally distributes the length of the two zero vectors during each switching period, while DPWM eliminates one zero vector and uses the sum of the two, as it can be seen in Figure 4.6. Finally, the signal selector block can be used in order to choose between the different modulation techniques.

4.4.3 GRID SYNCHRONIZATION

In order to inject electrical energy in a controlled manner into an AC grid, it is essential to synchronize the amplitude, phase, and frequency of the grid voltage. In practice, the electrical grid voltage is rich in harmonics, due to nonlinear loads, oscillations, etc., which makes the phase, frequency, and amplitude detection a challenging task. Several algorithms have been proposed in the literature for synchronization, from which phase-locked loop (PLL) gives the best performance [47]. The PLL is not a new technology; the analog implementation of a local oscillator, which is synchronized to a frequency component from a signal, has been successfully used in telecommunications for many decades. A PLL consists of three main blocks: phase detector (PD), loop filter (LF), and a voltage-controlled oscillator (VCO) as shown in the block diagram in Figure 4.7. Inside the PD block, the input signal is multiplied with the VCO output voltage, which mathematically can be expressed as

$$V_i = A_i \sin(\omega t + \theta); \quad V' = \cos(\omega' t + \theta')$$

$$V_{PD} = V_i \cdot V' = \frac{A_i k_{pd}}{2} \left(\begin{array}{l} \sin((\omega - \omega')t + \theta - \theta') \\ + \cos((\omega + \omega')t + \theta + \theta') \end{array} \right) \quad (4.1)$$

where
    ω and ω' are the actual and estimated angular velocity of the grid voltage
    θ and θ' are the actual and estimated phase of the grid voltage
    V_i and V' are the measured and estimated grid voltages

The output signal V_{PD} consists of a high-frequency component (sum of the two frequencies) and a low-frequency component (difference of the two frequencies). The duty of LF is to cancel out the high-frequency component, and in the ideal case, only the first sine term from (4.1) remains. Thus, the output of the LF block is a quasi-DC component. This has a positive value when the estimated frequency, given by the VCO block, is above and negative when the estimated frequency is below the frequency of the measured grid voltage. In this way, the estimated frequency can be controlled in order to match with the frequency of the measured grid voltage.

A better PLL performance can be achieved by using a quadrature PLL with quadratic signal generator (QSG), where the quadrature of the tracked frequency component is also created as shown in Figure 4.7b [48]. The second-order generalized integrator (SOGI) has the advantage compared to the other QSGs, which works also as a notch filter, rejecting any unwanted frequency components [49]. However, when the grid frequency varies, the SOGI has also to adapt to it, which can be done by using a frequency-locked loop (FLL), which is presented at the bottom of Figure 4.7b [50].

For three-phase systems, the quadrature signal can be obtained by applying the Clarke transformation; thus, the quadrature PLL can be adopted directly. This method is called synchronous reference

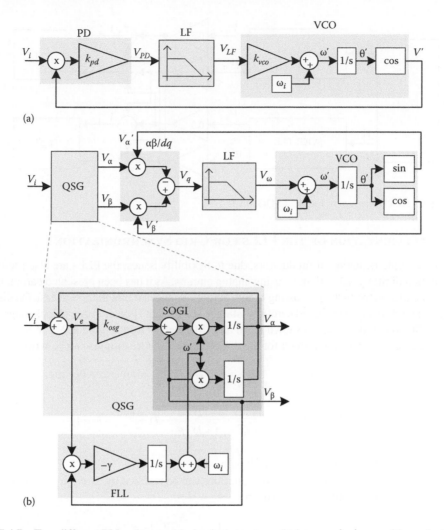

FIGURE 4.7 Two different PLL schemes: (a) classical structure, (b) improved scheme with quadrature feedback and quadrature signal generation with FLL.

frame (SRF) PLL [51]. In practice, the three-phase grid voltages are not balanced, especially during grid faults. Another concept for synchronization is based on the idea that any unbalanced three-phase system can be decomposed as a sum of symmetrical positive-, negative-, and zero-sequence components [52]. The $\alpha\beta$-components of the positive and negative sequence can be expressed by

$$\begin{bmatrix} V_\alpha'^+ \\ V_\beta'^+ \\ V_\alpha'^- \\ V_\beta'^- \end{bmatrix} = \frac{1}{2} \begin{bmatrix} 1 & -q & 0 & 0 \\ q & 1 & 0 & 0 \\ 0 & 0 & -q & 1 \\ 0 & 0 & 1 & q \end{bmatrix} \begin{bmatrix} V_\alpha' \\ V_\beta' \\ V_\alpha' \\ V_\beta' \end{bmatrix} \tag{4.2}$$

where q is the in-quadrature operator. With the SOGI-FLL presented in Figure 4.7, the quadrature of the V_α and V_β can be obtained. By using Equation 4.2, the $\alpha\beta$-components of the positive and negative sequence can be obtained, as shown in Figure 4.8, which can be transformed back to *abc* coordinates with the inverse Clarke transformation.

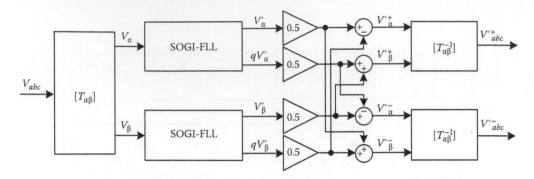

FIGURE 4.8 Block diagram of the dual SOGI-FLL.

4.4.4 IMPLEMENTATION OF THE PLLS FOR GRID SYNCHRONIZATION

Similar to the implementation of modulators, due to flexibility issues, the PLLs are also preferred for implementation digitally rather than by using analog circuits. As it has been presented earlier, the main element of the PLL is the SOGI consisting of two parallel-connected integrators. In case the discretization of the integrators is not designed carefully, stability issues can appear. From a block diagram point of view, the SOGI can be drawn as shown in (a) and (b) schemes from Figure 4.9. The SOGI transfer function in the continuous time domain for the direct and quadrature outputs can be written as

$$H_d(s) = \frac{V_\alpha}{V_e k_{OSG}} = \frac{s\omega}{s^2 + \omega^2} = \frac{D}{1 + D \cdot F}; \quad D = \frac{\omega}{s}; \quad F = \frac{\omega}{s};$$

$$H_q(s) = \frac{V_\beta}{V_e k_{OSG}} = \frac{\omega^2}{s^2 + \omega^2} = \frac{D}{1 + D \cdot F}; \quad D = \frac{\omega^2}{s^2}; \quad F = 1;$$

(4.3)

where D is the direct while F is the feedback path.

A typical way to discretize those transfer functions can be realized by using first-order integrators like forward Euler, backward Euler, or Tustin:

$$\frac{1}{s} \cong T_s \frac{z^{-1}}{1 - z^{-1}} \cong T_s \frac{1}{1 - z^{-1}} \cong \frac{T_s}{2} \frac{1 + z^{-1}}{1 - z^{-1}};$$

(4.4)

where T_s is the sampling frequency.

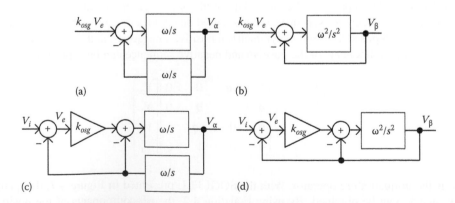

FIGURE 4.9 Block diagram representation of the SOGI in Figure 4.8 for (a) alpha component and (b) beta component and SOGI-OSG for (c) alpha and (d) beta component of the measured grid voltage.

When the forward Euler discretization method is used for both the direct path and feedback path, an algebraic loop is created. When the backward Euler method is used, the delay on the feedback path will be doubled; thus, the oscillator goes into instability. When SOGI is implemented in the continuous domain for a simulation software, such as MATLAB®/Simulink, and the discretization is performed by the simulator, the problems mentioned previously can appear. In order to obtain a stable implementation for SOGI the forward Euler implementation should be used for the direct path and the backward Euler for the feedback path or vice versa. The difference in the two ways of implementation consists of a sample delay in $H_d(z)$:

$$H_d(z) = \frac{\left(\omega T_s - \omega T_s z^{-1}\right)z^{-1}}{1 + \left(\omega^2 T_s^2 - 2\right)z^{-1} + z^{-2}}; \quad H_q(z) = \frac{\omega^2 T_s^2 z^{-1}}{1 + \left(\omega^2 T_s^2 - 2\right)z^{-1} + z^{-2}} \tag{4.5}$$

Due to the fact that V_α is a feedback signal for the orthogonal signal generator (OSG) (as it can be seen in (c) from Figure 4.9), in order to avoid the algebraic loop, a sample delay has to be present. Thus, it is preferable to use forward Euler for the discretization of the direct axes and backward Euler for the feedback axes. The perpendicularity between the alpha and beta components of the measured grid voltage is not ensured with forward and backward Euler implementation, due to the different delays on the direct and feedback path (half sampling period delay appears). This becomes an issue when the PLL has to run with a low sampling rate.

Another commonly used discretization for SOGI is to use the Tustin method for both direct and feedback path. However, in order to obtain the transfer function, this method requires more calculations. It has the advantage compared to the forward and backward Euler method, which ensures the perpendicularity between the alpha and beta components. By substituting the Tustin discretization from (4.4) into (4.3), the discrete transfer function becomes

$$H_d(z) = \frac{2\omega T_s - 2\omega T_s z^{-2}}{\left(\omega^2 T_s^2 + 4\right) + \left(2\omega^2 T_s^2 - 8\right)z^{-1} + \left(\omega^2 T_s^2 + 4\right)z^{-2}}$$

$$H_q(z) = \frac{\omega^2 T_s^2 + 2\omega^2 T_s^2 z^{-1} + \omega^2 T_s^2 z^{-2}}{\left(\omega^2 T_s^2 + 4\right) + \left(2\omega^2 T_s^2 - 8\right)z^{-1} + \left(\omega^2 T_s^2 + 4\right)z^{-2}} \tag{4.6}$$

By modifying the transfer function and substituting the parameters of the denominator with "a" and the parameters of the numerator with "b," the block diagram from Figure 4.10 can be created. As $H_d(z)$ has a term without delay (b0d from Figure 4.10), in order to avoid an algebraic loop in the feedback path of the OSG, a sampling delay has to be inserted (block shown with dashed lines in Figure 4.10). As it can be observed in both transfer functions ((4.5) and (4.6)), the denominators are identical; thus, they can be implemented by using only two state variables as shown in Figure 4.10.

The FLL block from the state point of view consists of one integrator, where forward Euler discretization is preferred to avoid the algebraic loops. It has to be noted here when FLL is used, the parameters of the SOGI have to be recalculated at each sampling, which might need extra calculation power from the digital implementation.

The implementation of the remaining three blocks of the PLL is straightforward: the phase detector block consists of only mathematical calculations, the LF typically is a proportional integrator (PI) controller where the integrator can be discretized with any of the methods mentioned earlier, and finally, the VCO can be implemented as a sine and a cosine function while the integrator is using backward or forward Euler.

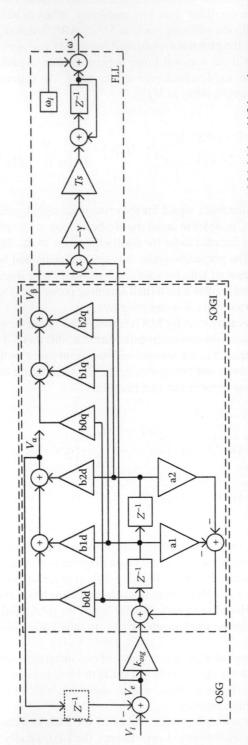

FIGURE 4.10 Digital implementation scheme for the OSG and the FLL where a1 and a2 are the denominator parameters; b0d, b1d, and b2d are the numerators of the direct path parameters; while b0q, b1q, and b2q are the quadrature parameters of the SOGI transfer functions.

4.4.5   CURRENT CONTROL

The two most used current control methods are the voltage-oriented SRF based on PI controllers and the stationary reference frame method based on resonant controllers [47]. As a first step, both of the methods convert the measured three-phase currents in two quadratic components (i_α, i_β) by using the Clarke transformation as it is shown in Figure 4.11. In the PI-based controller method, the α–β current components are transformed to DC values (Park transformation) by using the PLL phase angle θ estimation. The PI controller acts based on the difference between the d–q current references and the measured/calculated d–q currents. The reference d–q currents are obtained from the desired active and reactive power injection based on the following relation:

$$P^* = v_{gd}i_d + v_{gq}i_q$$
$$Q^* = -v_{gq}i_d + v_{gd}i_q$$

(4.7)

The active power on the d-axis current is adjusted by the V_{DC} controller in order to keep the DC voltage constant. After an inverse Park transformation, the α–β reference voltage components are obtained, which are the input signals for the modulator. The difference between using the resonant controllers compared to the PI-based method is that there is no need for the Park transformation; the control is applied directly on the α–β AC signals. The reference signals for the resonant controllers are obtained also from the active and reactive power injection based on the α–β equations:

$$P^* = v_{g\alpha}i_\alpha + v_{g\beta}i_\beta$$
$$Q^* = -v_{g\beta}i_\alpha + v_{g\alpha}i_\beta$$

(4.8)

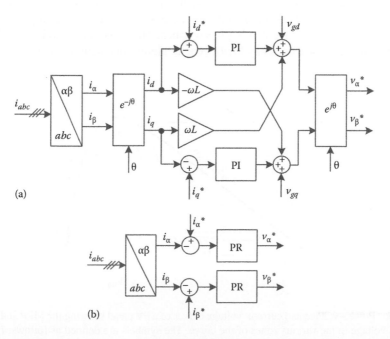

(a)

(b)

FIGURE 4.11   Current control schemes of three-phase grid-connected inverters: (a) voltage-oriented SRF using PI-based controller; (b) resonant controller–based method (PR).

4.4.6 IMPLEMENTATION OF THE CURRENT CONTROLLERS

From an implementation point of view, the PI controllers are simple to apply, and any of the discretization methods (backward Euler, forward Euler, or Tustin) presented earlier can be used. The implementation of the resonant controllers needs more attention, due to the two parallel-connected integrators. However, the resonant controller structure is identical with the SOGI implementation methods given for the SOGI in Section 4.4.4.

4.4.7 MAXIMUM POWER POINT TRACKING

MMPT is one of the essential functions in PV plants. Considering the nonlinear current–voltage characteristics of PV cells, the maximum power delivered by the PV array is at a well-defined operating point called maximum power point (MPP). MPPT is one of the key functions that every grid-connected PV inverter should have. There is a large amount of publications dealing with MPPT, and trackers in the majority of the commercial PV inverters are able to extract around 99% of the available power from the PV plant, over a wide irradiance and temperature range—at least in a steady state. An extensive overview of modern MPPT techniques has been presented in [53]. The most frequently applied MPPT algorithms are hill-climbing methods, such as the perturb and observe (P&O), and its alternate implementation (with identical behavior), the incremental conductance [54]. These methods are based on the fact that on the voltage–power characteristic (Figure 4.12), the variation of the power with respect to the voltage is positive ($dP/dV > 0$) in the left-hand side of the MPP, while it is negative ($dP/dV < 0$) on the right-hand side of MPP, as shown in Figure 4.12.

The main advantages of these methods are that they are generic, that is, they are suitable for any PV array, they do not require any information about the PV array, they work reasonably well in most conditions, and they are simple to implement in a digital controller with minimum computational demand.

However, the hill-climbing methods have some inherent shortcomings as they have limited tracking capability during varying conditions (e.g., windy day with fast moving clouds, see Figure 4.13), and in the case of partial shadow (see Figure 4.14), they may not find the global MPP, settling at a local maxima on the power–voltage curve. These shortcomings have inspired a large number of publications that aim at overcoming these issues. The most popular advanced MPPT methods in the literature are based on fuzzy logic, genetic algorithms, neural networks, particle swarm

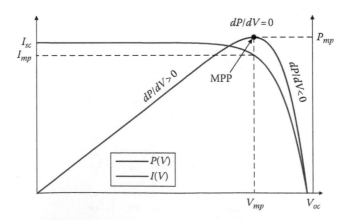

FIGURE 4.12   Power–voltage and current–voltage curves of a PV panel showing the MPP and the derivatives of power with voltage in the various zones of the curve. The symbols are defined as follows: I_{sc}, short-circuit current; V_{oc}, open-circuit voltage; I_{mp}, current at MPP; V_{mp}, voltage at MPP; P_{mp}, peak power.

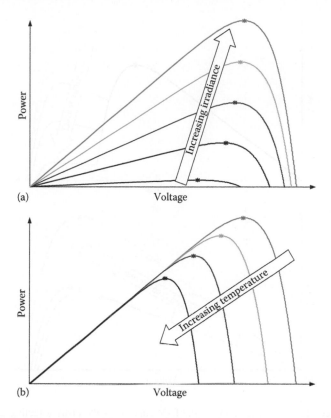

FIGURE 4.13 Variation of the power–voltage curve influenced by irradiance (a) and by temperature (b).

optimization, etc., as well as PV model based, which are theoretically well suited for tackling the nonlinear problem of MPPT [55, 56]. Indeed, good performances of these methods are reported; however, they share a common drawback in that they require detailed knowledge of the system they are applied to, which therefore decreases their general applicability.

The MPPT methods that are mostly preferred by the industry are the hybrid or enhanced hill-climbing ones, which have a P&O algorithm as their core, with an improvement/enhancement in conjunction, to overcome its limitations in variably and/or partially shaded conditions.

With the rise of large-scale PV plants as well as microinverters and DC modules, distributed MPPT systems have gained more attention (see Figure 4.15c) [57–60]. The main MPPT structures are outlined in Figure 4.15.

In Figure 4.15, the location/granularity of the MPP trackers in the PV systems is also determined by the use of converter system layout (see Figure 4.2). In fact, an important factor in choosing the PV plant converter layout is the desired level of resolution of the MPPT.

The central (or multistring) (a) and string (b) MPPT structures are the most common and can be considered as the traditional configuration. As mentioned before, module-level (or even submodule-level) MPPT is a promising structure, considering the potential for higher yield due to localized MPPT and condition monitoring.

Another possible MPPT structure is the hybrid solution, where only a part of the string is equipped with module-level MPPT [61]. This can be a cost-effective solution to minimize shading losses; however, they are more beneficial for, for example, rooftop installations. In this case (and for layout (c) in case a central MPPT is present), the interaction between the local and central MPP tracker has to be dealt with as discussed in [20, 62, 63].

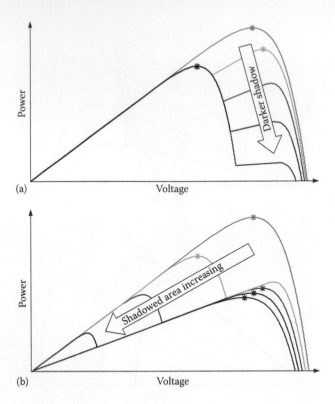

FIGURE 4.14 Effect of a partial shadow on the P–V curve of a PV array affecting 25% of its surface as changing from no shadow to 90% reduction of irradiance (a) and effect of a 50% intensity shadow as the shadowed area is increasing from nonshadowed to fully shadowed condition (b).

While the majority of the systems typically still use the traditional hill-climbing MPPT method for tracking, the focus has shifted toward the interaction of the parallel-operating trackers in a large PV system. Furthermore, with the appearance of new grid requirements, for example, power ramp limitation (PRL) and frequency-sensitive operation, the MPPT operation is no longer the standard operating mode of PV plants.

4.4.8 GRID INTEGRATION FUNCTIONS FOR PV SYSTEMS

Recently, with the increasing penetration of PV into the grid, as well as the increasing number of large PV plants connected to the MV network, grid integration and grid support requirements for large PV plants have been released or are under development. The European Network of Transmission Systems Operators for Electricity released grid codes where large-scale PV power plants (LPVPP) are required to have fault ride through, voltage support, and frequency-sensitive mode (FSM) capabilities [64]. Furthermore, LPVPPs have to adhere to the MV grid codes in providing reactive current injection during grid faults and maintaining connection during grid faults all the way down to 0 V grid voltage. While, in most cases, these requirements are for all generators, in a few countries (e.g., Germany, Spain), grid codes that are specific to plants without a synchronous generator have been published. An overview of grid codes for PV plants is provided in [65].

While many of the grid requirements PV plants must fulfill are not necessarily only PV specific, a few but general application for distributed generation functionalities/features that are typical to PV plants are discussed in the succeeding part.

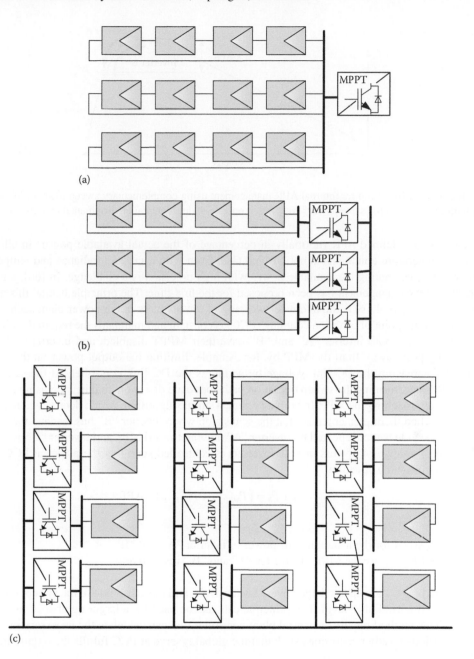

FIGURE 4.15 Main maximum power point tracking (MPPT) structures: (a) central, (b) string, (c) module level.

4.4.9 INTERNAL ACTIVE POWER RESERVE MANAGEMENT

Since the production of PV plants is determined by the available solar irradiance and therefore can only be controlled downwards, in order for a PV plant to be able to provide FSM operation and PRL, power reserve is necessary. This can be secured by either an external energy storage (e.g., battery storage system) or an internal reserve, where the PV plant operates below the MPP power level, and the curtailed power can be used as active power reserve (APR).

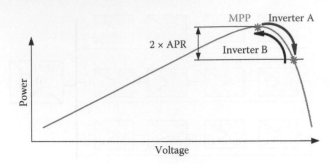

FIGURE 4.16 Illustration for internal APR management using complementary sweep method, in case of a plant with two CI with the same power level, which operate under identical environmental conditions.

Having a well-defined APR (normally in percentage of the actual available power) in all irradiance and temperature conditions is not a trivial task. Even with accurate irradiance and temperature measurements, estimating the available power with high precision is a challenge. In [66], a method for internal APR management has been proposed for the first time. The principle behind this method (illustrated in Figure 4.16) is the alternate sweeping of the inverters in the power plant such that the total power at the point of common coupling (PCC) is constant and maintains the required APR level.

In Figure 4.16, both inverter "A" and "B" have their MPPT disabled, and inverter "A" moves its operating point away from the MPP by, for example, limiting the output power on the AC side or directly increasing the DC-link voltage by acting as the DC-link controller. At the same time, inverter "B" is performing the sweep operation in the opposite direction, that is, increasing its output power, and approaching the MPP. In both inverters, a simple algorithm, which detects when the peak power is reached, is implemented so that the scans stop when inverter "B" has reached the MPP of its array. The cycle is then repeated by inverter "A" moving toward MPP and inverter "B" away from it. The rate of change of powers in both inverters is controlled such that the power at PCC (P_{PCC}) satisfies the following equation:

$$P_{PCC} = P_A + P_B = \left(P_{MPP\_A} + P_{MPP\_B} \right) - \text{APR} \tag{4.9}$$

where

P_A and P_B stand for the instantaneous power of inverter "A" and "B"

P_{MPP_A} and P_{MPP_B} are the MPP of inverter "A" and "B," respectively

P_{MPP_A} and P_{MPP_B} are updated upon every sweep cycle, and the amplitude of the sweep is adjusted according to the irradiance level. The principle can also be used for a larger number of inverters in the plant. This method requires a central plant controller that can calculate the necessary individual APRs in various weather conditions such that the global reserve at PCC fulfills the requirements.

4.4.10 ACTIVE POWER RAMP LIMITATION

In smaller networks, such as islands with high PV penetration, ramping obligations for PV plants have been introduced in the grid codes. This is done with the purpose of minimizing the effect of fast power fluctuations that could stress the electrical network (e.g., frequency variation, due to passing clouds). Although for the moment this characteristic is only for small/island networks, the German transmission system operator (TSO) has also imposed a 10%/min increasing power ramp limit, and it is expected that with the increasing penetration of PV, this requirement will be adopted by other TSOs as well [67].

Three main factors contribute to fulfilling the maximum ramping rate by PV plants. On one hand, the spatial distribution of large PV plants acts as a smoothing factor for the output power fluctuations caused by passing clouds [68–70]. Therefore, dispersing the generation capacity over a large geographical area can help mitigate the fast power fluctuations. On the other hand, in the case of a high PV generation density, reserves can be used to compensate for the plant output variability, either by auxiliary energy storage or by internal APR.

4.5 SUMMARY

The technology of three-phase PV inverters is maturing, and a few key topologies of success have emerged that dominate the market, with peak conversion efficiencies reaching 98.5%. The main grid interfacing, safety, and power extraction features are well understood and are standard elements of the grid-connected PV inverters. The exceptional success of PV energy technology in the last couple of decades has led to a maturity level where the focus is now shifting from grid interfacing issues to (large-scale) grid integration challenges. In many countries, grid parity has been achieved, and it is expected that the next challenge for (very) high penetration of PV will be related to the intermittency of the PV generation. As a consequence, APRs (both by internal and auxiliary storage systems) will keep increasing their significance.

REFERENCES

1. IEA-PVPS, Snapshot of global photovoltaic markets, *IEA PVPS T1-29:2016*, 2016.
2. J. W. Oliver Schäfer, *Global Market Outlook 2015–2019*, EPIA, Brussels, Belgium, 2015. http://helapco.gr/pdf/Global_Market_Outlook_2015_-2019_lr_v23.pdf.
3. DIN, DIN VDE V 0126-1-1:2013-08, in *Selbsttätige Schaltstelle zwischen einer netzparallelen Eigenerzeugungsanlage und dem öffentlichen Niederspannungsnetz*, ed, 2013.
4. V.-A.-N. 4105:2011, Generators connected to the low-voltage distribution network, in *Technical Requirements for the Connection to and Parallel Operation with Low-Voltage Distribution Networks*, ed. 2011.
5. B. I. Craciun, E. A. Man, V. A. Muresan, D. Sera, T. Kerekes, and R. Teodorescu, Improved voltage regulation strategies by PV inverters in LV rural networks, in *2012 Third IEEE International Symposium on Power Electronics for Distributed Generation Systems (PEDG)*, pp. 775–781, Aalborg, Denmark, 2012.
6. Z. Dehua, Z. Qian, A. Grishina, A. Amirahmadi, H. Haibing, J. Shen et al., A comparison of soft and hard-switching losses in three phase micro-inverters, in *2011 IEEE Energy Conversion Congress and Exposition (ECCE)*, pp. 1076–1082, Phoenix, AZ, 2011.
7. B. Johnson, P. Krein, Z. Zheming, and A. Lentine, A single-stage three-phase AC module for high-voltage photovoltaics, in *2012 27th Annual IEEE Applied Power Electronics Conference and Exposition (APEC)*, pp. 885–891, Orlando, FL, 2012.
8. D. Zhang, Q. Zhang, H. Hu, A. Grishina, J. Shen, and I. Batarseh, High efficiency current mode control for three-phase micro-inverters, in *2012 27th Annual IEEE Applied Power Electronics Conference and Exposition (APEC)*, pp. 892–897, Orlando, FL, 2012.
9. Enphase, Enphase microinverter datasheet, E. E. M. System, Ed., 2011.
10. W. Huai and F. Blaabjerg, Reliability of capacitors for DC-link applications in power electronic converters: An overview, *IEEE Transactions on Industry Applications*, 50, 3569–3578, 2014.
11. Z. Peng, W. Yang, X. Weidong, and L. Wenyuan, Reliability evaluation of grid-connected photovoltaic power systems, *IEEE Transactions on Sustainable Energy*, 3, 379–389, 2012.
12. C. Lin, A. Amirahmadi, Z. Qian, N. Kutkut, and I. Batarseh, Design and implementation of three-phase two-stage grid-connected module integrated converter, *IEEE Transactions on Power Electronics*, 29, 3881–3892, 2014.
13. A. Amirahmadi, H. Haibing, A. Grishina, Z. Qian, C. Lin, U. Somani et al., Hybrid ZVS BCM current controlled three-phase microinverter, *IEEE Transactions on Power Electronics*, 29, 2124–2134, 2014.
14. Q. Li and P. Wolfs, A review of the single phase photovoltaic module integrated converter topologies with three different DC link configurations, *IEEE Transactions on Power Electronics*, 23, 1320–1333, 2008.
15. G. Mehlmann, F. Schirmer, M. Zeuß, and G. Herold, DC/DC converters as linkages between photovoltaic plants and module integrated multilevel-inverters, in *International Conference on Renewable Energies and Power Quality*, Las Palmas de Gran Canaria, Spain, 2010.
16. C. Olalla, D. Clement, M. Rodriguez, and D. Maksimovic, Architectures and control of submodule integrated DC–DC converters for photovoltaic applications, *IEEE Transactions on Power Electronics*, 28, 2980–2997, 2013.

17. R. C. N. Pilawa-Podgurski, Architectures and circuits for low-voltage energy conversion and applications in renewable energy and power management, Department of Electrical Engineering and Computer Science, Massachusetts Institute of Technology, Cambridge, MA, 2012.

18. D. Shmilovitz and Y. Levron, Distributed maximum power point tracking in photovoltaic systems–emerging architectures and control methods, *AUTOMATIKA: casopis za automatiku, mjerenje, elektroniku, racunarstvo i komunikacije*, 53, 142–155, 2012.

19. G. G. R. Walker, J. Xue, and P. Sernia, PV string per-module maximum power point enabling converters, in *The Australasian Universities Power Engineering Conference*, Christchurch, New Zealand, p. 6, 2003.

20. R. C. N. Pilawa-Podgurski and D. J. Perreault, Submodule integrated distributed maximum power point tracking for solar photovoltaic applications, *IEEE Transactions on Power Electronics*, 28, 2957–2967, 2013.

21. SolarEdge, Solaredge power optimizer: Module embedded solution, 2013, http://www.solaredge.com/sites/default/files/se-pb-csi-datasheet.pdf. last accessed October 24, 2016.

22. R. Enne, M. Nikolic, and H. Zimmermann, Dynamic integrated MPP tracker in 0.35 µm CMOS, *IEEE Transactions on Power Electronics*, 28, 2886–2894, 2013.

23. S. Bernet and T. Bruckner, Open-loop and closed-loop control method for a three-point converter with active clamped switches, and apparatus for this purpose, ed: Google Patents US6697274 B2, US 10/388,022, 2003.

24. J. Ketterer, H. D. I. Schmidt, and C. D. Siedle, Inverter for transforming a DC voltage into an AC current or an AC voltage, ed: Google Patents EP1369985 B1, EP20030009882, 2009.

25. P. Knaup, Inverter, ed: Google Patents WO2009039888 A1, PCT/EP2007/060266, 2007.

26. M. D. Victor, F. Greizer, S. Bremicker, and U. Hübler, Method of converting a DC voltage of a DC source, in particular of a photovoltaic DC source, in an AC voltage, ed: Google Patents EP1626494 A3, EP20050011807, 2007.

27. R. Gonzalez, J. Lopez, P. Sanchis, and L. Marroyo, Transformerless inverter for single-phase photovoltaic systems, *IEEE Transactions on Power Electronics*, 22, 693–697, 2007.

28. T. Kerekes, M. Liserre, R. Teodorescu, C. Klumpner, and M. Sumner, Evaluation of three-phase transformerless photovoltaic inverter topologies, *IEEE Transactions on Power Electronics*, 24, 2202–2211, 2009.

29. C. Pham, R. Teodorescu, T. Kerekes, and L. Mathe, High efficiency battery converter with SiC devices for residential PV systems, in *2013 15th European Conference on Power Electronics and Applications (EPE)*, pp. 1–10, Lille, France, 2013.

30. J. Holtz, M. Holtgen, and J. O. Krah, A space vector modulator for the high-switching frequency control of three-level SiC inverters, *IEEE Transactions on Power Electronics*, 29, 2618–2626, 2014.

31. N. Oswald, P. Anthony, N. McNeill, and B. H. Stark, An experimental investigation of the tradeoff between switching losses and EMI generation with hard-switched all-Si, Si-SiC, and all-SiC device combinations, *IEEE Transactions on Power Electronics*, 29, 2393–2407, 2014.

32. A. Nabae, I. Takahashi, and H. Akagi, A new neutral-point-clamped PWM inverter, *IEEE Transactions on Industry Applications*, IA-17, 518–523, 1981.

33. M. Schweizer and J. W. Kolar, Design and implementation of a highly efficient three-level T-type converter for low-voltage applications, *IEEE Transactions on Power Electronics*, 28, 899–907, 2013.

34. S. Debnath, Q. Jiangchao, B. Bahrani, M. Saeedifard, and P. Barbosa, Operation, control, and applications of the modular multilevel converter: A review, *IEEE Transactions on Power Electronics*, 30, 37–53, 2015.

35. R. Teodorescu, E.-P. Eni, L. Mathe, and P. Rodriguez, Modular multilevel converter control strategy with fault tolerance, in *International Conference on Renewable Energies and Power Quality (ICREPQ'13)*, Bilbao, Spain, 2013.

36. Emerson, Utility scale solar inverters, 2013, http://www.emersonindustrial.com/en-EN/documentcenter/ControlTechniques/Brochures/photovoltaic_brochure.pdf, last accessed April 2016.

37. G. Gohil, R. Maheshwari, L. Bede, T. Kerekes, R. Teodorescu, M. Liserre et al., Modified discontinuous PWM for size reduction of the circulating current filter in parallel interleaved converters, *IEEE Transactions on Power Electronics*, 30, 3457–3470, 2014.

38. G. D. Holmes and T. A. Lipo, *Pulse Width Modulation for Power Converters: Principles and Practice*. Hoboken, NJ: John Wiley & Sons, Inc., 2003.

39. A. M. Hava, R. J. Kerkman, and T. A. Lipo, Simple analytical and graphical methods for carrier-based PWM-VSI drives, *IEEE Transactions on Power Electronics*, 14, 49–61, 1999.

40. C. C. Hou, C. C. Shih, P. T. Cheng, and A. M. Hava, Common-mode voltage reduction pulsewidth modulation techniques for three-phase grid-connected converters, *IEEE Transactions on Power Electronics*, 28, 1971–1979, 2013.

41. A. M. Hava and E. Un, A high-performance PWM algorithm for common-mode voltage reduction in three-phase voltage source inverters, *IEEE Transactions on Power Electronics*, 26, 1998–2008, 2011.
42. L. Mathe, H. Cornean, D. Sera, P. O. Rasmussen, and J. K. Pedersen, Unified analytical equation for theoretical determination of the harmonic components of modern PWM strategies, in *IECON 2011—37th Annual Conference on IEEE Industrial Electronics Society*, pp. 1648–1653, Melbourne, VIC, 2011.
43. L. Mathe, F. Lungeanu, D. Sera, P. O. Rasmussen, and J. K. Pedersen, Spread spectrum modulation by using asymmetric-carrier random PWM, *IEEE Transactions on Industrial Electronics*, 59, 3710–3718, 2012.
44. B. P. McGrath and D. G. Holmes, A comparison of multicarrier PWM strategies for cascaded and neutral point clamped multilevel inverters, in *2000 IEEE 31st Annual Power Electronics Specialists Conference, 2000 (PESC 00)*, Vol. 2, pp. 674–679, 2000.
45. A. Lewicki, Z. Krzeminski, and H. Abu-Rub, Space-vector pulsewidth modulation for three-level NPC converter with the neutral point voltage control, *IEEE Transactions on Industrial Electronics*, 58, 5076–5086, 2011.
46. J. Pou, J. Zaragoza, S. Ceballos, M. Saeedifard, and D. Boroyevich, A carrier-based PWM strategy with zero-sequence voltage injection for a three-level neutral-point-clamped converter, *IEEE Transactions on Power Electronics*, 27, 642–651, 2012.
47. R. Teodorescu, M. Liserre, and P. Rodriguez, Eds., *Grid Converters for Photovoltaic and Wind Power Systems*. Chichester, West Sussex, U.K.: Wiley-IEEE Press, 2011.
48. M. K. Ghartemani, S. A. Khajehoddin, P. K. Jain, and A. Bakhshai, Problems of startup and phase jumps in PLL systems, *IEEE Transactions on Power Electronics*, 27, 1830–1838, 2012.
49. M. Ciobotaru, R. Teodorescu, and F. Blaabjerg, A new single-phase PLL structure based on second order generalized integrator, in *37th IEEE Power Electronics Specialists Conference, 2006 (PESC '06)*, pp. 1–6, 2006.
50. I. Carugati, P. Donato, S. Maestri, D. Carrica, and M. Benedetti, Frequency adaptive PLL for polluted single-phase grids, *IEEE Transactions on Power Electronics*, 27, 2396–2404, 2012.
51. C. Se-Kyo, A phase tracking system for three phase utility interface inverters, *IEEE Transactions on Power Electronics*, 15, 431–438, 2000.
52. P. Rodriguez, A. Luna, R. S. Munoz-Aguilar, I. Etxeberria-Otadui, R. Teodorescu, and F. Blaabjerg, A stationary reference frame grid synchronization system for three-phase grid-connected power converters under adverse grid conditions, *IEEE Transactions on Power Electronics*, 27, 99–112, 2012.
53. N. Femia, G. Petrone, G. Spagnuolo, and M. Vitelli, *Power Electronics and Control Techniques for Maximum Energy Harvesting in Photovoltaic Systems*. London, U.K.: CRC Press, 2012.
54. D. Sera, L. Mathe, T. Kerekes, S. V. Spataru, and R. Teodorescu, On the perturb-and-observe and incremental conductance MPPT methods for PV systems, *IEEE Journal of Photovoltaics*, 3, 1070–1078, July 2013.
55. L. V. Hartmann, M. A. Vitorino, M. B. R. Correa, and A. M. N. Lima, Combining model-based and heuristic techniques for fast tracking the maximum-power point of photovoltaic systems, *IEEE Transactions on Power Electronics*, 28, 2875–2885, 2013.
56. C. Chian-Song, T-S fuzzy maximum power point tracking control of solar power generation systems, *IEEE Transactions on Energy Conversion*, 25, 1123–1132, 2010.
57. M. Miyatake, M. Veerachary, F. Toriumi, N. Fujii, and H. Ko, Maximum power point tracking of multiple photovoltaic arrays: A PSO approach, *IEEE Transactions on Aerospace and Electronic Systems*, 47, 367–380, 2011.
58. G. Petrone, G. Spagnuolo, and M. Vitelli, An analog technique for distributed MPPT PV applications, *IEEE Transactions on Industrial Electronics*, 59, 4713–4722, 2012.
59. R. Alonso, E. Roman, A. Sanz, V. E. M. Santos, and P. Ibanez, Analysis of inverter-voltage influence on distributed MPPT architecture performance, *IEEE Transactions on Industrial Electronics*, 59, 3900–3907, 2012.
60. N. Femia, G. Lisi, G. Petrone, G. Spagnuolo, and M. Vitelli, Distributed maximum power point tracking of photovoltaic arrays: Novel approach and system analysis, *IEEE Transactions on Industrial Electronics*, 55, 2610–2621, July 2008.
61. D. Sera, L. Mathe, and F. Blaabjerg, Hierarchical control of PV strings with module integrated converters in presence of a central MPPT, in *IEEE Energy Conversion Congress and Exposition ECCE 2014*, Pittsburgh, PA, 2014.
62. M. Vitelli, On the necessity of joint adoption of both distributed maximum power point tracking and central maximum power point tracking in PV systems, *Progress in Photovoltaics: Research and Applications*, 22, 283–299, 2012.

63. G. Petrone, G. Spagnuolo, and M. Vitelli, Distributed maximum power point tracking: Challenges and commercial solutions, *Automatika*, 53, 128–141, April–June 2012.

64. B.-I. Craciun, T. Kerekes, D. Sera, R. Teodorescu, and A. Udaya, Active power reserves evaluation in large scale PVPPs, in *International Workshop on Integration of Solar Power into Power System*, London, U.K., 2013.

65. B. I. Craciun, T. Kerekes, D. Sera, and R. Teodorescu, Overview of recent grid codes for PV power integration, in *2012 13th International Conference on Optimization of Electrical and Electronic Equipment (OPTIM)*, pp. 959–965, Brasov, Romania, 2012.

66. B.-I. Craciun, S. Spataru, T. Kerekes, D. Sera, and R. Teodorescu, Internal active power reserve management in large-scale PV power plants, in *International Workshop on Integration of Solar Power into Power Systems SIW2014*, Berlin, Germany, 2014.

67. B.-I. Craciun, T. Kerekes, D. Sera, R. Teodorescu, and U. D. Annakkage, Power ramp limitation capabilities of large PV power plants with active power reserves, *IEEE Transactions on Sustainable Energy*, Vol. PP, no. 99, pp. 1–1, 2014.

68. N. Kakimoto, H. Satoh, S. Takayama, and K. Nakamura, Ramp-rate control of photovoltaic generator with electric double-layer capacitor, *IEEE Transactions on Energy Conversion*, 24, 465–473, 2009.

69. J. Marcos, L. Marroyo, E. Lorenzo, D. Alvira, and E. Izco, From irradiance to output power fluctuations: The PV plant as a low pass filter, *Progress in Photovoltaics: Research and Applications*, 19, 505–510, 2011.

70. B. I. Craciun, S. Spataru, T. Kerekes, D. Sera, and R. Teodorescu, Power ramp limitation and frequency support in large scale PVPPs without storage, in *2013 IEEE 39th Photovoltaic Specialists Conference (PVSC)*, pp. 2354–2359, Tampa, FL, 2013.

5 Overview of Maximum Power Point Tracking Techniques for Photovoltaic Energy Production Systems

Eftichios Koutroulis and Frede Blaabjerg

CONTENTS

Abstract 91
5.1 Introduction 92
5.2 Operation and Modeling of PV Modules and Arrays 95
5.3 MPPT Methods for PV Arrays Operating under Uniform Solar Irradiation Conditions 96
     5.3.1 Constant-Voltage and Constant-Current MPPT 97
     5.3.2 Perturbation and Observation MPPT 97
     5.3.3 Incremental-Conductance MPPT 99
     5.3.4 Model-Based MPPT 100
     5.3.5 Artificial Intelligence–Based MPPT 101
     5.3.6 Single-Sensor MPPT 102
     5.3.7 MPPT Methods Based on Numerical Optimization Algorithms 103
     5.3.8 Ripple Correlation Control (RCC) MPPT 105
     5.3.9 Extremum Seeking Control (ESC) MPPT 106
     5.3.10 MPPT Based on Sliding-Mode Control 107
     5.3.11 Comparison of MPPT Methods for Uniform Solar Irradiation Conditions 107
5.4 MPPT Methods for PV Arrays Operating under Nonuniform Solar Irradiation Conditions 110
     5.4.1 PV Array Reconfiguration 110
     5.4.2 Evolutionary MPPT Algorithms 110
     5.4.3 MPPT Methods Based on Numerical Optimization Algorithms 112
     5.4.4 Stochastic and Chaos-Based MPPT Algorithms 113
     5.4.5 Distributed MPPT 113
     5.4.6 Other Global MPPT Methods 116
     5.4.7 Comparison of Global MPPT Methods for Nonuniform Solar Irradiation
            Conditions 118
5.5 Simulation Examples 119
5.6 Summary 122
References 125

ABSTRACT

A substantial growth of the installed photovoltaic (PV) systems capacity has occurred around the world during the last decade, thus enhancing the availability of electric energy in an environmentally friendly way. The maximum power point tracking (MPPT) technique enables to maximize the energy production of PV sources, despite the stochastically varying solar irradiation and ambient temperature conditions. Thereby, the overall efficiency of the PV energy production system is increased. Numerous techniques have been presented during the last decades for implementing the MPPT process in a PV system. This chapter provides an overview of the operating principles of these techniques, which are suited for either uniform or nonuniform solar irradiation conditions. The operational characteristics and implementation requirements of these MPPT methods are also analyzed in order to demonstrate their performance features.

5.1 INTRODUCTION

Motivated by the concerns on environmental protection (sustainability) and energy availability, the installation of photovoltaic (PV) energy production systems has been increased substantially during the last years. The falling prices of PV modules and more efficient power conversion have assisted in that direction by enhancing the economic viability of the installed PV systems. More than 40 GW of new PV capacity was installed across the world during 2014, thus achieving a worldwide cumulative installed capacity of 178 GW during that year [1].

A basic block diagram of a PV energy production system is shown in Figure 5.1, with a PV array comprising a number of PV modules, a power converter, and also a control unit. The PV source is connected to a DC/DC or DC/AC power converter, respectively, interfacing the PV-generated power to the load, which typically is connected to the electric grid, or operates in a stand-alone mode (e.g., using a battery bank) [2, 3]. The pulse width modulation (PWM) controller of the control unit is responsible for producing the appropriate control signals (with an adjustable duty cycle), which drive the power switches (e.g., metal-oxide-semiconductor field-effect transistors (MOSFETs), insulated-gate bipolar transistors (IGBTs)) of the power converter. Its operation is based on measurements of the input/output voltage and current and of internal reference signals of the power converter.

Examples of the power–voltage characteristics of a PV array under various atmospheric conditions are illustrated in Figure 5.2a for the case that the same amount of solar irradiation is incident on all PV modules of the PV array [2]. It is observed that the power–voltage curves exhibit a unique point where the power produced by the PV module is maximized (i.e., maximum power point, MPP). However, in the case that the solar irradiation, which is incident on one or more of the PV modules, is different (e.g., due to dust, shading caused by buildings or trees), then the power–voltage characteristic of the PV array is distorted, exhibiting one or more local MPPs (see Figure 5.2b) [4]. Among them, the operating point where the output power is maximized corresponds to the global

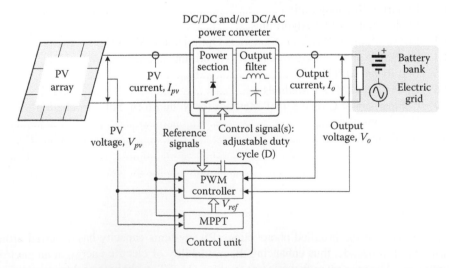

FIGURE 5.1 A block diagram of a PV energy production system including a maximum power point tracking (MPPT) unit to harvest maximum power.

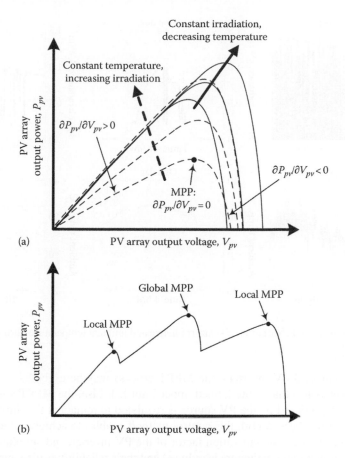

FIGURE 5.2 Examples of the power–voltage characteristics of a PV array: (a) under uniform solar irradiation conditions and (b) under partial shading conditions (different MPPs).

MPP of the PV array. However, the power generated by the PV array at the global MPP is less than the sum of the power values produced by the individual PV modules, when operating at their respective MPPs. The number and position of the local MPPs on the power–voltage curve of the PV array depend on both the configuration (i.e., connection in series and/or parallel) of the PV modules in the PV array and the time-varying form of the shading pattern on the surface of the PV modules.

As shown in Figure 5.3, the solar irradiance and ambient temperature conditions exhibit a stochastic variation during a year, a day, and an hour, respectively. During these operating conditions, the location of the MPPs on the power–voltage curve of the PV array varies accordingly. Thus, an appropriate operation is incorporated into the control unit of the PV energy production system, as also shown in Figure 5.1, for continuously adjusting the operation of the power converter under the stochastically changing weather conditions, such that the operating point of the PV array, which is determined by its output voltage and current, always corresponds to the global MPP. This process is termed as maximum power point tracking (MPPT). Employing an MPPT process is indispensable to every PV energy production system in order to ensure that the available PV energy production is optimally exploited, thus maximizing the energy production and therefore reducing the cost of the energy generated. Depending on the type of PV application, the MPPT process operates by controlling the power converter of the PV system based on measurements of the PV array output voltage and current, and it is appropriately integrated into the energy management algorithm, which is executed by the control unit. For example, in PV systems containing a battery energy storage unit, the battery charging control is also performed to protect the batteries from overcharging [2]. Also, in

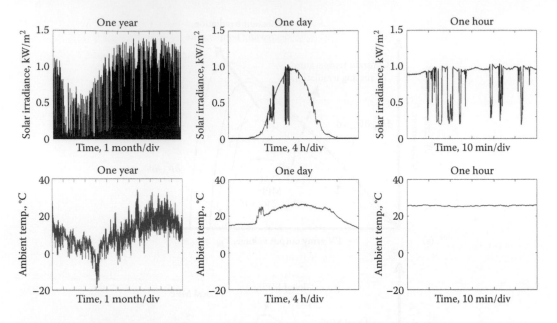

FIGURE 5.3 Examples of the variation of solar irradiance and ambient temperature during a year, a day, and an hour, respectively.

the case of grid-connected PV inverters, the MPPT process may be executed only as long as the PV-generated power is less than a predefined upper limit [5]. Else, the MPPT algorithm is deactivated and the power produced by the PV source is regulated to remain at that limit. By controlling the feed-in power to the electric grid, this control method enables to achieve a better utilization of the electric grid and increase the utilization factor of the PV inverter, and simultaneously, the thermal loading of its power semiconductors is reduced and their reliability is also increased.

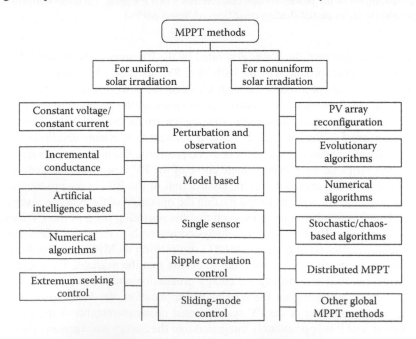

FIGURE 5.4 An overview of MPPT methods for PV arrays operating under uniform or nonuniform solar irradiation conditions.

Various methods have been presented in the research literature for performing the MPPT process under either uniform incident solar irradiation or partial shading conditions. Each of these methods is based on different operating principles and exhibits different hardware implementation requirements and performance [6–14]. In this chapter, the operation and modeling of PV modules and arrays are initially described. Then, the operating principles of the existing MPPT methods, which are suited for either uniform or nonuniform solar irradiation conditions, are analyzed with a primary focus on the MPPT techniques developed during the last years. An overview of these MPPT methods is presented in Figure 5.4. Their implementation requirements and operational characteristics are also analyzed in order to demonstrate their performance features, thus assisting the designers of PV power processing systems to select the MPPT technique, which would be most suitable for incorporation in their specific PV application.

5.2 OPERATION AND MODELING OF PV MODULES AND ARRAYS

The elementary structural units of a PV source are the solar cells, which operate according to the PV effect [15]. The photons of the incident solar irradiation are absorbed by the semiconducting material of the solar cell, exciting the generation of electron–hole pairs, which results in the flow of electric current when the solar cell is connected to an electric circuit. Multiple solar cells are connected electrically in series and parallel, thus forming a PV module. A PV array consists of multiple PV modules, also connected in series and parallel in order to comply with the voltage and power level requirements of the PV system. Currently, the PV modules that are available in the market are constructed using materials such as multicrystalline silicon and monocrystalline silicon, as well as by employing thin-film technologies based on cadmium telluride (CdTe), copper indium gallium selenide (CIGS), and amorphous silicon [16]. The multijunction solar cells are expensive, but they have the ability to operate under a high level of solar irradiation intensity with high efficiency; thus, they are mostly used in space and high-concentration PV applications. Also, new solar cell technologies such as the dye-sensitized and organic cells are under development [17] and seem to be improving year by year. A comparison of the primary commercially available PV module technologies in terms of efficiency and degradation rate of power production is presented in Table 5.1 [18, 19].

Various models have been presented in the scientific literature for describing the current–voltage characteristic of a PV source in order to estimate its performance under real operating conditions [20]. Due to its simplicity and ability to provide sufficient accuracy for a wide variety of studies and applications, the single-diode, five-parameter model is widely adopted (see Figure 5.5a). According to this model, the output current, I_{pv}, and power, P_{pv}, of a PV source are given by the following equations [21]:

$$I_{pv} = I_{ph} - I_s \cdot \left(e^{\frac{q \cdot \left(V_{pv} + I_{pv} \cdot R_s \right)}{N_s \cdot n \cdot k \cdot T}} - 1 \right) - \frac{V_{pv} + I_{pv} \cdot R_s}{R_p} \tag{5.1}$$

$$P_{pv} = V_{pv} \cdot I_{pv} \tag{5.2}$$

TABLE 5.1

Comparison of the Commercially Available Photovoltaic Module Technologies

	Multicrystalline Silicon	Monocrystalline Silicon	CdTe	CIGS	Amorphous Silicon
Average efficiency (%)	14.1–15.4	14.8–17.0	11.9–12.7	11.7–12.0	7.0–10.2
Degradation rate of power production (%/year)	0.61–0.64	0.36–0.47	0.40–3.33	0.96–1.44	0.87–0.96

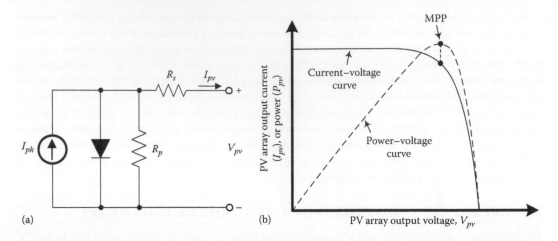

FIGURE 5.5   The single-diode model of a PV module/array: (a) the equivalent circuit and (b) the corresponding current–voltage and power–voltage characteristics.

where

V_{pv} is the output voltage of the PV source

I_{ph} is the photocurrent

I_s is the reverse saturation current

q is the electron charge ($q = 1.602176565 \cdot 10^{-19}$ C)

n is the ideality factor of the solar cells

k is the Boltzmann constant ($k = 1.3806488 \cdot 10^{-23}$ J/K)

N_s is the number of solar cells connected in series

T (K) is the (absolute) temperature of the solar cells

R_s, R_p are the series and parallel, respectively, resistance of the PV source

The values of I_{ph} and T in (5.1) depend on the solar irradiation and ambient temperature mission profile [15] (e.g., see Figure 5.3). The impact of R_p is usually neglected, while due to the small value of R_s, the short-circuit current of the PV module/array, $I_{sc} = I_{pv}|_{V_{pv}=0}$, is approximately equal to I_{ph}. The short-circuit current I_{sc} depends on the solar irradiance, which is incident on the surface of the PV source. The open-circuit voltage of the PV source, V_{oc}, is derived by setting $I_{pv} = 0$ in (5.1) and its value is affected significantly by the temperature of the solar cells. The current–voltage and power–voltage characteristics of a PV module/array are shown in Figure 5.5b. The location of the MPP on these curves is also illustrated in Figure 5.5b. When the meteorological conditions vary (like shown in Figure 5.3), the shape of the current–voltage and power–voltage characteristics is also modified, according to (5.1) and (5.2), respectively, and the position of the MPP changes (see Figure 5.2a).

The PV module/array model described previously can be used in simulation studies either for evaluating the performance of a PV system under uniform or nonuniform solar irradiance at the individual PV modules of the PV source, given the meteorological conditions at the installation site (e.g., Figure 5.2), or for implementing an MPPT method, as described next.

5.3   MPPT METHODS FOR PV ARRAYS OPERATING UNDER UNIFORM SOLAR IRRADIATION CONDITIONS

This class of MPPT techniques is suited for application in cases where the PV modules of the PV source operate under uniform solar irradiation conditions. In such a case, the power–voltage characteristic of the PV source exhibits a unique MPP. However, due to the short- and long-term variability of solar irradiation and ambient temperature (see Figure 5.3), the position of the MPP will be changed accordingly. Thus, the application of an MPPT control algorithm is required, which is capable of guaranteeing fast convergence to the continuously moving MPP of the PV source in order to optimize the energy production of the PV system. The operating principles of alternative techniques, which belong to this class of MPPT methods (see Figure 5.4), along with a comparison of their operational characteristics, are presented next.

5.3.1 CONSTANT-VOLTAGE AND CONSTANT-CURRENT MPPT

The constant-voltage (also referred to as fractional open-circuit voltage) MPPT technique is based on the assumption that the ratio of the MPP voltage to the open-circuit voltage of a PV module remains relatively constant at 70%–85% [22, 23]. Thus, by periodically disconnecting the power converter (see Figure 5.1) from the PV array, the output current of the PV array is set to zero and the resulting open-circuit voltage is measured. In the constant-current (or fractional short-circuit current) MPPT method, a similar approach is adopted [24]. In this case, the MPPT process is based on the assumption that the MPP power is proportional to the short-circuit current, which is measured by periodically setting the PV module/array under short-circuit conditions, through a power switch. In both the constant-voltage and constant-current MPPT methods, the corresponding MPP voltage is calculated by the control unit according to the measurements of the open-circuit voltage and short-circuit current, respectively, and then the power converter is regulated to operate at that point.

The constant-voltage and constant-current MPPT methods require only one sensor for their implementation (i.e., a voltage and current sensor, respectively), but the periodic interruption of the PV source operation for measuring the open-circuit voltage/short-circuit current results in power loss. In both of these methods, the accuracy of tracking the MPP is affected by the accuracy of knowing the value of the proportionality factors between the open-circuit voltage and short-circuit current, respectively, with the corresponding values at the MPP for the specific PV module/array used in each installation, as well as their variations with temperature and aging.

5.3.2 PERTURBATION AND OBSERVATION MPPT

The perturbation and observation (P&O) MPPT method is based on the property that the derivative of the power–voltage characteristic of the PV module/array is positive on the left side and negative on the right side (see Figure 5.2a), while at the MPP, it holds that

$$\frac{\partial P_{pv}}{\partial V_{pv}} = 0 \tag{5.3}$$

where P_{pv} and V_{pv} are the output power and voltage, respectively, of the PV module/array.

During the execution of the P&O MPPT process, the output voltage and current of the PV module/array are periodically sampled at consecutive sampling steps in order to calculate the corresponding output power and the power derivative with voltage. The MPPT process is performed by adjusting the reference signal of the power converter PWM controller (see Figure 5.1), V_{ref}, based on the sign of $\frac{\partial P_{pv}}{\partial V_{pv}}$, according to the following equation:

$$V_{ref}(k) = V_{ref}(k-1) + \alpha \cdot \text{sign}\left(\frac{\partial P_{pv}}{\partial V_{pv}}(k)\right) \tag{5.4}$$

where k, $k-1$ are consecutive time steps, $\alpha > 0$ is a constant determining the speed of convergence to the MPP, and the function sign$(\cdot)$ is defined as follows:

$$\text{sign}(x) = \begin{cases} 1 & \text{if } x > 0 \\ -1 & \text{if } x < 0 \end{cases} \tag{5.5}$$

The PV module/array output voltage is regulated to the desired value, which is determined by V_{ref} according to (5.4), using either a proportional integral (PI) or, for example, a fuzzy logic controller. The latter has the advantage of providing a better response under dynamic conditions [25]. Under steady-state conditions, the operating point of the PV module/array oscillates around the MPP with an amplitude determined by the value of α in (5.4). Increasing the perturbation step enables to converge faster to the MPP under changing solar irradiation and/or ambient temperature conditions but increases the steady-state oscillations around the MPP, thus resulting in power loss.

An MPPT system based on the P&O method can be developed by either implementing (5.4) in the form of an algorithm executed by a microcontroller or digital signal processing (DSP) unit or using mixed-signal circuits. A flowchart of the P&O MPPT algorithm based on the procedure proposed in [26], which can be executed by a microcontroller or DSP device of the control unit, is presented in Figure 5.6. The process shown in Figure 5.6 is executed iteratively until the value of the gradient $\partial P_{pv}/\partial V_{pv}$ drops below a predefined threshold, indicating that convergence close to the MPP has been achieved with the desired accuracy.

A methodology for the design of the control unit such that the P&O MPPT process operates with the optimal values of step size and perturbation period is proposed in [27]. The optimal perturbation period is calculated in [28] for adapting to the time-varying meteorological conditions, using a field-programmable gate array (FPGA) control unit, which executes the P&O-based MPPT process. An algorithm for dynamically adapting the perturbation size according to the solar irradiation conditions is presented in [29] for increasing the response speed of the P&O algorithm and reducing the steady-state oscillation around the MPP. The short-circuit current of the PV source is estimated in [30] during the execution of the P&O algorithm by applying the measured values of

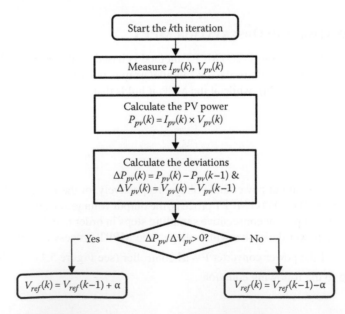

FIGURE 5.6 A flowchart of the algorithm implementing the P&O MPPT process based on the procedure proposed in Hua et al. (1998).

the current and voltage in the single-diode model of the PV modules. The resulting value is used to detect whether a variation of the PV source output power is due to a change of the solar irradiation conditions or the MPPT process itself.

The P&O method is characterized by its operational and implementation simplicity. However, it exhibits a slow convergence speed under varying solar irradiation conditions and its performance may also be affected by system noise.

5.3.3 INCREMENTAL-CONDUCTANCE MPPT

At the MPP of the PV source, it holds that

$$\frac{\partial P_{pv}}{\partial V_{pv}} = 0 \Rightarrow \frac{\partial \left(I_{pv} \cdot V_{pv} \right)}{\partial V_{pv}} = I_{pv} + \frac{\partial I_{pv}}{\partial V_{pv}} V_{pv} = 0 \Rightarrow \frac{\partial I_{pv}}{\partial V_{pv}} = -\frac{I_{pv}}{V_{pv}} \tag{5.6}$$

where I_{pv} is the output current of the PV array.

Due to the shape of the current–voltage characteristic of the PV module/array in Figure 5.5b, the value of $\frac{\partial I_{pv}}{\partial V_{pv}}$ is higher than $-\frac{I_{pv}}{V_{pv}}$ at the left side of the MPP and lower than $-\frac{I_{pv}}{V_{pv}}$ at its right side. The incremental-conductance (InC) MPPT technique operates by measuring the PV module/array output voltage and current and comparing the value of $\frac{\partial I_{pv}}{\partial V_{pv}}$ with $-\frac{I_{pv}}{V_{pv}}$. Then, the power converter is controlled based on the result of this comparison, according to the flowchart illustrated in Figure 5.7, which is based on the procedure presented in [31]. Similar to the P&O process, the execution of the algorithm shown in Figure 5.7 is iteratively repeated until the difference between $\frac{\partial I_{pv}}{\partial V_{pv}}$ and $-\frac{I_{pv}}{V_{pv}}$ is less than a predefined value, which indicates that the MPP has been tracked with an acceptable accuracy.

Alternatively, the InC method may be implemented by controlling the power converter according to the sign of $I_{pv} + \frac{\partial I_{pv}}{\partial V_{pv}} V_{pv}$ such that its value is adjusted to zero as dictated by (5.6).

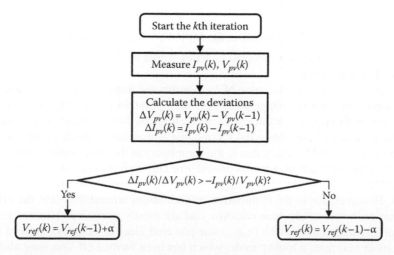

FIGURE 5.7 A flowchart of the InC MPPT algorithm based on the procedure presented in Elgendy et al. (2013).

Although the InC and P&O MPPT methods are based on the same operating principle, the former is implemented by using the individual measurements of the PV array output voltage and current, thus not requiring the computation of the corresponding output power. A variation of the InC algorithm, employing a dynamic adaptation of the step size during the tracking process, is proposed in [32].

In [33], it is demonstrated through experimental testing that the P&O and InC MPPT methods exhibit similar performance under both static and dynamic conditions.

5.3.4 MODEL-BASED MPPT

The operation of model-based MPPT methods is based on measuring the PV module/array output voltage and current at multiple operating points [34]. Using the resulting measurements, the parameters I_{ph}, I_s, V_T, and R_s, respectively, of the single-diode model of the PV source, which has been described in Section 5.2, are initially estimated (the shunt resistance R_p is neglected). Then, (5.1) is used to calculate the voltage and current of the PV source at the operating point, where the derivative of power with respect to the voltage is equal to zero (i.e., MPP) by applying numerical techniques (e.g., Newton–Raphson method). A similar approach has also been employed in [35], where successive measurements of the PV module output voltage are iteratively applied in a simplified empirical mathematical model of the PV module, until a convergence to the MPP has been achieved.

In [36], analytical equations are derived, which enable to calculate the PV module current and voltage at the MPP, as follows:

$$I_m = I_{1m} - \frac{V_{1m}}{R_p} \tag{5.7}$$

$$V_m = V_{1m} - I_m R_s \tag{5.8}$$

where

$$V_{1m} = n\alpha V_T \left[W\left(\frac{I_{ph} \cdot \exp(1)}{I_s} \right) - 1 \right] \tag{5.9}$$

and $I_m = I_{pv}|_{MPP}$ is the PV module current at the MPP, $V_m = V_{pv}|_{MPP}$ is the PV module voltage at the MPP, n is the number of PV cells connected in series within the PV module, α is the quality factor, and $W(\cdot)$ is the Lambert function.

In order to apply this method, the value of I_{ph} is estimated from (5.1) using measurements of the PV module output current and voltage at an operating point away from the open-circuit voltage. When using this technique, the accuracy of predicting the MPP voltage and current is highly affected by the accuracy of estimating the PV module temperature, which affects the values of V_T and I_s applied in (5.1) and (5.9). Also, due to the complexity of the computations required for calculating the MPP voltage or current, a microcontroller or DSP unit is required for the implementation of such an MPPT scheme, while additionally, the response speed of the MPP control algorithm is relatively low. However, due to the elimination of oscillations around the MPP, the MPPT units of this type achieve a better steady-state response and are mostly attractive in cases of continuously changing solar irradiation conditions (e.g., solar-powered electric vehicles). Instead of solving a set of equations in real time, a lookup table, which has been formed off-line, may also be used for calculating the MPP voltage [37], but this method is also characterized by computational complexity and requires knowledge of the operational characteristics of the PV source. In [38], the output of an

MPPT subsystem operating according to the InC method is added to the output of a model-based MPP tracker, thus forming a hybrid MPPT controller.

The model-based MPPT techniques have the advantage of not disconnecting the PV source during the execution of the MPPT process. The accuracy of the model-based MPPT method is affected by the accuracy of the single-diode model of the PV source, as well as by the aging of the PV modules, which results in the modification of the values of the PV module operating parameters during the PV system operational lifetime period.

5.3.5 ARTIFICIAL INTELLIGENCE–BASED MPPT

Artificial intelligence techniques, such as neural networks and fuzzy logic, have also been applied for performing the MPPT process. In the former case, measurements of solar irradiation and ambient temperature are fed into an artificial neural network (ANN) and the corresponding optimal value of the DC/DC power converter duty cycle is estimated, as shown in the diagram of Figure 5.8a, which is based on the structure presented in [39]. In order to obtain accurate results, the ANN will need to have been trained using a large amount of measurements prior to its real-time operation in the MPPT control unit [40], which is a disadvantage.

The controllers based on fuzzy logic have the ability to calculate the value of the power converter control signal (e.g., duty cycle) for achieving operation at the MPP using measurements of an error signal, e (e.g., $e = \dfrac{\partial P_{pv}}{\partial I_{pv}}$, $e = \dfrac{\partial P_{pv}}{\partial V_{pv}}$, or $e = \dfrac{\partial I_{pv}}{\partial V_{pv}} + \dfrac{I_{pv}}{V_{pv}}$), and its variation with time (i.e., Δe) [41, 42]. The structure of an MPPT scheme, which is employing a fuzzy logic controller based on the method proposed in [42], is presented in Figure 5.8b. The values of e and Δe are assigned by the fuzzy logic–based controller to linguistic variables such as "negative big", "positive small", etc. and the appropriate membership functions are applied. Based on the values resulting from this transformation, a lookup table that contains the desired control rules is used to calculate the output of the controller in the form of alternative linguistic variables, which then

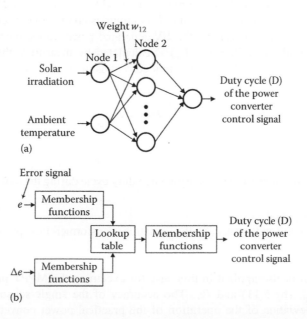

FIGURE 5.8 The structure of artificial intelligence–based techniques for MPPT: (a) ANN based on the architecture presented in Charfi et al. (2014) and (b) fuzzy logic controller based on the method proposed in Adly et al. (2012).

are combined through the corresponding membership functions into a numerical value (defuzzification stage), thus producing the duty cycle of the control signal driving the power converter such that the MPP is tracked. The fuzzy logic controllers have the advantage of not requiring knowledge of the exact model of the system under control. However, in order to obtain an effective performance, expert knowledge is required for forming the membership functions and rule sets. Thus, optimization algorithms such as genetic algorithms (GAs) and ant colony optimization have been applied for tuning the operational parameters of fuzzy logic controllers [42], while in [40] the structure of an ANN is exploited for that purpose.

5.3.6 SINGLE-SENSOR MPPT

The implementation of P&O and InC methods requires measurement of the PV module/array output current. The accuracy of current measurements is affected by the current sensor bandwidth and switching ripple imposed on the PV source output current due to the switching operation of the power converter. Additionally, the use of a current sensor increases the cost and power consumption of the MPPT control unit.

As analyzed in [43], the output power of the PV module/array is given by

$$P_{pv} = V_{pv} \cdot I_{pv} = V_{pv} \cdot \frac{V_{pv}}{R_{in}} = k \cdot V_{pv}^2 \qquad (5.10)$$

where R_{in} is the input resistance of the power converter, which is a function of the duty cycle of the control signal driving the power converter, and $k = \dfrac{1}{R_{in}}$.

The value of V_{pv} in (5.10) also depends on R_{in}. Thus, by modifying the control-signal duty cycle, this will affect the resulting operating values of both V_{pv} and P_{pv}. The power produced by the PV source can be calculated by applying in (5.10) the measurements of the PV module/array output voltage, thus avoiding the direct measurement of the corresponding output current. Using the calculated value of P_{pv}, the P&O algorithm may be applied for executing the MPPT process.

In [44], a flyback inverter operating in the discontinuous conduction mode is connected at the output of the PV module for interfacing the PV-generated power to the electric grid. The output power of the PV module (i.e., $P_{pv} = V_{pv} \cdot I_{pv}$) is calculated by measuring the PV module output voltage and also calculating the PV source output current using the following equation (assuming a lossless power converter):

$$I_{pv} = \frac{1}{4} \cdot \frac{D_{max}^2 \cdot T_s}{L_m} \cdot V_{pv} \qquad (5.11)$$

where

D_{max} is the maximum value of the primary-switch duty cycle during the half period of the electric grid voltage

T_s is the switching period

L_m is the magnetizing inductance of the isolation transformer incorporated into the flyback inverter circuit

A P&O algorithm is also applied in this case for executing the MPPT process using the calculated values of I_{pv} (by 5.11) and P_{pv}. The accuracy of the single-sensor MPPT approaches is affected by the deviation of the operation of the practical power converter circuit from that predicted by the theoretical Equations 5.10 and 5.11, respectively, due to the tolerance of the electric/electronic components values, circuit parasitics, etc. The MPPT accuracy of this method can be improved if the MPPT control unit is modified such that the deviation mentioned earlier is compensated by employing a suitable model of the power converter, but the complexity of the control unit would also be increased in such case.

5.3.7 MPPT Methods Based on Numerical Optimization Algorithms

A simple approach for deriving the position of a PV source MPP is to apply an exhaustive search process, where the entire power–voltage characteristic is sequentially scanned. By measuring and comparing the power production levels at the individual operating points that the PV source is set to operate at during the power–voltage curve scanning, the MPP position can be detected. Since this process requires a large number of search steps to be executed, which results in power loss until the tracking process has been accomplished, various MPPT algorithms based on numerical optimization techniques have been applied in order to detect the position of the MPP on the power–voltage curve of the PV array with reduced search steps.

A golden section search algorithm has been employed in [45], where the MPPT process is performed by iteratively narrowing the range of the PV output voltage values where the MPP resides. For each search range $\left[V_{\min}, V_{\max}\right]$ (initially it holds that $V_{\min} = 0$ and $V_{\max} = V_{oc}$), the output power of the PV source is measured at two operating points of the PV source, where the values of the PV source output voltage (i.e., parameter V_{pv} in Figures 5.1, 5.2, and 5.5b), $V_{pv,1}$ and $V_{pv,2}$, respectively, are given by

$$V_{pv,1} = V_{\max} - r \cdot \left(V_{\max} - V_{\min}\right) \tag{5.12}$$

$$V_{pv,2} = V_{\min} + r \cdot \left(V_{\max} - V_{\min}\right) \tag{5.13}$$

where $r = 0.618$ so that $V_{pv,1}$, $V_{pv,2}$ are placed symmetrically within $\left[V_{\min}, V_{\max}\right]$ and, also, $V_{pv,2}$ is placed at a position where the ratio of its distances from $V_{pv,1}$ and $V_{\max}$, respectively, is equal to the ratio of distances of $V_{pv,1}$ from $V_{\min}$ and $V_{\max}$, respectively.

Then, the PV module/array output power is measured at $V_{pv,1}$ and $V_{pv,2}$. If the output power at $V_{pv,1}$ is higher than that at $V_{pv,2}$, then it is set that $V_{\max} = V_{pv,2}$, or else it is set that $V_{\min} = V_{pv,1}$. This process is repeated until the distance between $V_{\min}$ and $V_{\max}$ is smaller than a predefined value.

In [46], a multistage MPPT process is presented, which comprises a combination of the P&O, golden section search, and InC algorithms. A flowchart of this process, which is based on the method proposed in [46], is depicted in Figure 5.9. Initially, the P&O algorithm is applied with a large perturbation step in order to quickly converge close to the MPP. Then, the golden section search algorithm is applied for accurately and quickly detecting the MPP, and finally, the InC algorithm is executed for ensuring the operation at the MPP in steady state and for triggering the initiation of a new search process in case that a large deviation from the MPP is detected (i.e., when $\partial P_{pv}/\partial V_{pv} > \varepsilon$, where ε is a preset threshold) due to changing environmental conditions.

An iterative approach, where the search window is progressively modified, is also performed in the linear iteration algorithm presented in [47]. However, in that case, the new search range at each iteration of the algorithm is calculated based on the power slope of the abscissa on the power–voltage characteristic of the point, which is defined as the intersection of the tangent lines at $V_{\min}$ and $V_{\max}$ (i.e., point Q in Figure 5.10, which is based on the procedure proposed in [47]). If the gradient at point Q is positive, then Q is set as the new lower limit of the search range, or else it will be the upper limit.

In the parabolic prediction MPPT algorithm [48], the power–voltage curve of the PV source, $P_{pv}\left(V_{pv}\right)$, is approximated by a parabolic curve, $Q\left(V_{pv}\right)$ (in Watt), which is given by

$$Q\left(V_{pv}\right) = P_{pv}\left(V_0\right)\frac{\left(V_{pv} - V_1\right)\cdot\left(V_{pv} - V_2\right)}{\Delta V_{01} \cdot \Delta V_{02}} + P_{pv}\left(V_1\right)\frac{\left(V_{pv} - V_0\right)\cdot\left(V_{pv} - V_2\right)}{\Delta V_{10} \cdot \Delta V_{12}} + P_{pv}\left(V_2\right)\frac{\left(V_{pv} - V_0\right)\cdot\left(V_{pv} - V_1\right)}{\Delta V_{20} \cdot \Delta V_{21}} \tag{5.14}$$

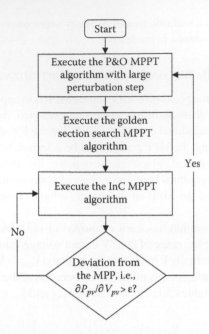

FIGURE 5.9 A flowchart of a multistage MPPT process, comprising the P&O, golden section search, and InC algorithms, based on the procedure proposed in Shao et al. (2014).

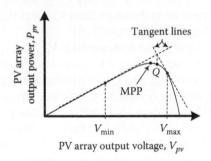

FIGURE 5.10 The operating principle of the linear iteration process, using numerical optimization algorithm for MPPT, based on the procedure proposed in Xu et al. (2014).

where V_i is the output voltage of the PV source (i.e., parameter V_{pv} in Figures 5.1, 5.2, and 5.5b) at the ith operating point, $\Delta V_{ij} = V_i - V_j$, and $i, j = 0, 1, 2$.

During the execution of the MPPT process, the output power and voltage of the PV source are measured at three operating points (e.g., A, B, and C in Figure 5.11, which is based on the procedure proposed in [48]) and the corresponding parabolic curve is calculated using (5.14). The resulting parabolic curve is used to estimate the MPP location (i.e., D in Figure 5.11), which deviates from the real MPP of the PV source depicted in Figure 5.11. Similarly, a new parabolic curve is calculated at the next iteration of the algorithm using the three operating points, which produce the highest values of power (i.e., B, D, and C in Figure 5.11), resulting in operation at E, which is closer to the MPP than point D. This process is repeated until the power deviation of the MPPs calculated at two successive iterations is less than a predefined level.

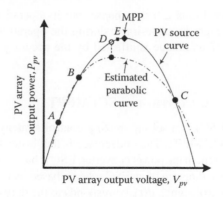

FIGURE 5.11 The parabolic prediction MPPT algorithm based on the procedure proposed in Pai et al. (2011).

Although effective in deriving the MPP of the PV source, these techniques exhibit higher implementation complexity as compared to the simpler algorithms, such as the P&O and InC MPPT methods.

5.3.8 RIPPLE CORRELATION CONTROL (RCC) MPPT

In order to avoid employing a derivative for performing the MPPT process, the gradient of the power–voltage curve, $\dfrac{\partial P_{pv}}{\partial V_{pv}}$, employed in the P&O technique for detecting the direction toward which the MPP resides is replaced in the ripple correlation control (RCC) MPPT method by the following correlation function [49]:

$$c(t) = \frac{\partial P_{pv}}{\partial t} \cdot \frac{\partial V_{pv}}{\partial t} \tag{5.15}$$

In case a DC/DC converter is used to interface the PV-generated energy to the load, the duty cycle, $d(t)$, of the power converter at time t, is adjusted according to the following control law:

$$d(t) = k \cdot \int_0^t \text{sign}(c_{lp}(\tau)) d\tau \tag{5.16}$$

where k is a constant and $c_{lp}(t)$ is the result of low-pass filtering the correlation function $c(t)$ given by (5.15).

In order to simplify the hardware implementation of the MPPT system, the values of $\dfrac{\partial P_{pv}}{\partial t}$ and $\dfrac{\partial V_{pv}}{\partial t}$ in (5.15) are calculated by measuring the AC disturbances (i.e., ripples) at the operating point of the PV source, which are due to the high-frequency switching operation of the power converter. The derivatives are measured using high-pass filters with a cutoff frequency higher than the ripple frequency (i.e., switching frequency) [50]. In [51], the PWM dithering technique is applied for increasing the resolution of the power converter PWM control signal. The resulting ripple in the output current and voltage of the PV source, which is due to the dithering process, is then exploited for applying the RCC MPPT method.

Targeting to increase the accuracy of the MPP tracking process, a variation of the RCC method is proposed in [52]. In this technique, the phase displacement of the PV source output voltage and current ripples is monitored. A change in this phase displacement indicates that the peak of the current ripple of the PV source has reached the MPP. Then, the DC component of the PV source output current is regulated at the value of the detected MPP.

The RCC MPPT method exhibits a fast response, but its operation is based on the existence of switching ripples, which might be undesirable during the operation of power converters. Also, the performance of this MPPT technique is affected by the accuracy of the measurements of the correlation function, $c(t)$.

5.3.9 EXTREMUM SEEKING CONTROL (ESC) MPPT

Extremum seeking control (ESC) is a self-optimizing control strategy [53], which operates based on a similar principle with RCC MPPT. Their difference is that instead of using the high-frequency switching ripple, which is inherent in the power converter, ESC is based on the injection of a sinusoidal perturbation [54–57]. The block diagram of an ESC scheme based on the method presented in [57] is shown in Figure 5.12. The control signal, $d(t)$, corresponds to the duty cycle of the power converter. The PV module/array output power, $P_{pv}(t)$, is passed through a high-pass filter and demodulated. The resulting signal (i.e., $f(t)$ in Figure 5.12a) has a positive sign in the case that the operating point is on the left side of the power–voltage curve in Figure 5.2a, since the perturbation and the PV source output power signals are in phase in that case, or else its sign is negative (Figure 5.12b). The control signal is produced by integrating $f(t)$ and then adding the perturbation $\alpha \cdot \sin(\omega t)$.

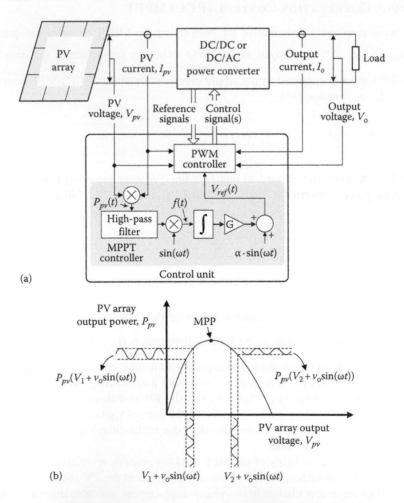

FIGURE 5.12 ESC-based PV MPPT based on the method presented in Malek et al. (2014): (a) a block diagram of the MPPT controller and (b) the resulting operating points on the power–voltage curve of the PV source.

The ESC-based MPPT process has the disadvantage that for its implementation in a PV power processing system, the development of a relatively complex control circuit is required.

5.3.10 MPPT Based on Sliding-Mode Control

In a sliding-mode control for MPPT, the output voltage of the PV source and the current of the power converter inductor comprise a set of state variables. A switching surface is defined using these state variables as follows [58]:

$$S\left(v_{pv}, i_{in}\right) = c_1 \cdot i_{in} - c_2 \cdot v_{pv} + V_{ref} \tag{5.17}$$

where

i_{in} is the current of the power converter inductor
c_1 and c_2 are positive constants
V_{ref} is an adjustable control signal

A block diagram of a sliding-mode control MPPT system based on the method proposed in [58] is illustrated in Figure 5.13. During operation, the value of $S\left(v_{pv}, i_{in}\right)$ is evaluated; in the case that $S\left(v_{pv}, i_{in}\right) > 0$, transistor T_1 is turned off and the energy is transferred towards the load, or else T_1 is turned on such that the energy is stored in the input inductor L. The value of V_{ref} is adjusted by a P&O MPPT algorithm such that the PV source operates at the MPP. In order to accelerate the convergence to the MPP, the values of c_1 and c_2 in (5.17) are selected such that the operating points $\left[v_{pv}, i_{in}\right]$, which are defined by $S\left(v_{pv}, i_{in}\right) = 0$, match to the locus of the PV source MPPs under various solar irradiation conditions, with the minimum possible deviation. Thus, for the implementation of this MPPT technique, knowledge of the PV source operational characteristics is required, which is a disadvantage.

As demonstrated in [58], compared to the MPPT based on the PWM principle, sliding-mode control for MPPT provides a faster response under dynamic conditions.

5.3.11 Comparison of MPPT Methods for Uniform Solar Irradiation Conditions

A comparison of the operational characteristics of the MPPT methods for uniform solar irradiation conditions is presented in Table 5.2. As analyzed in Sections 5.3.2 through 5.3.3, the P&O and InC methods are characterized by implementation simplicity and exhibit equivalent static and dynamic

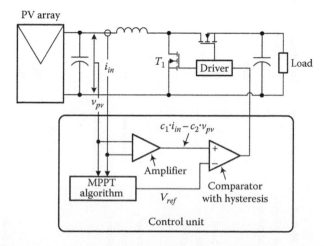

FIGURE 5.13 A block diagram of an MPPT system employing sliding-mode control, based on the method proposed in Levron et al. (2013).

TABLE 5.2

Comparison of the Operational Characteristics of the MPPT Methods for Uniform Solar Irradiation Conditions

MPPT Method	Sampling Rate	Complexity	Robustness		Efficiency	Sampled Parameters	System or Expert Knowledge Required	Constant-Power Operation
			To External Disturbances	To Aging of the PV Modules				
Constant voltage/constant current	High	Very low	Very high	Low	Low	PV voltage or current	High	Difficult
P&O	Low	Low	High	High	High	PV voltage and current	Low	Easy
InC	Low	Low	High	High	High	PV voltage and current	Low	Easy
Model based	Low	High	Very high	Low	Low	PV voltage and current	Very high	Easy
Artificial intelligence based	Low	Very high	Very high	ANN: low Fuzzy logic: high	Low	Irradiation and temperature PV current and voltage	Very high	Difficult Easy
Single sensor	Low	Low	High	High	Low	PV voltage	High	Difficult
Numerical optimization	Multistage and parabolic prediction: low Others: high	High	Multistage and parabolic prediction: high Others: low	High	Multistage and parabolic prediction: high Others: low	PV voltage and current	Low	Easy
RCC	Low	High	Low	Low	High	PV voltage and current	High	Easy
ESC	Low	High	High	High	High	PV voltage and current	High	Easy
Sliding-mode control	Low	Low	Very high	Low	Very high	PV voltage and current	High	Easy

performance. Although their operation can be affected by external disturbances (e.g., system noise, short-term or rapidly changing meteorological conditions, etc.), they are able to recover and move in the correct direction, where the MPP is located as soon as the disturbance has subsided. The constant-voltage, constant-current, model-based MPPT and artificial intelligence–based methods are more robust compared to the P&O and InC methods, since they are less affected by external disturbances. However, their efficiency is lower due to the periodic interruption of the PV source for measuring the open-circuit voltage/short-circuit current of the PV source. The resulting efficiency is further reduced in case accurate knowledge of the PV source operational parameters, which is required for their implementation, is not available. The single-sensor MPPT approach comprises a P&O MPPT method, thus exhibiting equivalent robustness to external disturbances with the P&O approach. However, its efficiency is lower because of the deviation of the power converter operation, which is predicted using a theoretical model, from the actual performance obtained under practical operating conditions. This is due to the tolerance of the electric/electronic components, values, circuit parasitics, etc. In the numerical optimization MPPT algorithms (except the multistage and parabolic prediction MPPT methods), a scan process is periodically repeated in order to detect possible changes of the MPP position, which results in efficiency reduction due to the associated power loss until convergence to the MPP has been achieved. The numerical optimization MPPT algorithms do not require a significant system knowledge for their application, but their implementation complexity is higher than that of the P&O and InC methods. Among the numerical optimization MPPT algorithms, the multistage and parabolic prediction MPPT methods exhibit similar performance to the P&O and InC methods. The robustness of the remaining numerical optimization MPPT algorithms is affected by external disturbances, since they are not able to recover from possible error estimations, which are performed due to the decisions taken during each iteration, until the next scan process is reinitiated. Due to the exploitation of the inherent, low-amplitude switching ripples of the power converter for performing the MPPT process, the robustness of the RCC MPPT technique may easily be affected by the impact of external disturbances on the accuracy of calculating the PV power–voltage correlation function. Additionally, appropriate codesign of the power converter and MPPT control system is required for the implementation of the RCC MPPT method, thus requiring system knowledge availablility. The control-circuit complexity of the RCC and ESC MPPT techniques is relatively high. A better robustness to external disturbances is obtained using the ESC method compared to the RCC-based MPPT approach since its operation is based on the injection of perturbation signals, rather than using the inherent, low-amplitude switching ripples of the power converter. However, detailed knowledge of the PV system operational characteristics is still required by the ESC method for tuning the operational parameters of the MPPT control loop. Both the RCC and ESC MPPT methods operate by employing a continuously operating feedback loop; thus, their efficiency is not affected by periodic disruptions of the PV source operation. The MPPT method based on sliding-mode control requires knowledge of the PV source operational characteristics. Since the PWM generator is replaced by a sliding-mode controller and a P&O MPPT process is also performed during its execution, the complexity of the corresponding control circuit is similar to that of the P&O MPPT process. However, a better efficiency and robustness to external disturbances may be obtained under dynamic conditions, compared to the PWM-based P&O MPPT method, due to the faster response of the sliding-mode MPPT controller.

The constant-voltage, constant-current, model-based MPPT and ANN-based and sliding-mode control methods operate based on the knowledge of the PV source operational characteristics. Thus, their accuracy is highly affected by the PV module aging, unless the drift of the PV source electrical characteristics with time is compensated through a suitable model, which may be difficult to derive and would increase the complexity of the control unit.

In contrast to the rest of the MPPT methods, which perform the MPPT process through a continuously operating feedback loop, the periodic reinitialization of the MPPT process required in the constant-voltage, constant-current, and numerical optimization techniques (except the multistage and parabolic prediction MPPT methods) imposes the need to apply a high sampling rate in order to be able to quickly detect the MPP position changes.

All MPPT methods presented in this section are suitable to accommodate a constant-power-mode control scheme [5], except the constant-voltage, constant-current, ANN-based and single-sensor MPPT techniques, since, due to the types of sampled parameters, which are employed in these methods, they do not comprise the sensors required to facilitate the measurement of the PV output power.

5.4 MPPT METHODS FOR PV ARRAYS OPERATING UNDER NONUNIFORM SOLAR IRRADIATION CONDITIONS

When the individual modules of the PV array receive unequal amount of solar irradiation, the power–voltage characteristic of the PV source exhibits multiple MPPs as their positions change frequently under the influence of stochastically varying meteorological conditions. In such circumstances, the target of an MPPT process is to derive, among the individual local MPPs of the PV source, the global MPP where the overall power production of the PV array is optimized. Multiple alternative techniques have been developed in the past, which are suited for application under nonuniform solar irradiation conditions (see Figure 5.4), and their operating principles are described and compared in the following.

5.4.1 PV Array Reconfiguration

In order to increase the power, which, for example, is supplied to a constant resistive load by a PV array operating under partial shading conditions, the use of a matrix of power switches has been proposed in [59]. Using this matrix, the connections between the PV cells/modules are dynamically modified, such that the PV strings comprise PV cells/modules operating under similar solar irradiation conditions.

The PV array reconfiguration method has the disadvantages of higher implementation complexity and cost due to the high number of power switches required but increases the energy production of the PV array. According to [6], since the power–voltage curve of the PV array after reconfiguration may still exhibit local MPPs, a power converter executing one of the global MPPT algorithms presented in the following should be connected at the output of the PV source, in order to optimize the generated power.

5.4.2 Evolutionary MPPT Algorithms

In this class of MPPT techniques, the MPPT process is treated as an optimization problem, where the optimal value of the decision variable is calculated in real time, such that the objective function, which corresponds to the power–voltage curve of the PV source, is optimized. Thus, various alternative evolutionary optimization algorithms, which in some cases have been inspired from biological and natural processes, have been applied for that purpose. A generalized flowchart of an evolutionary algorithm for implementing an MPPT process is shown in Figure 5.14a. Initially, the designer specifies the values of the optimization algorithm operational parameters, which define the speed and accuracy of convergence to the global optimum solution. During the execution of the optimization/MPPT process, multiple sets of values of the decision variable are produced in a way defined by the operating principle of the specific optimization algorithm, which has been employed. By appropriately controlling the power converter, the PV source is set to operate at alternative operating points corresponding to each of these sets (e.g., PV array output voltage levels $V_1 - V_k$ in the power–voltage curve shown in Figure 5.14b). At each operating point, the power generated by the PV source is measured (i.e., the objective function of the optimization problem is evaluated) and compared to the power produced at other operating points (e.g., PV array output power levels $P_1 - P_k$ in Figure 5.14b). This process is iteratively repeated, until a convergence criterion has been satisfied, which indicates that the position of the global MPP on the power–voltage curve has been derived

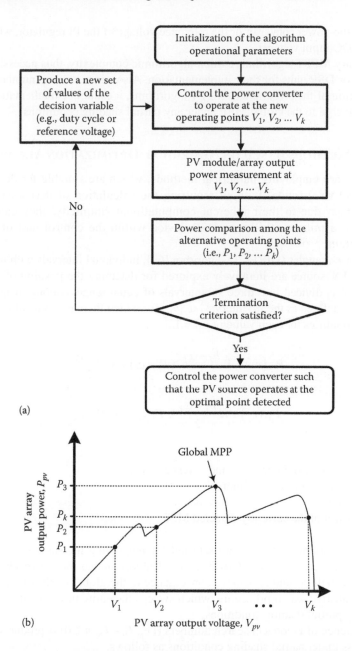

(a)

(b)

FIGURE 5.14 Operation of partially shaded modules: (a) a generalized flowchart of an evolutionary algorithm for implementing an MPPT process and (b) the operation of an evolutionary algorithm during the execution of the MPPT process.

(e.g., operating point $\left[V_3, P_3\right]$ in Figure 5.14b). Then, the PV source is set to operate at the optimal operating point derived during the execution of the optimization algorithm.

Alternative evolutionary algorithms and their variations have been applied for performing the MPPT process, such as GAs [60], differential evolution (DE) [61], particle swarm optimization (PSO) [62–64], the firefly algorithm [65], and the artificial bee colony algorithm [66]. A hybrid MPPT technique, which is a combination of the P&O and PSO algorithms, is proposed in [67], while the PSO and DE techniques are combined in [68] for performing the MPPT process. The decision variable employed during the application of the previously mentioned algorithms is either the duty cycle of the power converter or the reference voltage of the PI regulator, which controls the power converter DC input voltage.

The evolutionary MPPT algorithms exhibit algorithmic complexity, thus necessitating the use of microcontrollers or DSP units for their implementation. Also, since random numbers are employed during the execution of the evolutionary MPPT algorithms, it cannot be mathematically guaranteed that they will converge to the global MPP under any partial shading conditions.

5.4.3   MPPT Methods Based on Numerical Optimization Algorithms

These algorithms are employing numerical methods, which are suitable for deriving the maximum of an objective function, without requiring the calculation of derivatives of the objective function. Thus, due to their inherent computational simplicity, they can effectively be implemented into a microcontroller or DSP device within the control unit of the PV energy management system.

In the *dividing rectangles* (DIRECT) algorithm [69], individual intervals within the output voltage range of the PV source are iteratively explored for detecting the position of the global MPP. Each such interval is divided into three subintervals of equal range. Among them, the potentially optimal interval is defined as the jth interval $\left[\alpha_j, b_j\right]$ for which there exists a value $K > 0$ satisfying the following inequalities for each value of i $(i = 1,...,3)$:

$$P_{pv}\left(c_j\right) + K\frac{b_j - \alpha_j}{2} \geq P_{pv}\left(c_i\right) + K\frac{b_i - \alpha_i}{2} \tag{5.18}$$

$$P_{pv}\left(c_j\right) + K\frac{b_j - \alpha_j}{2} \geq P_{pv,\max} + \varepsilon \cdot P_{pv,\max} \tag{5.19}$$

where

  c_j, c_i are the midpoints of intervals j and i, respectively
  α_i, b_i are the end points of the ith interval
  $\varepsilon > 0$ is a constant
  $P_{pv,\max}$ is the power at the currently detected MPP

Only potentially optimal intervals are selected for reapplying the same dividing process until the global MPP is detected [70]. The DIRECT MPPT algorithm has the drawback of not being able to guarantee that it will be able to achieve convergence to the global MPP with a fewer number of steps than an exhaustive search procedure, which sweeps the entire power–voltage curve of the PV source, under any partial shading conditions.

In [71], a sequence of Fibonacci search numbers (i.e., $F_0 - F_n$, $n \geq 0$) is produced for performing the MPPT process under partial shading conditions as follows:

$$\begin{aligned} F_0 &= 0 \\ F_1 &= 1 \\ F_n &= F_{n-2} + F_{n-1} \quad (n \geq 2) \end{aligned} \tag{5.20}$$

In the ith iteration of the search process, the control signal of a boost-type DC/DC converter, connected to the PV source, is adjusted to the values c_1^i and $c_2^i\left(c_1^i < c_2^i\right)$, which lie within the search interval $\left(c_3^i c, c_4^i\right)$. The distance α_i between c_1^i, c_3^i and c_2^i, c_4^i and the distance b_i between c_1^i and c_2^i are given by

$$\begin{aligned} \alpha_i &= F_{n+1}, \quad b_i = F_n \\ \alpha_{i+1} &= F_n, \quad b_{i+1} = F_{n-1} \end{aligned} \tag{5.21}$$

The next search interval is decided by comparing the PV output power at c_1^i and c_2^i as follows: if $P_{pv}\left(c_1^i\right) < P_{pv}\left(c_2^i\right)$, then $c_3^{i+1} = c_1^i$ and $c_4^{i+1} = c_4^i$, or else $c_3^{i+1} = c_3^i$ and $c_4^{i+1} = c_2^i$. This search process is continued until the variable n of F_n is reduced to zero or the distances between c_3^i, c_4^i and $P_{pv}\left(c_3^i\right)$, $P_{pv}\left(c_4^i\right)$, respectively, drop below predefined thresholds. The MPPT algorithm based on the Fibonacci sequence does not guarantee convergence to the global MPP.

5.4.4 STOCHASTIC AND CHAOS-BASED MPPT ALGORITHMS

Random search method has been applied in [72] for deriving the global MPP of a PV array with partial shading. Using this approach, the duty cycle of a DC/DC power converter is iteratively modified using random numbers, such that it progressively moves toward values that operate the PV source at points providing a higher output power.

The chaotic-search global MPPT process presented in [73] is based on two recursive functions (i.e., dual carrier) in order to perform iterative fragmentations of the PV array power–voltage characteristic. For that purpose, sequences of numbers are generated through the use of appropriate functions, which correspond to alternative operating points on the power–voltage characteristic of the PV array. By measuring the power generated by the PV array at these positions, the global MPP is detected. In [74], the global MPP tracking process for flexible PV modules, which also exhibit local MPPs on their power–voltage curves, is performed using a combination of the dual-carrier (i.e., using two recursive functions) chaotic-search and PSO optimization algorithms.

Due to their operational complexity, a microcontroller- or DSP-based control unit is required for executing these global MPPT algorithms.

5.4.5 DISTRIBUTED MPPT

In the case that the PV source, which is connected to the power converter of the PV energy production system shown in Figure 5.1, comprises strings of series-connected PV modules, then a bypass diode is connected in antiparallel with each PV module in order to conduct the string current in cases of partial shading conditions. In contrast to this design approach, in the distributed MPPT (DMPPT) architectures, a separate DC/DC power converter is connected at the output of each PV module of the PV array.

In the current equalization DMPPT topology, the DC/DC converter connected at the output of each PV module is power supplied by the DC bus of the PV string. A diagram of this topology based on the architecture proposed in [75] is depicted in Figure 5.15. Under partial shading conditions, the nth PV module produces a current equal to $I_{pv,n}$ and the corresponding DC/DC converter is controlled to supply an additional current, which is equal to $I_s - I_{pv,n}$, such that the total string current is equal to I_s. At the same time, the output voltage of each PV module is regulated such that operation at its MPP is ensured [76]. The position of MPP is different for each PV module, depending on the geometry of the shading pattern on the PV array. Each DC/DC converter is required to supply only the equalization current, thus operating at a low power level with relatively low power losses.

In order to enable multiple strings, each employing the current equalization topology described previously, to be connected in parallel without forcing their PV modules to operate away from their MPPs, the current equalization topology presented earlier has been extended to the shunt-series compensation topology. A diagram of this topology based on the architecture proposed in [77] is shown in Figure 5.16. In this architecture, a current-compensating DC/DC converter with MPPT controller is connected in parallel with each PV module and a voltage-compensating DC/DC converter is connected in series with each PV string, which balances the deviation of the total voltage produced by parallel-connected strings, thus enabling the individual PV modules to operate at their own MPPs.

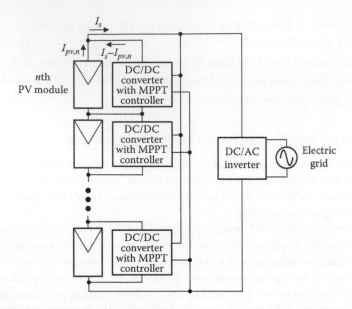

FIGURE 5.15 A diagram of the current equalization Distributed MPPT (DMPPT) topology based on the architecture proposed in Sharma et al. (2012).

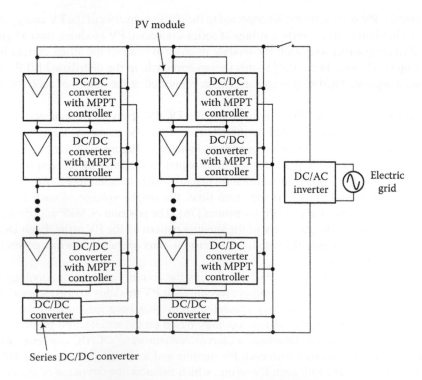

FIGURE 5.16 A diagram of the shunt-series compensation DMPPT topology based on the architecture proposed in Sharma et al. (2014).

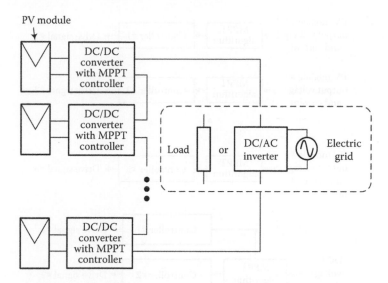

FIGURE 5.17 DMPPT topology where the outputs of the DC/DC converters are connected in series, based on the methods presented in Sharma et al. 2012.

An alternative DMPPT topology based on the methods presented in [78–80] is illustrated in Figure 5.17. In this case, the PV strings are formed by connecting in series the outputs of the DC/DC converters, which are connected at the output of each PV module. Each DC/DC converter processes the entire power generated by the corresponding PV module and executes the MPPT process for that individual PV module.

Alternative DMPPT control schemes based on the methods presented in [79] are illustrated in Figure 5.18. The MPPT process may be performed by either executing the MPPT process (e.g., P&O) at each DC/DC converter separately or measuring the total power of the DC bus and then sending the appropriate control signal to each DC/DC converter. In the latter case, the power losses of the individual power converters are also taken into account in the MPPT process.

The diagram of an architecture employing a triggering circuit in parallel with each PV module of the PV string, together with an energy recovery unit across the PV string, based on the design method proposed in [81], is depicted in Figure 5.19. The triggering circuit measures the voltage developed across the bypass diode. When this voltage exhibits a low negative value, indicating that the corresponding bypass diode conducts current, thus a partial shading condition has evolved, the energy recovery circuit is activated in order to bypass that diode. In this case, part of the current of the less shaded PV modules is diverted into the energy recovery circuit, thus maintaining the current of all PV modules at the same value, without requiring the activation of the bypass diodes of the shaded PV modules. The resulting power–voltage curve of the PV string exhibits a single MPP, without local MPPs, which is tracked by the MPPT unit of a central DC/AC inverter.

Solar irradiance mismatch conditions may also arise among the individual solar cells of a PV module. In the case that access to individual groups of solar cells (i.e., submodules) is provided within the junction box of the PV module, then the DMPPT techniques mentioned earlier may also be applied at the submodule level in order to further enhance the PV system energy production [82].

The DMPPT approach has the advantage that the total available MPP power of the PV array is increased. However, compared to the PV system topology where a single central power converter is used for processing the energy generated by the entire PV array, the implementation complexity of the DMPPT architectures is higher due to the requirement of having to install a separate DC/DC converter at each PV module of the PV source.

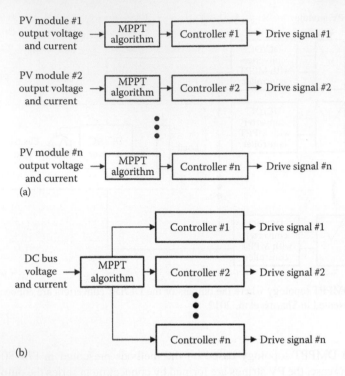

(a)

(b)

FIGURE 5.18 Alternative DMPPT control schemes based on the methods presented in Ramos-Paja et al. (2013): (a) MPPT at the DC/DC converter level and (b) centralized MPPT.

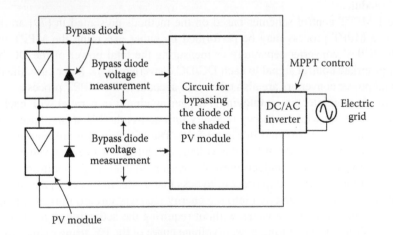

FIGURE 5.19 Diagram of a DMPPT topology with an energy recovery circuit in parallel with each PV module based on the design method proposed in Ramli et al. (2014).

5.4.6 OTHER GLOBAL MPPT METHODS

Scanning of the power–voltage curve is performed in [83] by varying the duty cycle of a DC/DC power converter (see Figure 5.1) in order to detect the position of the global MPP either during initial operation or during varying atmospheric conditions. Then, a fuzzy logic–based MPPT process is applied in order to track short-term changes of the global MPP. A similar two-stage process is applied in [84]. Targeting to restrict the voltage range to be scanned, thus reducing the time required to accomplish the scan process and the associated power loss, the voltage windows explored during the scanning process are continuously updated in [85] based on the geometry of the power–voltage curve of the PV array under partial shading conditions. The calculation of the voltage windows is performed using the value of the open-circuit voltage of the PV modules comprising the PV source, which must be known prior to the application of this global MPPT method.

In [4], a buck-type DC/DC converter is controlled such that it operates as an adjustable constant-power load of the PV array, thus avoiding operation at local MPPs of the PV power–voltage curve. A diagram of the operation in adjustable constant-power mode for avoiding convergence to local MPPs during the global MPPT process based on the method proposed in [4] is depicted in Figure 5.20. Although the periodic scanning of the PV source power–voltage curve is also performed by this global MPPT method, the number of search steps, which must be executed for detecting the position of the global MPP, is lower than those required by an exhaustive search process. Additionally, this technique does not require knowledge of either the PV source configuration or the individual PV modules electrical characteristics.

A thermal imaging camera is used in [86] for acquiring thermal images of a PV array containing partially shaded PV modules. These images are then analyzed using a model of the PV array in order to estimate the voltage corresponding to the global MPP. Although the implementation cost of this MPPT method is relatively high due to the thermal imaging camera employed, the operation of the PV system is not disturbed for tracking different operating points during the global MPP detection process.

In [87], the global MPPT process is based on the assumption that the local MPPs occur at voltages that are an integer multiple of about $0.8 \cdot V_{oc}$, where V_{oc} is the open-circuit voltage of the PV modules, which comprise the PV source. Thus, the PV array voltage is modified at steps of $0.8 \cdot V_{oc}$ and at each step the position of the local MPP is derived using an InC MPPT algorithm. The power levels of the local MPPs are compared, and among them, the global MPP corresponds to that providing the maximum amount of power. The InC MPPT algorithm is also used to maintain operation at the previously detected global MPP until a variation of incident solar irradiation is detected, which ignites a new execution of the global MPP tracking process. The application of this MPPT technique requires knowledge of the value of V_{oc} and its sensitivity to the ambient temperature and solar irradiation.

An ANN is trained in [88] for producing the location of the global MPP of a PV array under various solar irradiation conditions. The ANN training is performed by using the power–voltage curves of the PV source, which are produced by using the single- or two-diode model of the PV modules. Thus, this MPPT method requires knowledge of the PV source operational characteristics in order to perform the ANN training process. During the global MPP process, measurements of incident

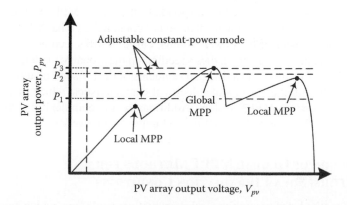

FIGURE 5.20 Diagram of the operation in adjustable constant-power mode for avoiding convergence to local MPPs during the global MPPT process based on the method proposed in Koutroulis et al. (2012).

solar irradiation on each PV module of the PV array are fed in the off-line-trained ANN, which then produces an estimation of the approximate position of the global MPP. This information is used as an initial operating point by a P&O MPPT algorithm for deriving the true global MPP.

In [89], the current–voltage curve of the PV source is initially traced by disconnecting it from the power converter and connecting it to a parallel RC circuit containing a discharged capacitor. This action causes the PV source voltage to sweep in the range of $0 - V_{oc}$. The current and voltage measurements acquired during that tracing process are used to move the PV source operating point to a region close to the global MPP. Then, a P&O algorithm is applied for converging to the global MPP. This technique has the disadvantage that during the execution of the current–voltage curve tracing process, the PV source power is not transferred to the PV system load.

In an alternative MPPT method, the total output voltage range of the PV source (i.e., $0 - V_{oc}$) is divided into intervals and the corresponding output power is measured at each interval [90]. The P&O MPPT method with variable step size (i.e., reducing the step size as the MPP is approached) is then applied for tracking the global MPP at the particular interval, which gave the highest power measurement. The number of intervals is selected to be higher than the number of PV modules connected in series in the PV source, thus requiring the knowledge of the PV source operational characteristics.

The output voltage of each PV module of the PV string is measured in [91] for calculating the number, q, of different levels of solar irradiation, $G_j \left(G_1 < G_2 < \cdots < G_q \right)$, which are received by the PV string and the corresponding number of PV modules receiving each such level, $M_j \left(j = 1,...,q \right)$. Then, a P&O MPPT process is applied to the entire PV string, which is initiated at the open-circuit voltage of the PV string, V_{oc}, as well as at each of the distinct PV string voltage levels defined by the following equation for $j = 1,...,q - 1$:

$$V_{pv,j} = 0.85 \cdot \left(1 - \frac{\sum_{i=1}^{j} M_i}{N} \right) \cdot V_{oc} \tag{5.22}$$

where N is the total number of PV modules of the PV string.

By comparing the power produced at the individual MPPs tracked by the P&O algorithm, the position of the global MPP is derived. However, the number of voltage sensors and accompanying signal-conditioning circuits is significantly high when applying this technique in PV sources composed of strings with a high number of PV modules, thus increasing the complexity and cost of the control unit.

In [92], the ESC MPPT method is sequentially applied at individual segments of the voltage range of the PV array in order to detect the positions of local MPPs. Among them, the local MPP where the maximum power is produced corresponds to the global MPP. A similar process is also applied in [93], but in this case the segments that do not contain the global MPP are identified using information about the gradient of the power–voltage curve. By this technique, convergence to the (local) MPP of these segments is avoided, thus speeding up the global MPP detection process. In the ESC-based global MPPT method, the individual segments of the PV source output voltage range are selected using the values of the PV module electrical characteristics, which must be known when developing the corresponding MPPT system. Additionally, the complexity of the control circuit, which implements the ESC-based global MPPT technique, is relatively high.

5.4.7 COMPARISON OF GLOBAL MPPT METHODS FOR NONUNIFORM SOLAR IRRADIATION CONDITIONS

The global MPPT methods for nonuniform solar irradiation conditions, which have been described previously, are compared in terms of their operational characteristics in Table 5.3. With the exception of the DMPPT techniques, the remaining global MPPT algorithms, which have been presented in this section, require periodic reinitialization of their execution. This is indispensable in order to be able to detect a possible displacement of the global MPP position, due to a change of either the meteorological conditions or the shading pattern on the surface of the PV array (e.g., change of the shadow shape due to sun movement). This results in power loss until the MPPT algorithm converges to the new global MPP and also imposes the need of a high sampling rate in order to be able to quickly detect the global MPP changes. The application of the PV array reconfiguration method requires knowledge of the configuration of the PV source but provides a high efficiency, since the PV system is able to produce more power than at the global MPP without reconfiguration. Due to their inherent randomness, the evolutionary, stochastic, and chaos-based MPPT algorithms do not guarantee convergence to the global MPP under any partial shading conditions. Thus, they exhibit a lower energy yield. However, they have the advantage of not requiring detailed knowledge of the PV system operational characteristics. The global MPPT methods that are based on numerical optimization algorithms either do not guarantee convergence to the global MPP or convergence can be accomplished after a large number of search steps, both resulting in a reduction of the PV-generated energy. The robustness of the PV array reconfiguration, evolutionary, stochastic, and chaos- and numerical optimization–based MPPT algorithms may easily be degraded by external disturbances affecting the correctness of the decisions taken during the global MPP tracking process with respect to the direction toward which the global MPP resides. In such a case, possible wrong estimations may only be recovered at the next reexecution of the corresponding MPPT algorithm. The DMPPT techniques require knowledge of the PV source configuration but they are able to extract the maximum possible energy from the PV source at the cost of a significantly higher hardware complexity. Most of the other global MPPT algorithms, which have been presented in Section 5.4.6, require knowledge of the operational characteristics of the PV source. Also, since a search process is applied periodically during their execution in order to detect possible changes of the global MPP position (e.g., due to a change of meteorological conditions), the power produced by the PV source is reduced and also a high sampling rate is required. Their performance in terms of the remaining metrics considered in Table 5.3 depends on the specific technique applied in each case.

The evolutionary, stochastic, and chaos- and numerical optimization–based MPPT methods, as well as some of the global MPPT approaches presented in Section 5.4.6, do not operate by using information about the electrical characteristics of the PV source, so their accuracy is not affected by the PV module aging. The robustness of DMPPT architectures to the PV module aging depends on the type of method that has been employed for performing the MPPT process.

Among the global MPPT methods presented earlier, the PV array reconfiguration approach is the least suitable for operation in combination with a constant-power-mode controller [5], due to the low power adjustment resolution achieved by controlling the configuration of the PV modules within a PV array, instead of directly controlling a power converter.

5.5 SIMULATION EXAMPLES

The MATLAB®/Simulink® model shown in Figure 5.21 (filename: "MPPT.mdl") simulates the operation of the P&O MPPT process applied to a buck-type DC/DC converter with a constant output voltage. The buck converter is assumed to operate in the continuous conduction mode and its model has been implemented in an embedded MATLAB® function. Also, an ideal PV array has been modeled as a controlled current source, with an output current calculated through an embedded MATLAB® function using the single-diode model, according to [15]. A series resistance is connected to the output of the controlled current source, in order to form a nonideal PV array. The parallel resistance of the PV array is assumed negligible. The P&O MPPT process has been implemented in the corresponding embedded MATLAB® function, based on [2]. Initially, the PV array open-circuit voltage, short-circuit current, number of solar cells in series, solar cell temperature, and ideality factor, as well as the P&O MPPT process perturbation step and the DC/DC converter output voltage, must be set to the desired values. The "scope" of the MATLAB®/Simulink® model provides

TABLE 5.3

Comparison of the Operational Characteristics of the Global MPPT Methods for Nonuniform Solar Irradiation Conditions

MPPT Method	Sampling Rate	Complexity	Robustness		Efficiency	Sampled Parameters	System or Expert Knowledge	Constant-Power Operation
			To External Disturbances	To Aging of the PV Modules				
PV array reconfiguration	High	High	Low	Low	High	PV voltage and current	High	Very difficult
Evolutionary MPPT algorithms	High	High	Low	High	Low	PV voltage and current	Low	Easy
Numerical optimization	High	Low	Low	High	Low	PV voltage and current	Low	Easy
Stochastic/chaos-based algorithms	High	High	Low	High	Low	PV voltage and current	Low	Easy
DMPPT	Low	Very high	High	Depends on the type of MPPT method employed	Very high	PV voltage and current	High	Easy
Other global MPPT methods	High	Method dependent	Method dependent	Method dependent	Low	PV voltage and current	High in most cases	Easy in most cases

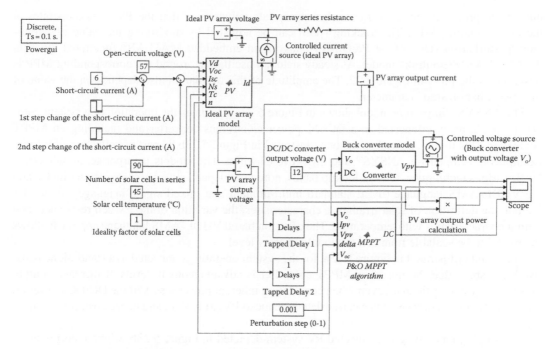

FIGURE 5.21 A MATLAB®/Simulink® model of the P&O MPPT process applied to a buck-type DC/DC converter with a constant output voltage.

plots of the variations in the PV array output current, voltage, and power, as well as the DC/DC converter duty cycle, which arise when the P&O MPPT process is executed and step changes of the PV array short-circuit current also occur.

An example of the PV array output current, voltage, and power, as well as the DC/DC converter duty cycle, during the operation of the P&O MPPT process, is shown in Figure 5.22. Three different levels of short-circuit current are applied at time instants $t = 0$, 15, and 25 s, respectively. In each of these cases, the PV array output current, voltage, and power and the DC/DC converter

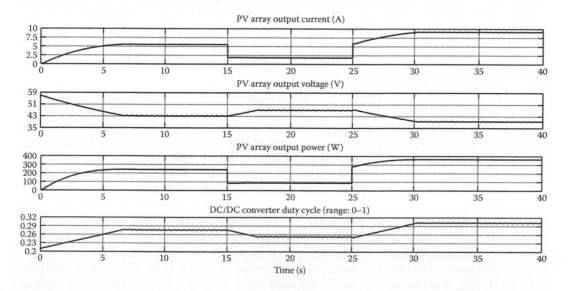

FIGURE 5.22 Simulation results of the PV system operation when applying the P&O MPPT process.

duty cycle progressively converge to the appropriate values such that the PV source operates at the corresponding MPP. The tracking speed can be adjusted by modifying the value of parameter "perturbation step" of the P&O MPPT algorithm embedded MATLAB® functional block in the MATLAB®/Simulink® model. At steady state, an oscillation around the corresponding MPP is observed in the plots in Figure 5.22. The amplitude of this oscillation also depends on the value of the "perturbation step" parameter.

The ANSYS Simplorer® model shown in Figure 5.23a simulates the operation of a system comprising a PV array connected to a DC/DC power electronics converter and including an MPPT controller. This model, as well as the one illustrated in Figure 5.23b, is part of collection of tutorial examples developed for ANSYS®, Inc. by Steve Chwirka. The models incorporate, at subsystem level, various techniques, such as lookup tables, nonlinear dependent sources, equation blocks, and the VHDL-AMS language (IEEE standard multi-domain and mixed signal language). Examples include the definition of solar irradiance characteristics, the variation of the PV cell resistance, and output characteristics with temperature. The equation based VHDL-AMS language approach allows the model to be scalable from cell to module to array level.

The results illustrated in Figure 5.23a, for the system operating connected to a stand-alone resistive load, show that the use of an MPPT controller is advantageous in terms of increased output power and system efficiency, even when considering inherent power losses of the DC/DC converter. At the highest system level, this can translate in reduced PV array size, and hence cost, for specified rated power.

The single-phase AC grid connected PV system depicted in Figure 5.23b, which incorporates a DC/AC power electronics converter, provides additional functionality, such that during the day the load can be supplied in combination by the PV array and the grid and any excess PV power is fed to the grid. Through the integration of a battery, further capability can be added, such that part or all of the PV energy is stored for local use during sun shading or at night, resulting in a reduction of the energy demand from the grid.

5.6 SUMMARY

The power–voltage curves of PV modules/arrays exhibit a point where the PV-generated power is optimized. Under uniform solar irradiation conditions, this point is unique, while in the case that different amounts of solar irradiation are incident on the individual PV modules of the PV array, then multiple local MPPs may also exist. Thus, the control unit of the PV energy conversion system must execute an MPPT process in order to operate the PV source at the point where the generated power is maximized. This process enables to optimally exploit the installed PV capacity, thus increasing the energy conversion efficiency of the overall PV system and, simultaneously, enhancing the economic benefit obtained during the PV system lifetime.

The P&O and DMPPT methods have already been incorporated into commercial PV power converters [94, 95]. However, a wide variety of methods have additionally been proposed in the scientific literature during the last years for performing the MPPT process in a PV system. Aiming to assist the designers of PV power processing systems to select the most suitable MPPT method, the operational characteristics and implementation requirements of these techniques have been analyzed in this chapter.

In case the PV system has to operate under nonuniform solar irradiation conditions, the MPPT methods that have been developed for PV arrays operating under uniform solar irradiation conditions should not be applied, since they do not guarantee that the global MPP will be derived.

The operational complexity of each MPPT method affects the implementation cost of the corresponding control unit. Digital control units are most frequently employed in the modern PV power processing systems. The MPPT techniques relying on the execution of an optimization algorithm (e.g., P&O, InC, evolutionary algorithms) are more easily integrated into such devices, compared to the techniques that require the addition of specialized analog and/or digital control circuits for

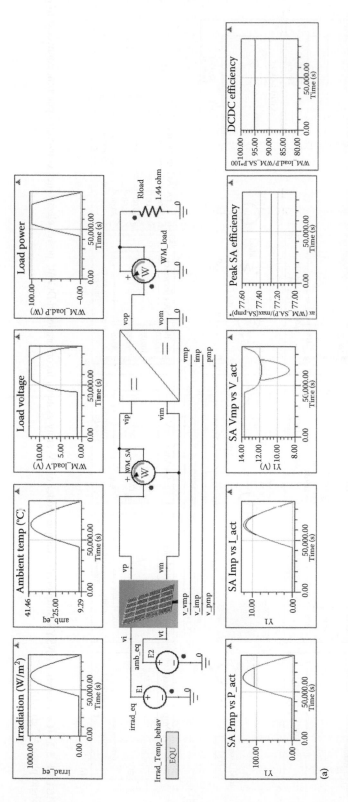

FIGURE 5.23 ANSYS Simplorer® example model of PV array systems incorporating a DC/DC power electrics converter with an MMPT controller and operating on a stand-alone restive load (a).

(Continued)

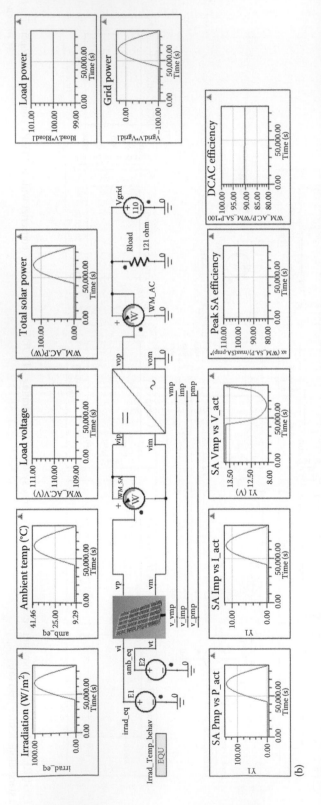

FIGURE 5.23 (*Continued*) ANSYS Simplorer® example model of PV array systems incorporating a DC/AC power electronics converter operating connected to a single-phase AC grid (b).

their operation. Thus, the economic burden imposed on the cost of the total PV power processing interface is minimized and its flexibility to adapt to different operating conditions (e.g., installation in alternative sites) is increased. Also, the MPPT methods, which require knowledge of one or more operational characteristics of the PV source regarding either its configuration (e.g., number of PV modules connected in series) or its operational parameters (e.g., temperature coefficients, open-circuit voltage, etc.), are not suitable for incorporation into commercial PV power management products. In such cases, the specifications of the target PV source are not known during the design and manufacturing stages of the power converter and associated control unit but will be defined by the PV system designer considering the specific target application requirements. Furthermore, the accuracy of the MPPT methods, which operate by using the values of electrical parameters of the PV source, is affected by (1) the uncertainty of the estimated electrical parameters values for the specific PV modules employed in each PV installation (e.g., due to measurement errors during an experimental characterization process, deviation of the actual operating characteristics from the corresponding datasheet information, etc.); (2) the electrical parameter deviation among the individual PV modules supplied by a manufacturer, due to nonidealities of the manufacturing process; and (3) the drift of the PV module electrical characteristics during the operational lifetime period of a PV system (e.g., 25 years).

Each MPPT method also exhibits a different speed in deriving the MPP of the PV source. However, the intensity of the quantitative impact of this feature on the energy production performance of a PV system depends on the magnitude and duration of the short-term variability of solar irradiation and ambient temperature at each particular installation site under consideration. Thus, in order to select an MPPT method for incorporation in a PV energy management system, its performance should be evaluated in terms of the total energy produced by the PV source under both static and dynamic operating conditions, considering the time-varying profile of the meteorological conditions that prevail during the year at the installation site of interest. Testing procedures, which are suitable for evaluating and comparing the performance of MPPT algorithms, are presented in [96, 97]. As analyzed in [98], the reliability of the software and hardware components of the MPPT control unit also affects the energy production of the PV system; thus, it must also be considered during the performance evaluation process of an MPPT method.

REFERENCES

1. Solar Power Europe, Global market outlook for solar power/2015–2019, 2015, available online at: www.solarpowereurope.org. Accessed 2 December, 2015.
2. Koutroulis, E., Kalaitzakis, K., and Voulgaris, N.C. Development of a microcontroller-based, photovoltaic maximum power point tracking control system, *IEEE Transactions on Power Electronics*, 16(1), 46–54, 2001.
3. Kjaer, S.B., Pedersen, J.K., and Blaabjerg, F. A review of single-phase grid-connected inverters for photovoltaic modules, *IEEE Transactions on Industry Applications*, 41(5), 1292–1306, 2005.
4. Koutroulis, E. and Blaabjerg, F. A new technique for tracking the global maximum power point of PV arrays operating under partial-shading conditions, *IEEE Journal of Photovoltaics*, 2(2), 184–190, 2012.
5. Yang, Y., Wang, H., Blaabjerg, F., and Kerekes, T. A hybrid power control concept for PV inverters with reduced thermal loading, *IEEE Transactions on Power Electronics*, 29(12), 6271–6275, 2014.
6. Bastidas-Rodriguez, J.D., Franco, E., Petrone, G., Andrés Ramos-Paja, C., and Spagnuolo, G. Maximum power point tracking architectures for photovoltaic systems in mismatching conditions: A review, *IET Power Electronics*, 7(6), 1396–1413, 2014.
7. Esram, T. and Chapman, P.L. Comparison of photovoltaic array maximum power point tracking techniques, *IEEE Transactions on Energy Conversion*, 22(2), 439–449, 2007.
8. Ali, A.N.A., Saied, M.H., Mostafa, M.Z., and Abdel-Moneim, T.M. A survey of maximum PPT techniques of PV systems, *Proceedings of 2012 IEEE EnergyTech*, pp. 1–17, Cleveland, OH, May 29–31, 2012.
9. Subudhi, B. and Pradhan, R. A comparative study on maximum power point tracking techniques for photovoltaic power systems, *IEEE Transactions on Sustainable Energy*, 4(1), 89–98, 2013.
10. Liu, Y.-H., Chen, J.-H., and Huang, J.-W. A review of maximum power point tracking techniques for use in partially shaded conditions, *Renewable and Sustainable Energy Reviews*, 41, 436–453, 2015.

11. Ishaque, K. and Salam, Z. A review of maximum power point tracking techniques of PV system for uniform insolation and partial shading condition, *Renewable and Sustainable Energy Reviews*, 19, 475–488, 2013.
12. Eltawil, M.A. and Zhao, Z. MPPT techniques for photovoltaic applications, *Renewable and Sustainable Energy Reviews*, 25, 793–813, 2013.
13. Liu, L., Meng, X., and Liu, C. A review of maximum power point tracking methods of PV power system at uniform and partial shading, *Renewable and Sustainable Energy Reviews*, 53, 1500–1507, 2016.
14. Lyden, S. and Haque, M.E. Maximum power point tracking techniques for photovoltaic systems: A comprehensive review and comparative analysis, *Renewable and Sustainable Energy Reviews*, 52, 1504–1518, 2015.
15. Lorenzo, E. *Solar Electricity: Engineering of Photovoltaic Systems*, 1st edn., Seville, Spain: Progensa, 1994, ISBN: 84-86505-55-0.
16. Kaa, G., Rezaei, J., Kamp, L., and Winter, A. Photovoltaic technology selection: A fuzzy MCDM approach, *Renewable and Sustainable Energy Reviews*, 32, 662–670, 2014.
17. International Energy Agency, *Technology Roadmap: Solar Photovoltaic Energy*, 2014 edn., available online at: www.iea.org.
18. Jordan, D.C. and Kurtz S.R. Photovoltaic degradation rates—An analytical review, *Progress in Photovoltaics: Research and Applications*, 21(1), 12–29, 2013.
19. de Wild-Scholten, M.J. Energy payback time and carbon footprint of commercial photovoltaic systems, *Solar Energy Materials and Solar Cells*, 119, 296–305, 2013.
20. Laudani, A., Fulginei, F.R., and Salvini, A. Identification of the one-diode model for photovoltaic modules from datasheet values, *Solar Energy*, 108, 432–446, 2014.
21. Cristaldi, L., Faifer, M., Rossi, M., and Toscani, S. An improved model-based maximum power point tracker for photovoltaic panels, *IEEE Transactions on Instrumentation and Measurement*, 63(1), 63–71, 2014.
22. Tauseef, M. and Nowicki, E. A simple and cost effective maximum power point tracker for PV arrays employing a novel constant voltage technique, *Proceedings of 25th IEEE Canadian Conference on Electrical & Computer Engineering (CCECE)*, pp. 1–4, Montreal, Quebec, Canada, April 29–May 2, 2012.
23. Ramasamy, A. and Vanitha, N.S. Maximum power tracking for PV generating system using novel optimized fractional order open circuit voltage-FOINC method, *Proceedings of 2014 International Conference on Computer Communication and Informatics (ICCCI)*, pp. 1–6, Coimbatore, India, January 3–5, 2014.
24. Di, X., Yundong, M., and Qianhong, C. A global maximum power point tracking method based on interval short-circuit current, *Proceedings of 16th European Conference on Power Electronics and Applications (EPE'14-ECCE Europe)*, pp. 1–8, Lappeenranta, Finland, August 26–28, 2014.
25. El Khateb, A., Abd Rahim, N., Selvaraj, J., and Uddin, M.N. Fuzzy-logic-controller-based SEPIC converter for maximum power point tracking, *IEEE Transactions on Industry Applications*, 50(4), 2349–2358, 2014.
26. Hua, C., Lin, J., and Shen, C. Implementation of a DSP-controlled photovoltaic system with peak power tracking, *IEEE Transactions on Industrial Electronics*, 45(1), 99–107, 1998.
27. Femia, N., Petrone, G., Spagnuolo, G., and Vitelli, M., A technique for improving P&O MPPT performances of double-stage grid-connected photovoltaic systems, *IEEE Transactions on Industrial Electronics*, 56(11), 4473–4482, 2009.
28. Ricco, M., Manganiello, P., Petrone, G., Monmasson, E., and Spagnuolo, G. FPGA-based implementation of an adaptive P&O MPPT controller for PV applications, *Proceedings of IEEE 23rd International Symposium on Industrial Electronics (ISIE)*, pp. 1876–1881, Istanbul, Turkey, June 1–4, 2014.
29. Kollimalla, S.K. and Mishra, M.K. Variable perturbation size adaptive P&O MPPT algorithm for sudden changes in irradiance, *IEEE Transactions on Sustainable Energy*, 5(3), 718–728, 2014.
30. Huynh, D.C., Nguyen, T.A.T., Dunnigan, M.W., and Mueller, M.A. Maximum power point tracking of solar photovoltaic panels using advanced perturbation and observation algorithm, *Proceedings of Eighth IEEE Conference on Industrial Electronics and Applications (ICIEA)*, pp. 864–869, Melbourne, Victoria, Australia, June 19–21, 2013.
31. Elgendy, M.A., Zahawi, B., and Atkinson, D.J. Assessment of the incremental conductance maximum power point tracking algorithm, *IEEE Transactions on Sustainable Energy*, 4(1), 108–117, 2013.
32. Rahman, N.H.A., Omar, A.M., and Saat, E.H.M., A modification of variable step size INC MPPT in PV system, *Proceedigns of IEEE Seventh International Power Engineering and Optimization Conference (PEOCO)*, pp. 340–345, Langkawi, Malaysia, June 3–4, 2013.

33. Sera, D., Mathe, L., Kerekes, T., Spataru, S.V., and Teodorescu, R. On the perturb-and-observe and incremental conductance MPPT methods for PV systems, *IEEE Journal of Photovoltaics*, 3(3), 1070–1078, 2013.

34. Blanes, J.M., Toledo, F.J., Montero, S., and Garrigós, A. In-site real-time photovoltaic I–V curves and maximum power point estimator, *IEEE Transactions on Power Electronics*, 28(3), 1234–1240, 2013.

35. Raj, J.S.C.M. and Jeyakumar, A.E. A novel maximum power point tracking technique for photovoltaic module based on power plane analysis of characteristics, *IEEE Transactions on Industrial Electronics*, 61(9), 4734–4745, 2014.

36. Farivar, G., Asaei, B., and Mehrnami, S. An analytical solution for tracking photovoltaic module MPP, *IEEE Journal of Photovoltaics*, 3(3), 1053–1061, 2013.

37. Bhatnagar, P. and Nema, R.K. Maximum power point tracking control techniques: State-of-the-art in photovoltaic applications, *Renewable and Sustainable Energy Reviews*, 23, 224–241, 2013.

38. Hartmann, L.V., Vitorino, M.A., Correa, M.B.R., and Lima, A.M.N. Combining model-based and heuristic techniques for fast tracking the maximum-power point of photovoltaic systems, *IEEE Transactions on Power Electronics*, 28(6), 2875–2885, 2013.

39. Charfi, S. and Chaabene, M. A comparative study of MPPT techniques for PV systems, *Proceedings of Fifth International Renewable Energy Congress (IREC)*, pp. 1–6, Hammamet, Tunisia, March 25–27, 2014.

40. Afghoul, H., Krim, F., Chikouche, D., and Beddar, A. Tracking the maximum power from a PV panels using of neuro-fuzzy controller, *Proceedings of IEEE International Symposium on Industrial Electronics (ISIE)*, pp. 1–6, Taipei, Taiwan, May 28–31, 2013.

41. Sheraz, M. and Abido, M.A. An efficient approach for parameter estimation of PV model using DE and fuzzy based MPPT controller, *Proceedings of IEEE Conference on Evolving and Adaptive Intelligent Systems (EAIS)*, pp. 1–5, Linz, Austria, June 2–4, 2014.

42. Adly, M. and Besheer, A.H. An optimized fuzzy maximum power point tracker for stand alone photovoltaic systems: Ant colony approach, *Proceedings of Seventh IEEE Conference on Industrial Electronics and Applications (ICIEA)*, pp. 113–119, Singapore, July 18–20, 2012.

43. Veerachary, M. and Shinoy, K.S. V²-based power tracking for nonlinear PV sources, *IEE Proceedings: Electric Power Applications*, 152(5), 1263–1270, 2005.

44. Choi, B.-Y., Jang, J.-W., Kim, Y.-H., Ji, Y.-H., Jung, Y.-C., and Won, C.-Y. Current sensorless MPPT using photovoltaic AC module-type flyback inverter, *Proceedings of IEEE International Symposium on Industrial Electronics (ISIE)*, pp. 1–6, Taipei, Taiwan, May 28–31, 2013.

45. Agrawal, J. and Aware, M. Golden section search (GSS) algorithm for maximum power point tracking in photovoltaic system, *Proceedings of IEEE Fifth India International Conference on Power Electronics (IICPE)*, pp. 1–6, Delhi, India, December 6–8, 2012.

46. Shao, R., Wei, R., and Chang, L. A multi-stage MPPT algorithm for PV systems based on golden section search method, *Proceedings of 29th Annual IEEE Applied Power Electronics Conference and Exposition (APEC)*, pp. 676–683, Fort Worth, TX, March 16–20, 2014.

47. Xu, W., Mu, C., and Jin, J. Novel linear iteration maximum power point tracking algorithm for photovoltaic power generation, *IEEE Transactions on Applied Superconductivity*, 24(5), 1–6, 2014.

48. Pai, F.-S., Chao, R.-M., Ko, S.H., and Lee, T.-S. Performance evaluation of parabolic prediction to maximum power point tracking for PV array, *IEEE Transactions on Sustainable Energy*, 2(1), 60–68, 2011.

49. Carraro, M., Costabeber, A., and Zigliotto, M., Convergence analysis and tuning of ripple correlation based MPPT: A sliding mode approach, *Proceedings of 15th European Conference on Power Electronics and Applications (EPE)*, pp. 1–10, Lille, France, September 2–6, 2013.

50. Esram, T., Kimball, J.W., Krein, P.T., Chapman, P.L., and Midya, P. Dynamic maximum power point tracking of photovoltaic arrays using ripple correlation control, *IEEE Transactions on Power Electronics*, 21(5), 1282–1291, 2006.

51. Barth, C.B. and Pilawa-Podgurski, R.C.N. Dithering digital ripple correlation control with digitally-assisted windowed sensing for solar photovoltaic MPPT, *Proceedings of 29th Annual IEEE Applied Power Electronics Conference and Exposition (APEC)*, pp. 1738–1746, Fort Worth, TX, March 16–20, 2014.

52. Moo, C. and Wu, G. Maximum power point tracking with ripple current orientation for photovoltaic applications, *IEEE Journal of Emerging and Selected Topics in Power Electronics*, 2(4), 842–848, 2014.

53. Li, X., Li, Y., and Seem, J.E. Maximum power point tracking for photovoltaic system using adaptive extremum seeking control, *IEEE Transactions on Control Systems Technology*, 21(6), 2315–2322, 2013.

54. Xiao, W., Elnosh, A., Khadkikar, V., and Zeineldin, H. Overview of maximum power point tracking technologies for photovoltaic power systems, *Proceedings of 37th Annual Conference of the IEEE Industrial Electronics Society (IECON 2011)*, pp. 3900–3905, Melbourne, Victoria, Australia, November 7–10, 2011.

55. Zazo, H., Leyva, R., and del Castillo, E. MPPT based on Newton-like extremum seeking control, *Proceedings of IEEE International Symposium on Industrial Electronics (ISIE)*, pp. 1040–1045, Hangzhou, China, May 28–31, 2012.

56. Bazzi, A.M. and Krein, P.T. Ripple correlation control: An extremum seeking control perspective for real-time optimization, *IEEE Transactions on Power Electronics*, 29(2), 988–995, 2014.

57. Malek, H. and Chen, Y. A single-stage three-phase grid-connected photovoltaic system with fractional order MPPT, *Proceedings of 29th Annual IEEE Applied Power Electronics Conference and Exposition (APEC)*, pp. 1793–1798, Fort Worth, TX, March 16–20, 2014.

58. Levron, Y. and Shmilovitz, D. Maximum power point tracking employing sliding mode control, *IEEE Transactions on Circuits and Systems I: Regular Papers*, 60(3), 724–732, 2013.

59. Cheng, Z., Pang, Z., Liu, Y., and Xue, P. An adaptive solar photovoltaic array reconfiguration method based on fuzzy control, *Proceedings of Eighth World Congress on Intelligent Control and Automation (WCICA)*, pp. 176–181, Jinan, China, July 7–9, 2010.

60. Daraban, S., Petreus, D., and Morel, C. A novel global MPPT based on genetic algorithms for photovoltaic systems under the influence of partial shading, *Proceedings of 39th Annual Conference of the IEEE Industrial Electronics Society (IECON)*, pp. 1490–1495, Vienna, Austria, November 10–13, 2013.

61. Taheri, H., Salam, Z., Ishaque, K., and Syafaruddin A novel maximum power point tracking control of photovoltaic system under partial and rapidly fluctuating shadow conditions using differential evolution, *Proceedings of IEEE Symposium on Industrial Electronics & Applications (ISIEA)*, pp. 82–87, Penang, Malaysia, October 3–5, 2010.

62. Ishaque, K. and Salam, Z. A deterministic particle swarm optimization maximum power point tracker for photovoltaic system under partial shading condition, *IEEE Transactions on Industrial Electronics*, 60(8), 3195–3206, 2013.

63. Liu, Y.-H., Huang, S.-C., Huang, J.-W., and Liang, W.-C. A particle swarm optimization-based maximum power point tracking algorithm for PV systems operating under partially shaded conditions, *IEEE Transactions on Energy Conversion*, 27(4), 1027–1035, 2012.

64. Ting, T.O., Man, K.L., Guan, S.-U., Jeong, T.T., Seon, J.K., and Wong, P.W.H. Maximum power point tracking (MPPT) via weightless swarm algorithm (WSA) on cloudy days, *Proceedings of IEEE Asia Pacific Conference on Circuits and Systems (APCCAS)*, pp. 336–339, Kaohsiung, Taiwan, December 2–5, 2012.

65. Sundareswaran, K., Peddapati, S., and Palani, S. MPPT of PV systems under partial shaded conditions through a colony of flashing fireflies, *IEEE Transactions on Energy Conversion*, 29(2), 463–472, 2014.

66. Benyoucef, A.S., Chouder, A., Kara, K., Silvestre, S., and Sahed, O.A. Artificial bee colony based algorithm for maximum power point tracking (MPPT) for PV systems operating under partial shaded conditions, *Applied Soft Computing*, 32, 38–48, 2015.

67. Lian, K.L., Jhang, J.H., and Tian, I.S. A maximum power point tracking method based on perturb-and-observe combined with particle swarm optimization, *IEEE Journal of Photovoltaics*, 4(2), 626–633, 2014.

68. Seyedmahmoudian, M., Rahmani, R., Mekhilef, S., Maung Than Oo, A., Stojcevski, A., Tey Kok Soon, and Ghandhari, A.S. Simulation and hardware implementation of new maximum power point tracking technique for partially shaded PV system using hybrid DEPSO method, *IEEE Transactions on Sustainable Energy*, 6(3), 850–862, 2015.

69. Bidram, A., Davoudi, A., and Balog, R.S. Control and circuit techniques to mitigate partial shading effects in photovoltaic arrays, *IEEE Journal of Photovoltaics*, 2(4), 532–546, 2012.

70. Nguyen, T.L. and Low, K.-S. A global maximum power point tracking scheme employing DIRECT search algorithm for photovoltaic systems, *IEEE Transactions on Industrial Electronics*, 57(10), 3456–3467, 2010.

71. Ahmed, N.A. and Miyatake, M. A novel maximum power point tracking for photovoltaic applications under partially shaded insolation conditions, *Electric Power Systems Research*, 78(5), 777–784, 2008.

72. Sundareswaran, K., Peddapati, S., and Palani, S. Application of random search method for maximum power point tracking in partially shaded photovoltaic systems, *IET Renewable Power Generation*, 8(6), 670–678, 2014.

73. Zhou, L., Chen, Y., Guo, K., and Jia, F. New approach for MPPT control of photovoltaic system with mutative-scale dual-carrier chaotic search, *IEEE Transactions on Power Electronics*, 26(4), 1038–1048, 2011.

74. Konstantopoulos, C. and Koutroulis, E. Global maximum power point tracking of flexible photovoltaic modules, *IEEE Transactions on Power Electronics*, 29(6), 2817–2828, 2014.

75. Sharma, P., Peter, P.K., and Agarwal, V. Exact maximum power point tracking of partially shaded PV strings based on current equalization concept, *Proceedings of 38th IEEE Photovoltaic Specialists Conference (PVSC)*, pp. 1411–1416, Austin, TX, June 3–8, 2012.

76. Sharma, P. and Agarwal, V. Comparison of model based MPPT and exact MPPT for current equalization in partially shaded PV strings, *Proceedings of IEEE 39th Photovoltaic Specialists Conference (PVSC)*, pp. 2948–2952, Tampa, FL, June 16–21, 2013.

77. Sharma, P. and Agarwal, V. Maximum power extraction from a partially shaded PV array using shunt-series compensation, *IEEE Journal of Photovoltaics*, 4(4), 1128–1137, 2014.

78. Sharma, P. and Agarwal, V. Exact maximum power point tracking of grid-connected partially shaded PV source using current compensation concept, *IEEE Transactions on Power Electronics*, 29(9), 4684–4692, 2014.

79. Ramos-Paja, C.A., Spagnuolo, G., Petrone, G., Serna, S., and Trejos, A. A vectorial MPPT algorithm for distributed photovoltaic applications, *Proceedings of International Conference on Clean Electrical Power (ICCEP)*, pp. 48–51, Alghero, Italy, June 11–13, 2013.

80. Scarpetta, F., Liserre, M., and Mastromauro, R.A. Adaptive distributed MPPT algorithm for photovoltaic systems, *Proceedings of 38th Annual Conference of the IEEE Industrial Electronics Society (IECON)*, pp. 5708–5713, Montreal, Quebec, Canada, October 25–28, 2012.

81. Ramli, M.Z. and Salam, Z. A simple energy recovery scheme to harvest the energy from shaded photovoltaic modules during partial shading, *IEEE Transactions on Power Electronics*, 29(12), 6458–6471, 2014.

82. Qin, S., Cady, S.T., Dominguez-Garcia, A.D., and Pilawa-Podgurski, R.C.N. A distributed approach to maximum power point tracking for photovoltaic submodule differential power processing, *IEEE Transactions on Power Electronics*, 30(4), 2024–2040, 2015.

83. Alajmi, B.N., Ahmed, K.H., Finney, S.J., and Williams, B.W. A maximum power point tracking technique for partially shaded photovoltaic systems in microgrids, *IEEE Transactions on Industrial Electronics*, 60(4), 1596–1606, 2013.

84. Yeung, R.S.-C., Chung, H.S.-H., and Chuang, S.T.-H. A global MPPT algorithm for PV system under rapidly fluctuating irradiance, *Proceedings of 29th Annual IEEE Applied Power Electronics Conference and Exposition (APEC)*, pp. 662–668, Fort Worth, TX, March 16–20, 2014.

85. Boztepe, M., Guinjoan, F., Velasco-Quesada, G., Silvestre, S., Chouder, A., and Karatepe, E. Global MPPT scheme for photovoltaic string inverters based on restricted voltage window search algorithm, *IEEE Transactions on Industrial Electronics*, 61(7), 3302–3312, 2014.

86. Hu, Y., Cao, W., Wu, J., Ji, B., and Holliday, D. Thermography-based virtual MPPT scheme for improving PV energy efficiency under partial shading conditions, *IEEE Transactions on Power Electronics*, 29(11), 5667–5672, 2014.

87. Tey, K.S. and Mekhilef, S. Modified incremental conductance algorithm for photovoltaic system under partial shading conditions and load variation, *IEEE Transactions on Industrial Electronics*, 61(10), 5384–5392, 2014.

88. Jiang, L.L., Nayanasiri, D.R., Maskell, D.L., and Vilathgamuwa, D.M. A simple and efficient hybrid maximum power point tracking method for PV systems under partially shaded condition, *Proceedings of 39th Annual Conference of the IEEE Industrial Electronics Society (IECON)*, pp. 1513–1518, Vienna, Austria, November 10–13, 2013.

89. Bifaretti, S., Iacovone, V., Cina, L., and Buffone, E. Global MPPT method for partially shaded photovoltaic modules, *Proceedings of IEEE Energy Conversion Congress and Exposition (ECCE)*, pp. 4768–4775, Raleigh, NC, September 15–20, 2012.

90. Chen, J.-H., Cheng, Y.-S., Wang, S.-C., Huang, J.-W., and Liu, Y.-H. A novel global maximum power point tracking method for photovoltaic generation system operating under partially shaded condition, *Proceedings of International Power Electronics Conference (IPEC)*, pp. 3233–3238, Hiroshima, Japan, May 18–21, 2014.

91. Chen, K., Tian, S., Cheng, Y., and Bai, L. An improved MPPT controller for photovoltaic system under partial shading condition, *IEEE Transactions on Sustainable Energy*, 5(3), 978–985, 2014.

92. Lei, P., Li, Y., and Seem, J.E. Sequential ESC-based global MPPT control for photovoltaic array with variable shading, *IEEE Transactions on Sustainable Energy*, 2(3), 348–358, 2011.

93. Elnosh, A., Khadkikar, V., Xiao, W., and Kirtely, J.L. An improved extremum-seeking based MPPT for grid-connected PV systems with partial shading, *Proceedings of IEEE 23rd International Symposium on Industrial Electronics (ISIE)*, pp. 2548–2553, Istanbul, Turkey, June 1–4, 2014.

94. Hohm, D.P. and Ropp, M.E. Comparative study of maximum power point tracking algorithms, *Progress in Photovoltaics: Research and Applications*, 11(1), 47–62, 2003.

95. Petrone, G., Spagnuolo, G., and Vitelli, M. Distributed maximum power point tracking: Challenges and commercial solutions, *Automatika*, 53(2), 128–141, 2012.
96. Valentini, M., Raducu, A., Sera, D., and Teodorescu, R. PV inverter test setup for European efficiency, static and dynamic MPPT efficiency evaluation, *Proceedings of 11th International Conference on Optimization of Electrical and Electronic Equipment (OPTIM)*, pp. 433–438, Brasov, Romania, May 22–24, 2008.
97. Xiao, W., Zeineldin, H.H., and Zhang, P. Statistic and parallel testing procedure for evaluating maximum power point tracking algorithms of photovoltaic power systems, *IEEE Journal of Photovoltaics*, 3(3), 1062–1069, 2013.
98. Zhang, P., Li, W., Li, S., Wang, Y., and Xiao, W. Reliability assessment of photovoltaic power systems: Review of current status and future perspectives, *Applied Energy*, 104, 822–833, 2013.

6 Design of Residential Photovoltaic Systems

Tamás Kerekes, Dezső Séra, László Máthé,
and Kenn H.B. Frederiksen

CONTENTS

Abstract 131
6.1 Introduction about Worldwide PV Systems 131
6.2 Design Procedure for Residential PV Systems 132
    6.2.1 Grid-Connected or Stand-Alone Systems 132
    6.2.2 Load Pattern Evaluation (Cover the Load over 1-Hour/1-Year Periods) 133
    6.2.3 Solar Resource Evaluation 134
    6.2.4 PV Array Sizing (kWp) (Over- versus Undersizing) 134
    6.2.5 Choosing the String and Array Configuration 135
    6.2.6 Choosing the PV Inverter 136
    6.2.7 What Is the Yield and Performance Ratio? 138
6.3 Case Study for Designing a Residential PV System 141
    6.3.1 Methodology: Design Procedure 141
    6.3.2 Comments to the Design Differences for TF and Crystalline-Si Systems 147
    6.3.3 System Performance Monitoring 148
    6.3.4 Warranty of Modules 148
    6.3.5 Warranty of Inverters 148
    6.3.6 Warranty of BOS Components 148
    6.3.7 PV System Price 148
6.4 Summary 149
References 150

ABSTRACT

Renewable energy has become very important both worldwide and on the European market, mainly due to the decrease in the photovoltaic (PV) system cost (up to 75%) during the last decade. PV installations worldwide have reached 227 GW at the end of 2015 with a predicted extra 50 GW of new installation in 2016. Residential systems are a key element in the success story of PV rooftop installations. This chapter discusses the design of residential PV installations, with focus on rooftop grid-connected systems, which represent the vast majority of small-scale PV systems.

6.1 INTRODUCTION ABOUT WORLDWIDE PV SYSTEMS

The photovoltaic (PV) technology continues to increase its share in the global energy market, with an exceptionally fast growth in the last few decades, reaching a cumulative capacity of 227 GW by the end of 2015, with a predicted extra 50 GW of new installations for 2016 [1].

According to a report from SolarPower Europe (formerly European Photovoltaic Industry Association), the price of PV systems decreased by more than 75% in the last 10 years, making PV cost competitive with fossil fuel energy generation in several countries. Fueled by this strong cost reduction, the PV industry is transitioning from being driven by subsidies into a viable option for investment for both large power plants and residential installations on a pure cost competition basis [2].

Residential PV systems are a key element in the success story of PV rooftop installations, and large utility-scale plants share about 50% of new installations today.

This chapter discusses the design of residential PV installations, with focus on rooftop grid-connected systems, which represent the vast majority of small-scale PV systems.

Residential rooftop systems may be considered a special case when it comes to the sizing and design of the system. While in the case of ground-mounted systems (typical for utility-scale plants), the type of components, placement, and layout can all be optimized for the design criteria, typically to minimize the levelized cost of energy, in the case of residential rooftop systems, several factors may influence the design, such as the roof area, angle and shape, and easthetic requirements.

In the following sections, the general guidelines for the designing and sizing of residential PV systems are given. The example is for Denmark; nevertheless, the same methodology can of course be applied to other countries as well.

6.2 DESIGN PROCEDURE FOR RESIDENTIAL PV SYSTEMS

The task of designing PV systems is a very tricky process due to the fact that PV panels are still relatively expensive and energy production is very sensitive to atmospheric conditions and the physical location. In the case of ground-mounted PV systems, one can choose the optimum tilt angle and orientation, and often the physical size is the only limiting factor. In the case of residential PV systems, PV panels are usually mounted on the roof, which might not have the optimum angle or orientation. Besides these limitations, the size of the roof is fixed; therefore, several parameters are already fixed at the beginning of the design. Such design parameters or constraints will affect the following:

- The required annual energy production (AEP)
- The available budget for the installation
- Location-specific limitations: roof size, tilt, and orientation

The first task is to decide whether the PV system will be connected to the electrical grid. Afterward, the load pattern should be evaluated in order to estimate power and energy requirements. When these requirements have been defined, the PV cell technology can be chosen; the PV array can be sized for the required amount of power. Furthermore, the PV array needs to be configured to fit the specifications of the PV inverter. Finally, at the end of this chapter, the whole design procedure is evaluated through a case study, by using freely available design tools. The results are presented and discussed.

6.2.1 GRID-CONNECTED OR STAND-ALONE SYSTEMS

Residential PV systems can be divided into two major groups: grid connected or stand-alone. A PV system can be connected to the electrical grid when the house is connected to the low-voltage (LV) utility network; thus, the power network can be used to dump the surplus energy production. The PV system can be connected to the energy meter of the house, thereby increasing self-consumption. Another solution is to add an individual energy meter that measures the energy that the PV system produces, which is then accounted for separately. Both energy metering solutions are presented in Figure 6.1.

In both cases, the grid is used like a large battery, storing the surplus energy produced by the PV system. Thus, in such a case, during periods with high irradiation and low load conditions, the grid will be stressed, leading to voltage rise, due to the mainly resistive nature of the LV network. To mitigate the problem of voltage rise, PV inverters are nowadays required to support the grid with reactive power, thereby keeping the grid voltage within the limits imposed by the grid codes.

On the other hand, in houses in remote areas that are far away from any electrical network, a stand-alone PV system has to be installed. The complexity of the design will increase, since one

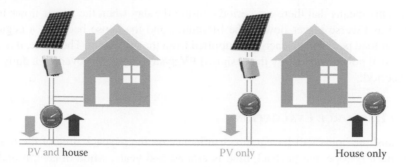

PV and house   PV only   House only

FIGURE 6.1 Energy metering of a residential grid-connected PV system, which can be connected to the same energy meter or have a separate meter for measuring PV production.

can install a battery pack to store the produced energy, which can be used later when extra energy is required. In this case, an extra converter is needed, which controls the state of charge of the battery and communicates this information to the PV system control.

To increase the penetration of PV and the self-consumption ratio, advanced storage solutions using Li-ion batteries can also be added to grid-connected PV systems. This trend can mainly be observed in Germany and Japan, where self-consumption is encouraged and has been subsidized since 2013. Up until now, more than 15,000 German households use PV systems that are combined with battery storage [3].

6.2.2 LOAD PATTERN EVALUATION (COVER THE LOAD OVER 1-HOUR/1-YEAR PERIODS)

In the case of residential homes, the average load profile shows two distinct peaks during the morning and evening hours, while it is much lower with a flat shape during the working hours on weekdays, as detailed in Figure 6.2.

Such a load profile does not match the PV energy production curves, which are shown with dotted lines in Figure 6.2. During a sunny day, production starts from sunrise, gradually increasing with the available sunshine, and peaks around noon, gradually decreasing as a function of irradiation

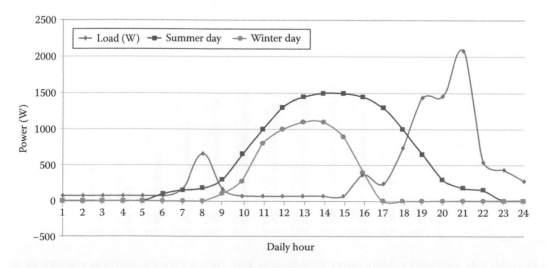

FIGURE 6.2 Typical residential load profile compared to the production of a 1.5 kWp PV system on a summer day and on a winter day on the Northern Hemisphere.

until sundown. This means that there are periods during the day when the power flows toward the grid (also called as reverse power flow in the literature) and the house becomes a negative load, while during peak load periods, the energy is supplied from the grid again. Therefore, it is important to decide how much power and energy the designed PV system needs to cover, on a daily, monthly, or even yearly period.

6.2.3 SOLAR RESOURCE EVALUATION

The available sunshine hours will have a direct influence on the payback time of the designed PV system. In southern Europe, the payback time is around 5–6 years, while in central and northern Europe, the payback time can reach periods up to 9–10 years or even longer, depending on the local energy price. This means that the PV system will produce the energy for "free" only after these years have passed.

If the location for the PV system is known, then there are several free tools online, for example, PVGIS (http://re.jrc.ec.europa.eu/pvgis/apps4/pvest.php [4]), that use the average irradiation data from satellite images of the last 10 years. Based on this, the monthly and yearly energy production data in kWh/m²/year/kWp for the chosen location, both for horizontal and for the optimum tilt angle, can be predicted, as shown in Figure 6.3.

The physical placement of the panels is also of importance. If the location is known, then a Sun chart can be plotted, where the elevation (position of the sun) is given for the whole year. Furthermore, if the height and distance to shading sources (trees, neighboring houses) are known, this information can also be added to the Sun chart, as shown with a black dashed line in Figure 6.4. Based on this, the periods when shadows will be casted over the PV system and the extent of their influence on the energy production can be estimated. Such a shaded period can be seen in Figure 6.4, where some trees have been considered as sources of shadow. As a consequence, on December 21 and January 21, the PV panels are permanently under shadow before noon.

6.2.4 PV ARRAY SIZING (KWP) (OVER- VERSUS UNDERSIZING)

With a good estimation of the required AEP and the data for the available kWh/m²/year/kWp, one can roughly estimate the size of the PV array for the different PV technologies. It is important to mention

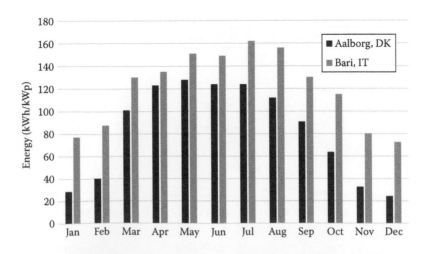

FIGURE 6.3 Estimated monthly energy production in kWh for a 1 kWp PV system at optimum tilt in Aalborg (Denmark) and Bari (Italy). (Based on data from PVGIS, Photovoltaic Geographical Information System—Interactive maps, available: http://photovoltaic-software.com/pvgis.php.)

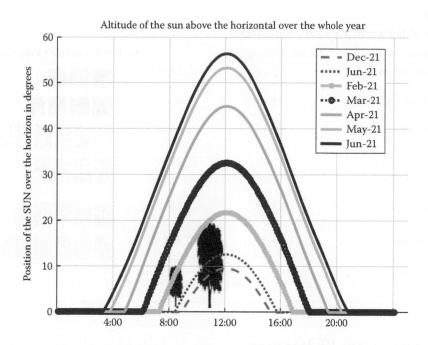

FIGURE 6.4 Sun chart for Aalborg location (latitude, 57; longitude, 9.5) showing also sources of shading during certain periods of the year represented by two trees.

that the different technologies have different cell efficiencies and price. Therefore, the cost, physical size, and AEP of the designed PV system will be influenced by the chosen PV cell technology.

There are three main technologies within the market today:

1. Monocrystalline silicone PV cells, which have the highest efficiency, but are the most expensive
2. Multicrystalline silicone PV cells, which offer a trade-off between efficiency and price
3. Thin-film (TF) PV cells, which tend to have the lowest efficiency, although price/m^2 might be the lowest

Depending on the design parameters, one can choose the technology that best fits the constraints. In this way, the PV system can be optimized for minimum investment cost or maximum energy production. Table 6.1 presents a comparison of the different PV cell technologies, based on panel efficiency and required surface for a power level of 1 kWp.

6.2.5 CHOOSING THE STRING AND ARRAY CONFIGURATION

Now that the solar cell technology and the kWp size for the PV array have been chosen, one will need to connect the individual PV panels into strings and arrays in order to achieve a certain maximum power point (MPP) voltage and current rating at standard test conditions (STC). STC is defined as 1000 W/m^2 irradiance intensity, 25 °C cell temperature, and an irradiance spectrum corresponding to an air mass ratio of 1.5; PV panel datasheet values are universally given in STC.

It is always recommended to recalculate the MPP voltage and current for extreme conditions for the PV panel temperature. For winter, this temperature is considered to be −10 °C while for summer it is +40 °C. Knowing the MPP voltage and current range during all conditions, one will be able to find a suitable PV inverter as it is presented in the next section.

TABLE 6.1

Efficiency of Different PV Cell Technologies

PV Cell Material	Module Efficiency (%)	Surface Needed for 1 kWp (m²)	
Monocrystalline silicon	15–18	5.5–6.5	
Polycrystalline silicon	13–16	6–8	
Micromorphous tandem cell (a-Si/ucSi)	8–10	10–12.5	
Thin-film copper indium/ gallium sulfur/diselenide (CI/GS/Se)	11–15	7–9	
Thin-film cadmium telluride (CdTe)	11–13	7.5–9	
Amorphous silicon	5–8	12.5–20	

Source: Remmers, K.-H., Inverter, storage and PV system technology, *Industry Guide 2014*, 2014.

6.2.6 CHOOSING THE PV INVERTER

A PV inverter will convert the DC current supplied by the PV panels into a grid-synchronized AC current that can be injected into the distribution network, as discussed in Chapter 4. Commercial PV inverters will make sure that the injected energy satisfies the grid codes and standards enforced by the distribution system operator.

PV inverters work with a wide range of power levels in a day. This is due to the fact that a PV array will deliver its kWp rating only during high irradiation conditions. When the cell temperature increases, or the irradiation decreases, the delivered power by the array drops as well. This is why the efficiency of PV inverters is expressed using a weighted formula that is adjusted to the specific location scenario [6]:

$$\eta_{EU} = 0.03\eta_{5\%} + 0.06\eta_{10\%} + 0.13\eta_{20\%} + 0.1\eta_{30\%} + 0.48\eta_{50\%} + 0.2\eta_{100\%} \quad (6.1)$$

$$\eta_{CEC} = 0.04\eta_{10\%} + 0.05\eta_{20\%} + 0.12\eta_{30\%} + 0.21\eta_{50\%} + 0.53\eta_{75\%} + 0.05\eta_{100\%} \quad (6.2)$$

where

    η_{EU} is the European weighted efficiency, tailored for locations in central and northern Europe

    η_{CEC} is the Californian efficiency, for locations with a lot of sunshine hours, like California and southern Europe

PV inverters are designed to achieve the highest possible efficiency over a wide power range, nevertheless, only at an optimum voltage range. This can be seen in Figure 6.5, which details the test measurements of a commercial PV inverter. The highest efficiency curve is achieved at 600 V input voltage over the whole power range of the inverter.

It is important to configure the PV array in such a way that best fits the PV inverter efficiency curve, MPP voltage range, and maximum allowable DC-link voltage, thereby eliminating any type of unnecessary losses and limitations and maximizing the AEP for the designed PV system.

As an example, a PV panel is chosen, for example, the RPP250 UP72 from REW Solartechnik with the datasheet values given in Table 6.2.

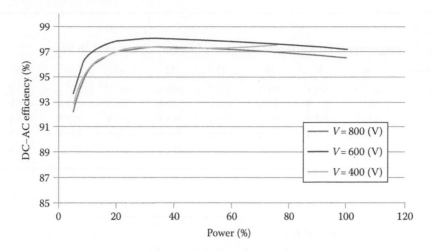

FIGURE 6.5 PV inverter efficiency at different input voltages and power levels.

TABLE 6.2
Datasheet Values for the Chosen PV Panel

Manufacturer	REW Solartechnik GmbH	
Type code	RPP250 UP72	
P_{MPP} at STC	250	W
V_{MPP} at STC	36.1	V
V_{MPP} (at 70 °C)	30.4	V
I_{MPP} at STC	7.49	A
V_{OC} at STC	43.9	V
V_{OC} (at −10 °C)	49.3	V
I_{SC} at STC	8.19	A
Length	1953	mm
Width	997	mm
Weight	15	kg
Module efficiency at STC	13	%
Temperature coefficient of I_{SC}	0.065	%/°C
Temperature coefficient of V_{OC}	−0.350	%/°C
Max system voltage	1000	V

Using the datasheet values of the chosen PV panel, a string can be configured. Such a string design case has a minimum of 8 panels in series and a maximum of 16. In this way, the PV string parameters can be calculated based on the datasheet values from Table 6.2.

Comparing the PV array configuration from Table 6.3 to the datasheet values of the PV inverter shown in Table 6.4, the sizing of the array and the choice of PV inverter can be obtained. This is an iterative process that should result in an optimum solution with the best match between the electrical specification for the PV array and the PV inverter.

In certain cases, the initial sizing of the PV array will lead to voltage and/or current ranges that will not fit the limits of the maximum power point tracker (MPPT) or the specifications of the PV inverter. For example, the maximum power point (MPP) voltage could be outside the range of the PV inverter specifications. In such a case, the MPPT algorithm of the PV inverter will limit

TABLE 6.3

PV String Configuration for Different Number of Panels Connected in Series

No. of Modules	8	9	10	11	12	13	14	15	16
V_{MPP} at STC	288 V	324 V	361 V	397 V	433 V	469 V	505 V	541 V	577 V
V_{OC} at STC	351 V	395 V	439 V	482 V	526 V	570 V	614 V	658 V	702 V
V_{MPP_min} at 70 °C	243 V	273 V	304 V	334 V	364 V	395 V	425 V	456 V	486 V
V_{OC_max} at −10 °C	394 V	443 V	492 V	542 V	591 V	640 V	689 V	739 V	788 V
V_{MPP} at −10 °C	288 V	325 V	361 V	397 V	433 V	469 V	505 V	541 V	577 V

TABLE 6.4

Datasheet Values for the Chosen PV Inverter

Max. DC power	3200 W
Max. input voltage	750 V
MPP voltage range	175–500 V
Rated input voltage	400 V
Min. input voltage	125 V
Initial input voltage	150 V
Max. input current	15 A
MPP inputs	2
Rated AC power	3000 W
Rated frequency	50 Hz ± 5 Hz
Max. output current	16 A
Power factor at rated power	1
Displacement power factor	±0.8
Max. efficiency	97%
EU weighted efficiency η_{EU}	96%

the output power according to the allowable voltage range. This means that the PV system will not operate at its maximum capability, resulting in a reduced AEP. Therefore, in such cases, it is recommended to configure a new PV array or choose another PV inverter that better fits the electrical parameters of the initial configuration of the PV array.

6.2.7 WHAT IS THE YIELD AND PERFORMANCE RATIO?

Finally, the PV system has been designed and configured for delivering the maximum energy in optimum conditions. To estimate the monthly and yearly energy production, a simple simulation model can be built, where all the components are modeled and real-life data are used for irradiation, cell temperature, availability of grid, etc. Usually, 1-minute averaged data are used for such simulations, but in certain cases, 1-second averaged data are recommended, to show a realistic scenario also for locations with fast-changing atmospheric conditions (fast cloud movements). In certain cases, it can happen that when data with a lower resolution are used, the peak irradiation values are filtered out and the converter is sized based on these averaged irradiation values, resulting in power limitations during high-irradiation periods and thereby resulting in energy loss. Such a scenario is shown in Figure 6.6, which presents sampled data with 5-minute averaged values in Figure 6.6a and data with 1-hour averaged values in Figure 6.6b, for a 48-hour period.

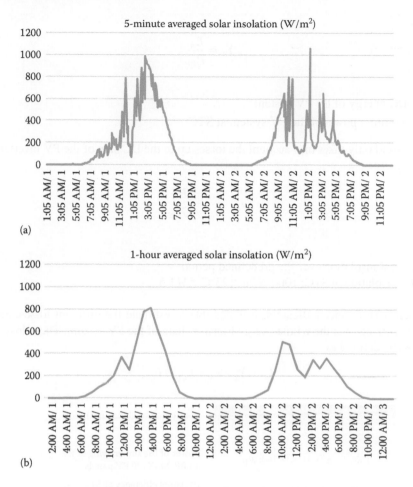

FIGURE 6.6 The same measured irradiation data for a 48-hour period in the same location, presented with different averaging periods: 5-minute averaged values (a) and 1-hour averaged values (b).

PV systems vary based on the location, PV technology, kWp size, PV array configuration, and also PV inverter type, just to mention a few. This means that direct power production or energy yield will not lead to the best criteria for comparing the performance of different PV systems. A simple way to measure the performance of a PV system is to compare the measured yearly energy yield to the same PV system under ideal conditions. The influence of the array size can be eliminated in case the produced energy is divided by the nominal power of the PV system at STC [7].

Using this idea, several performance factors can be defined.

Final yield:

$$Y_F = \frac{E_{usable}}{P_{G0}} \tag{6.3}$$

where
E_{usable} is the net energy output of the PV system
P_{G0} is the nominal power of the PV system at STC

The final yield includes all losses up to the point of connection to the AC electrical network.

Array yield:

$$Y_A = \frac{E_A}{P_{G0}} \tag{6.4}$$

where

E_A is the DC energy of the PV system

P_{G0} is the nominal power of the PV system at STC

The array yield only takes into account the losses up to the DC side of the PV inverter.

Reference yield:

$$Y_R = \frac{H_G}{G_0} = \frac{H_G}{G_{STC}} \tag{6.5}$$

where

H_G is the total insolation over the predefined period

G_{STC} is the insolation at STC: $1000 \ \text{W/m}^2; 25°C; AM 1.5$

Using such performance indicators, it is possible to estimate the different losses, presented in Figure 6.7, occurring over the whole energy transfer chain of the PV system. These losses could be generator losses, defined as

$$L_C = Y_R - Y_A = L_{CT} + L_{CM} \tag{6.6}$$

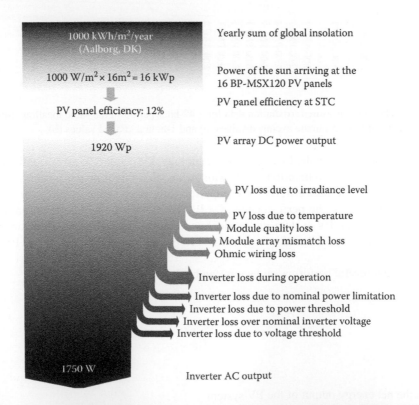

FIGURE 6.7 PV system losses, including all the sources for loss throughout the whole energy conversion chain, starting from the available energy from the solar insolation up till the point of connection to the distribution network of the AC grid.

where L_{CT} represents the thermal capture losses:

$$L_{CT} = P_{G0}\left[1 + c_T\left(T_C - T_{STC}\right)\right] \tag{6.7}$$

where
c_T is the thermal coefficient for power
T_C is the cell temperature
T_{STC} is the temperature at STC
L_{CM} represents nontemperature related losses (miscellaneous capture losses) that are due to

- String diodes and wiring
- Partial shading, dust, and snow on the panels
- Module mismatch
- PV inverter tracking errors (MPPT)
- Errors in irradiation measurement

There are also considered system losses that are defined as

$$L_{BOS} = Y_A - Y_F \tag{6.8}$$

which includes the total losses without the generator losses, like in the case of DC–AC conversion.

Now that all these yields and losses have been defined; the final performance indicator can be defined, which has been named as performance ratio (PR):

$$PR = \frac{Y_F}{Y_R} \tag{6.9}$$

giving a value of degree for approximation to the ideal case. Using this performance indicator, the performances of PV systems of different sizes, PV cell technologies, and PV inverters and located in different corners of the world can be compared.

6.3 CASE STUDY FOR DESIGNING A RESIDENTIAL PV SYSTEM

6.3.1 METHODOLOGY: DESIGN PROCEDURE

As it can be concluded from the previous sections, PV installations constitute a long-term and relatively expensive investment. Therefore, it is essential that an exact design of a certain PV installation is carried out. Figure 6.8 presents a simple flowchart for the design procedure of residential PV systems. In order to be able to design a PV installation, extensive and precise information of a wide range of parameters are required as described earlier in this chapter.

Besides all the technical aspects, there are also some financial issues to be considered for such an investment: payback time and internal rate of return. The financial calculations can be complicated and have many variables included. For small private PV systems on residential houses, the homeowners often look at the investment and the simple payback time in years, which can be calculated as the investment is divided by the yearly savings. Many homeowners wish to be self-supplied with energy, so the PV system can be dimensioned to cover the annual electricity consumption of the household. The average energy consumption in a typically Danish four-member family house is approximately 4300 kWh per year. Depending on the type of feed-in tariff for the produced solar energy, another threshold for the dimension could be the financial optimization of the system. If the selling price for the produced solar energy is less than the purchase price, many plants are designed to maximize the local consumption of the produced solar power.

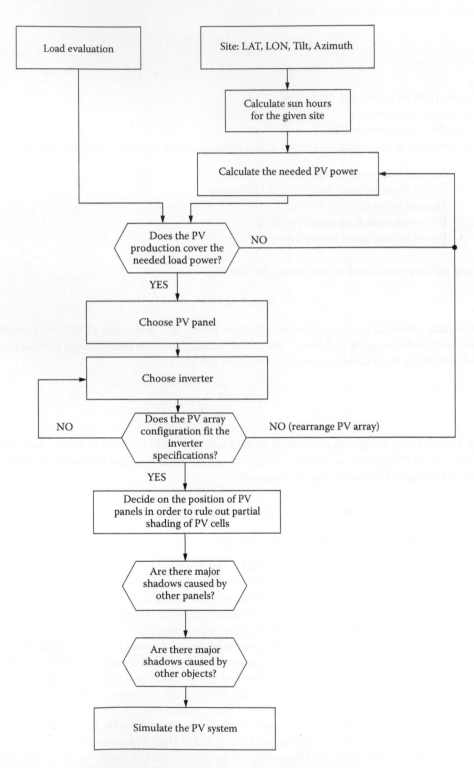

FIGURE 6.8 Simple flowchart for a residential PV system design.

TABLE 6.5

Deviation from Ideal Production Index in Percent of a PV Plant at Different Orientations and Tilt Angles

Production Index from a PV Plant at Different Orientations and Tilts (on the Northern Hemisphere: South = 0°, East = −90° and West = 90°)

Azimuth (°)	90	60	30	0	−30	−60	−90
Tilt (°)							
0	86%	86%	86%	86%	86%	86%	86%
15	84%	89%	93%	94%	93%	90%	85%
30	81%	90%	97%	99%	97%	91%	82%
45	77%	89%	97%	100%	98%	90%	79%
60	72%	85%	93%	97%	94%	86%	73%
75	65%	77%	86%	89%	86%	78%	66%
90	57%	67%	75%	77%	75%	68%	58%

As a rule of thumb, a PV plant in Denmark produces 950 kWh per year per installed kWp if all the optimum conditions are met. In Table 6.5, this corresponds to an installation at azimuth of 0 (the azimuth is related to the orientation of the panel and the azimuth angle is 0, if the panels are facing south on the Northern Hemisphere) and a module tilt of 45° (optimum for Denmark). If the plant is placed with an azimuth of −30° and a module tilt at 0°, the yearly production can roughly be estimated to be (950 kWh × 0.86) ≈ 817 kWh.

Designing the actual PV plant begins with choosing the module type and the amount of modules used in order to achieve the planned amount of kWh, which is followed by the choice of the PV inverter.

On the basis of this information, it will be possible, via a computer and relevant software, to simulate the yield of a certain PV installation with considerable accuracy, depending on the available data. There are independent software products that can be purchased from several different suppliers, which contain all the required information about PV modules and inverters on the market. The database within such software is regularly updated with new products. Many inverter manufacturers offer free software for system design. However, these are often restricted to the use of inverters from the inverter supplier, which offers the software tool. The planning of the system in the following text is done with a program supplied for free from the inverter manufacturer SMA and this gives a configuration solution with SMA inverters only [8].

The case discussed here presents the sizing of a PV system to supply approximately 2500 kWh a year. The system is placed close to the optimum conditions, so the system size can roughly be calculated to 2500 kWh/year divided by the factor 0.95 = 2631 Wp.

The planning is done with a 270 Wp PV module from a Trina-type TSM-270 DC05A.08. The technical data can be seen in Table 6.6.

Based on the rough size estimation, the number of modules to produce 2500 kWh per year can be calculated: 2631 Wp/270 Wp = 9.74, which if rounded up gives 10 modules.

The first step for the actual design in the program is to create the PV project where all general information about the project is entered:

- Project name
- Customer information
- Project location (for calculate the yield of the system)
- PV plant orientation and tilt
- Installation type (free standing, roof integrated, detached on the roof)

TABLE 6.6

Datasheet for the Chosen PV Module

Manufacturer	Trina Solar Energy	Cell Technology		Mono cSi
PV Module	TSM-270DC05A.08 (02/2014)	Certification		EU
Electric properties		**Temperature coefficients**		
Nominal power	270.00 Wp	MPP voltage	—	—
Performance tolerance	1.11%	Open-circuit voltage	−0.3200%/°C	−123.5 mV/°C
MPP voltage	30.80 V	Short-circuit current	0.0530%/°C	4.89 mA/°C
MPP current	8.77 A	**Degradation due to aging**		
Open-circuit voltage	38.60 V	Open-circuit voltage tolerance		0.00%
Short-circuit current	9.23 A	MPP voltage tolerance		0.00%
Permissible system voltage	1000 V	MPP current tolerance		0.00%
PV module efficiency (STC)	16.50%	Short-circuit current tolerance		0.00%
Grounding recommendation	No grounding	**Additional information**		
Mechanical properties		Current PV module		Yes
Number of cells in the PV module	60	Own PV module		No
Width	992 mm	Favorite		No
Length	1650 mm	**Comment**		
Weight	18.60 kg			

Source: S. GmbH, Sunny design web, 2015, available: http://www.sunnydesignweb.com/sdweb/#/Home.

After that, the type and numbers of the PV modules are chosen. In this design, there are 10 modules that give the following main plant specifications:

- 10×270 Wp = 2700 Wp
- Maximum open-circuit voltage from the system voltage: $V_{oc} = 35.8 \times 10 = 358$ V
- Maximum current in one string: $I_{sc} = 7.45$ A

When designing PV systems in cold regions, the maximum voltage has to be calculated for −10 °C, whereas the module data are given for +25 °C for most of the cases. It is important to consider the temperature coefficient of the module, as it can be read from Table 6.6 to be −0.32%/K. Therefore, the voltage rise in the system due to the low temperature can be calculated as $(25 + 10) \cdot 0.32 = 10.85\%$, which is automatically given by the software and is an important information based on which the PV inverter is chosen.

As seen in Table 6.7, the software returns the number of inverters that can be used for the chosen configuration. For the design presented in this chapter, the Sunny Boy 2.5-1VL-40 has been chosen. The inverter has a maximum output power of 2.5 kW, while the maximum input voltage is 600 V_{DC} with a maximum current of 10 A. As seen in Table 6.7, all the design parameters are within the limits of the inverter.

After the selection of the inverter, the software returns an overview showing the main technical data for the designed PV system. The annual yield is calculated to be 2558 kWh with a specific yield per year at 959 kWh/kWp. The inverter is a single-phase inverter, and therefore, the manufacturer

TABLE 6.7
Specifications Shown for the Chosen PV Inverter

SB 2.5-1VL-40			2.50 kW
Details	Peak power: 2.70 kW	Nominal power ratio: 98%	Energy usability factor: 100%

Performance	
Annual energy yield	2588.70 kWh
Spec. energy yield	959 kWh/kWp
Performance ratio	85.7 %
Line losses (in % of PV energy)	—

PV inverter parameter	Inverter	Input A
Max. DC power	2.65 kW	2.7 kW
Min. DC voltage	50 V	267 V
Typical PV voltage		288 V
Max. DC voltage (inverter)	600 V	431 V

Source: S. GmbH, Sunny design web, 2015, available: http://www.sunnydesignweb.com/sdweb/#/Home.

TABLE 6.8
Simple Overview of System Design Showing the Most Important Data for the Designed PV System

Total number of PV modules	10	Annual energy yield (approx.)	2588.70 kWh
Peak power	2.7 kWp	Energy usability factor	100%
Number of inverters	1	Performance ratio (approx.)	85.7%
Nominal AC power	2.5 kW	Spec. energy yield (approx.)	959 kWh/kWp
AC active power	2.5 kW	Line losses (in % of PV energy)	—
Active power ratio	92.6%	Unbalanced load	2.5 kVA

Source: S. GmbH, Sunny design web, 2015, available: http://www.sunnydesignweb.com/sdweb/#/Home.

highlights that there is 2.5 kVA of unbalanced "load" as seen in Table 6.8. Furthermore, the monthly energy productions will also be calculated, and for the presented example, these are given in Figure 6.9 and Table 6.9.

The software also allows the user to add and edit several standard parameters such as the cable type and length on both the DC and the AC side, so cable losses can be calculated and taken into account too. The basic formula behind the calculation is as follows:

$$P_{loss} = I^2 \cdot R \tag{6.10}$$

where

P_{loss} the power loss in the cable (W)

R is the resistance in the cable (Ω)

I is the nominal current in ampere (A)

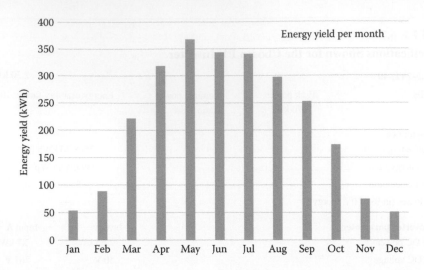

FIGURE 6.9 Energy production values shown for each month of the year calculated for the 10 panel installation given in Table 6.8.

TABLE 6.9

Calculated Total Energy and Performance Ratio for Each Month Separately for the 10 Panel Installations

Month	Energy Yield (kWh)	Performance Ratio (%)
Jan	54	82
Feb	89	85
Mar	222	88
Apr	319	88
May	368	86
Jun	344	85
Jul	341	85
Aug	298	85
Sep	253	86
Oct	174	86
Nov	75	84
Dec	51	82

Source: S. GmbH, Sunny design web, 2015, available: http://www.sunny designweb.com/sdweb/#/Home.

The resistance R of the cable can be calculated as

$$R = \delta \cdot \frac{l}{S} \tag{6.11}$$

where
  δ is the resistivity of the material conductor: 0.023 for copper and 0.037 for aluminum (ambient temperature = 25 °C) in $(\Omega \cdot mm^2/m)$
  l is the length of the cable in meters (m)
  S is the cross section of the cable in (mm^2)

A more detailed design software allows the user to build and parametrize custom modules, adjust the PV inverter parameters, calculate and adjust the different losses in the system, and also conduct yield studies, taking into account the module losses placed in shaded positions.

6.3.2 COMMENTS TO THE DESIGN DIFFERENCES FOR TF AND CRYSTALLINE-SI SYSTEMS

A residential PV system consists of several elements that can be categorized into four main topics:

1. *PV modules*—the PV modules on the roof that generate electricity
2. *Inverter*—the unit that converts the DC power from the modules to AC power usable in the house
3. *Balance of system (BOS) components*—cable, mounting structure, fuses, etc.
4. *Labor cost*—the cost for installing the system (electrician/carpenter)

TF system design is basically similar to cSi system design methodology; however, due to the lower efficiency and higher output voltage on TF modules, there are some design issues that occur more often when designing smaller systems using this new PV cell technology.

Furthermore, with TF modules, a PV system requires a larger area for the same kWp installation. This also means that there are more BOS components and the labor cost is also higher.

If the same rating of 2.7 kWp is used for the design of a PV system but with TF modules, there will be some differences in the final design. As an example, one can take the SF 150-S module from Solar Frontier. Its technical specifications are presented in Table 6.10. The SF 150-S module has a power output of 150 Wp per module so in this case, 18 modules are needed to reach the target of 2.7 kWp. Each module has an open-circuit voltage of 108 V_{DC}, and the maximum input voltage of the SMA Sunny Boy 2.5 inverter is 600 V_{DC}, which results in a maximum of 5 series-connected modules to the DC bus of the inverter. Since the target is 18 modules, one possible configuration is to create 6 parallel strings with 3 modules in series for this inverter. This is not

TABLE 6.10
Solar Frontier, Thin-Film Module Datasheet

S-Power Modules		SF 145-S	SF 150-S	SF 155-S
Nominal power	P_{max}	145 W	150 W	155 W
Power tolerance	+5 W/0 W			
Module efficiency	%	11.8%	12.2%	12.6%
Open-circuit voltage	V_{oc}	107.0 V	108.0 V	109.0 V
Short-circuit current	I_{sc}	2.20 A	2.20 A	2.20 A
Current at nominal power	I_{mpp}	1.80 A	1.85 A	1.88 A
Maximum system voltage	V_{DC}		1000 V	
Maximum reverse current			7 A	
Maximum fuse rating			4 A	
Temperature Characteristics				
Temperature coefficient of I_{sc}			+0.01%/K	
Temperature coefficient of V_{oc}			−0.30%/K	
Temperature coefficient of P_{max}			−0.31%/K	

an optimal configuration because more material is needed for the cabling. Furthermore, a lower input voltage results in high current, which at the end gives higher losses in the system. This means that another inverter should be used for this particular design case. Many inverters have a maximum input voltage of 1000 V_{DC}, and by choosing such an inverter, the PR of the designed system could be optimized.

6.3.3 System Performance Monitoring

When a PV system is installed and commissioned, it is important that the system is properly maintained during its lifetime, which is expected to be above 25 years. Many PV systems have been sold with the promise that no service or maintenance is needed. This is nearly true. However, faults might appear in the system that will need to be dealt with. Failure in the different components might happen during the early years of the PV system. Therefore, it is important to know how long the standard warranty period for the different components is.

6.3.4 Warranty of Modules

PV modules always come with a limited product warranty; this is typically a 10–15 years' repair, replacement, or refund remedy. The warranty covers defects in materials and workmanship under normal operating conditions.

There is also a standard in the PV business prescribing that the PV modules are delivered with a limited power warranty. This warranty guarantees that the customer with an output performance of the module in a period up to 30 years depends on the supplier conditions. The performance warranty typically includes a 10–15 years power guarantee of, for example, 90% of nominal output power and after 25–30 years of operation, 80% power output is guaranteed.

If the power output is below the guaranteed power values, the supplier has to replace such losses in power by either providing the customer additional PV modules to make up for such a loss, by providing monetary compensation equivalent to the cost of the additional PV modules, or by repairing or replacing the defective PV modules.

6.3.5 Warranty of Inverters

Today, inverters are delivered normally with a minimum of 5 years, warranty. The warranty covers defects in materials and workmanship under normal operating conditions. Most inverter suppliers offer 5–15 years, extra warranty, which can be purchased before installation or for some suppliers, up to 5 years after commission of the PV system.

6.3.6 Warranty of BOS Components

BOS components and the installation itself often come with a 10-year warranty.

When the PV plant is in operation, a minimum of surveillance is needed. Most new inverters today have a built-in Wi-Fi connectivity, so the inverter and the performance of the plant can be monitored through an app or over the Internet.

6.3.7 PV System Price

The price of PV systems has been decreasing since the start of their commercialization back in the 1980s. The chart in Figure 6.10 shows both the actual price for cSi and CdTe for the last 35 years and also the production learning curves when the production is increased. The actual price follows

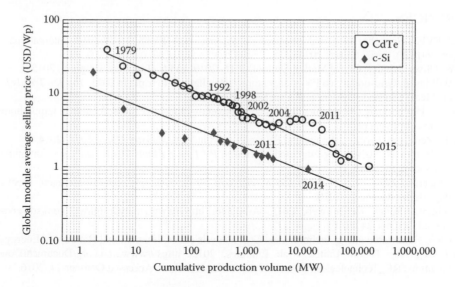

FIGURE 6.10 The global PV module price learning curve for cSi wafer-based and CdTe modules, 1979–2015. (Based on IRENA, Renewable energy technologies: Cost analysis series, International Renewable Energy Agency (IRENA), available: https://www.irena.org/DocumentDownloads/Publications/RE_Technologies_Cost_Analysis-SOLAR_PV.pdf2012.)

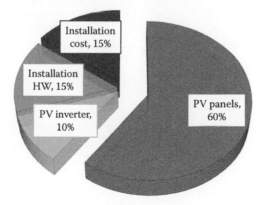

FIGURE 6.11 Cost distribution for grid-connected PV systems.

the learning curve and shows a 22% reduction in price each time the production is doubled. The pie chart in Figure 6.11 shows the cost distribution of the major components in the case of a PV system, where the "PV panels" take up 60% and "installation cost" and "installation hardware " take up 15% each, while the "PV inverter" only 10%.

6.4 SUMMARY

There has been a fall in the cost of not only PV modules; inverters, mounting structures, and other BOS components have also decreased in price and have become more efficient. In 2015, a turnkey residential PV system with a size of approximately 6 kWp costs around 85,000 DKK in Denmark. This corresponds to a price of approximately 14 DKK per Wp. Larger PV systems have a lower price per installed kWp, and prices are still expected to fall in the future.

REFERENCES

1. IEA-PVPS, Snapshot of global photovoltaic markets, IEA PVPS T1-29:2016, International Energy Agency (IEA), Paris, France, 2016.
2. J. W. Oliver Schäfer, Global Market Outlook 2015–2019, EPIA, Bruxelles, Belgium, 2015.
3. S. Wilkinson, Grid-connected PV energy storage installations to triple in 2015, *Top Ten Trends for 2015*, Solar Media Limited, London, U.K., pp. 9–10, 2015.
4. PVGIS, Photovoltaic Geographical Information System—Interactive maps. Available: http://photovoltaic-software.com/pvgis.php.
5. K.-H. Remmers, Inverter, storage and PV system technology, *Industry Guide 2014*, 2014.
6. K.-H. Remmers, Striving for increased efficiency, *PV Power Plants 2013*, Sunbeam Communications, Berlin, Germany, 2013.
7. H. Häberlin, *Photovoltaik: Strom aus Sonnenlicht für Verbundnetz und Inselanlagen*, Electrosuisse-Verlag, Berlin, Germany, 2010.
8. S. GmbH, Sunny design web, 2015. Available: http://www.sunnydesignweb.com/sdweb/#/Home. Accessed October 24, 2016.
9. IRENA, Renewable energy technologies: Cost analysis series, International Renewable Energy Agency (IRENA), Abu Dhabi, United Arab Emirates, 2012. https://www.irena.org/DocumentDownloads/Publications/RE_Technologies_Cost_Analysis-SOLAR_PV.pdf. Accessed October 24, 2016.

7 Small Wind Energy Systems

Marcelo Godoy Simões, Felix Alberto Farret,
and Frede Blaabjerg

CONTENTS

Abstract 151
7.1   Introduction 151
7.2   Generator Selection for Small-Scale Wind Energy Systems 152
7.3   Turbine Selection for Wind Energy 157
7.4   Self-Excited Induction Generators for Small Wind Energy Applications 159
7.5   Permanent Magnet Synchronous Generators for Small Wind Power Applications 162
7.6   Grid-Tied Small Wind Turbine Systems 168
7.7   Magnus Turbine–Based Wind Energy System 171
7.8   Summary 175
References 175
Further Reading 176

ABSTRACT

This chapter intends to serve as a brief guide when someone is considering the use of wind energy for small power applications. It is discussed that small wind energy systems act as the major energy source for residential or commercial applications, or how to make it part of a microgrid as a distributed generator. In this way, sources and loads are connected in such a way to behave as a renewable dispatch center. With this regard, non-critical loads might be curtailed or shed during times of energy shortfall or periods of high costs of energy production. If such a wind energy system is connected to the public distributor, it can serve as a backup system, as a non-interruptible power supply (with storage aggregation), provide low-voltage support, or give a clean surplus of energy transferred to the public network under economical and technological basis.

In this chapter, several factors are also considered when selecting a generator for a wind power plant, including capacity of the AC system, types of loads, availability of spare parts, voltage regulation, technical personal and cost. If several loads are likely inductive, such as phase-controlled converters, motors, and fluorescent lights, it is evaluated that synchronous generators or induction generators are an open-ended problem to be solved for the best choice.

7.1 INTRODUCTION

The application of small wind turbines for residential and commercial use depends on how a set of distributed generation and loads using controllers can operate in a suitable way, because in addition to variable demand, there is a random nature of the wind speed. Small wind turbines can supply electrical power for stand-alone applications, or grid connected or even connected to microgrids, that is, a group of generating sources and single or multiple end users for residential, industrial, commercial, rural, or public applications.

Alternative sources of energy for microgrid systems may include wind turbines, small hydroplants, photovoltaic panels, fuel cell stacks, geothermal energy, and small-scale renewable generators. In general, either the management of energy storage or the control of a dummy load and a centralized distribution control are necessary to ensure a power balance. The wind energy conversion has some simple foundations—the kinetic energy of the wind is converted into a mechanical shaft movement and then into electrical energy through a wind turbine coupled to an electrical generator. However, wind energy has some unavoidable constraints. Its real performance is affected by local conditions and random nature of the wind, nearby physical obstructions, power demand profiles, and several turbines-related factors, in addition to a possible deterioration of the performance due to aging.

A small wind energy system may be the major energy source for residential or commercial applications, or it can be part of a microgrid. All controlled sources and loads are interconnected in a manner that enables the devices to function as dispatch centers, and noncritical loads might be curtailed or shed during the times of energy shortfall or possible high costs of energy production. If such a wind energy system is connected to the public distributor, it can serve as a backup system and as a noninterruptible power supply (with storage aggregation), provides low-voltage support, or gives surplus of energy transferred to the public network at an economical price. In order to establish a wind-based grid-connected system, it is important to observe the following:

- Supply energy requirements of the present and future loads.
- Establish civil liabilities in case of accidents and financial losses due to shortage or low quality of energy.
- Negotiate collective conditions to interconnect the microgrid with the public network or with other sources of energy that are independent of wind resources.
- Establish performance criteria of power quality and reliability to reduce costs and guarantee an acceptable energy supply.

Good quality wind energy systems may require extensive data processing of the electrical network characteristics and analysis of how the power quality impacts the overall plant performance. The required data for assessing real-time performance of wind turbines are limited to three real-time variables: (1) the output power of the turbine (W), (2) the rotational speed of the turbine (rad/s), and (3) the wind speed (m/s). Data analysis would require instrumentation to obtain four parameters: (1) generator voltage, (2) load current, (3) distribution of the wind speed, and (4) wind turbine speed for every machine in the field. Chapters 8 through 10 cover the details for such design.

7.2 GENERATOR SELECTION FOR SMALL-SCALE WIND ENERGY SYSTEMS

Criteria must be established for selecting the electrical generators for small wind energy power plants. The level of active and reactive powers for a particular application is dependent on the variable-speed features of the generator. This analysis supports a wide set of other variables such as voltage tolerance, frequency, speed, output power, slip factor, required source of reactive power, and field excitation, among other parameters. In applications with variable speed, a DC link is usually used between the generator and the load and may decouple the generator operation from the grid.

There are other factors to consider when selecting a generator for an AC system, including capacity of the system, types of loads, availability of spare parts, voltage regulation, and cost. If several loads are likely inductive, such as phase-controlled converters, motors, and fluorescent lights, a synchronous generator could be a better choice than an induction generator (IG) for larger power applications. IGs cannot supply on their own a high start-up surge current required for starting other motor loads when operating in stand-alone mode. Therefore, selecting and sizing the generator is a highly technical decision and viability studies should be conducted, particularly using a power systems or power electronics software simulator. Figure 7.1 shows how to convert the low-speed (typically 10–30 rotations per minute [rpm]) high-torque turbine shaft power to electrical power. A gearbox is often used in small wind turbines, whereas multipole generator systems are custom made for multi-megawatt solutions. Between the grid and the generator, a power converter might be inserted in order to attain higher flexibility. The most typical machines to convert the turbine shaft mechanical power into electrical

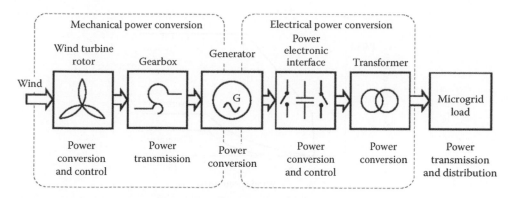

FIGURE 7.1   Power conversion stages in a modern wind turbine.

power are (1) permanent magnet synchronous generator (PMSG), (2) self-excited induction generator (SEIG), (3) squirrel cage induction generator (SCIG), and (4) doubly fed induction generator (DFIG).

For small power commercial applications (both onshore and offshore), the recommended solution is a multistage geared drive train with an IG [1–4]. The IG is a simple solution for a nonsynchronous direct connection to the grid, for which it is sufficient to guarantee the electrical rotation above the synchronous speed, that is, for $\omega_e > \omega_r$, in order to allow the mechanical power to be transferred to the electrical terminals.

The IGs have several other advantages: they are light, rugged, naturally protected against short circuit; usually, they can produce and distribute power in large scale for industrial applications, and they are cheaper than synchronous generators. The IG is more common for small power stand-alone applications, but it is also used from medium to high levels of power generation. The active power flow depends on the rated slip factor. A large slip factor decreases the power factor and a speed control or at least a speed limiter should always be implemented in the IG control system.

The IG is very often a standard three-phase induction motor, which is made to operate as a generator. Self-excitation capacitors are used for the voltage building-up process, particularly for smaller stand-alone systems, less than 15 kW mechanical shaft power rating. Requirements on constant frequency and voltage should not be demanding for stand-alone SEIG, but an electronic load control may improve the overall operation [4]. The efficiency of IGs depends on their size. However, a rough estimation for the SEIG efficiency from shaft to terminals is approximately 75% at full load to as low as 60% or less at light loads. At high speeds, the operating frequency for the IG is from 100 to 200 Hz, depending on the number of poles to maintain the required match of shaft angular speed to the machine terminals' electrical angular speed within a reasonable range [5, 6].

The primary energy source to drive the generator's shaft, the turbine type, number of poles, and electrical terminal characteristics of the generator determine the rated speed, commonly specified in rpm. Most electrical loads demand that generators should be driven at a speed that generates a steady power flow at a frequency of 50/60 Hz. The number of poles defines the necessary shaft speed of the turbine. In the United States, the electrical power grid frequency is 60 Hz. So it is common to have a two-pole generator demanding speeds as high as 3600 rpm, while for 50 Hz electrical networks, the machine runs typically at 3000 rpm; for a 60 Hz small wind power system, the 900 rpm eight-pole generator is often used in field applications. However, this range of 900–3600 rpm is still too high for practical use with small wind power. It is necessary to use speed multipliers (gearbox), making the whole system heavier, more expensive, more maintenance demanding, and relatively less efficient. The cost of the generating unit is more or less inversely proportional to the turbine speed and the type of primary energy. The lower the speed is, the larger the machine size is for such output power.

Currently, for medium and higher power applications, turbines have a variable-speed and variable-pitch control. In higher power applications, it is usual to adopt a DFIG with a multiple-stage gearbox. Many manufacturers have developed such a solution. The benefits of using DFIGs are as follows: (1) it is not necessary to have special sensors, (2) it is possible to achieve high rotor speeds,

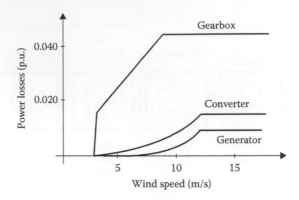

FIGURE 7.2 Losses in a typical wind turbine drive train.

and (3) it has a fractional power converter rating. In addition, there are other considerations to make the DFIG a proper generator solution for high power applications: possible minimization of reactive power needs at the stator side, and it makes the wind turbine technology independent of permanent magnets (PMs) from countries that control the business of rare-earth materials. However, a DFIG system has also the following disadvantages: (1) it needs usually a large, heavy, and noisy gearbox; (2) it gives heat dissipation because of gearbox friction; (3) gearbox maintenance procedures need to be done; (4) high torque peaks in the machine and large stator and rotor peak currents under grid fault conditions; (5) the brush–slip ring set to bring power to the rotor needs maintenance; (6) external synchronization by power converters is required between the stator and the grid to limit the start-up current (soft start); (7) detailed transient models and good knowledge of the DFIG parameters are required to make a correct estimate of occurring torques and speeds; and (8) when grid disturbances are present, a ride-through capability of the DFIG is required, and the control strategies may become more complex. Figure 7.2 shows the typical losses in a wind turbine system, composed of machine, converter, and gearbox. Because a machine generates very low power for wind speed less than 4 m/s, the off-shelf wind systems usually shut down the system, because it is only running for providing heat, and no real power is usually converted in the very low wind speed range.

Some wind turbine applications may also use switched reluctance (SR) generators; their operating frequency can be extremely high, in the range of 6 kHz at 60,000 rpm, requiring high-speed power switches at very high switching frequency rates. For example, the slip control of an IG, or even a scalar or a vector control, requires a precise measurement of speed in order to optimize the power. However, the control of SR generators requires very precise measurements of the rotor position involving high technological and expensive components. In the SR generator controller, rates of currents and voltages may result in high stress levels for the power electronic devices. On the other hand, the IG has a natural, well-regulated sinusoidal output that can be conditioned without using stressed electronic components [7–9]. In PM generators, the power rating of the converters has to cope with several complexities due to wide variation in the output voltage. The power electronic components must function at high stress levels.

For selecting the generator, it is also important to compare power outputs, operation hours, available technology, special needs of personnel, and cost. The power unit can be stationary or portable. Considerations about installation and maintenance must be made by qualified professionals, who will decide about additional accessories such as a protection cover against wear and tear of nature, protecting devices, a transfer switch, and a data logger [10–12].

Tables 7.1 through 7.4 list some general criteria to compare generators for small and medium power applications. These criteria are classified in electrical, mechanical, control, and constructive aspects, respectively. They can help decision-makers in selecting the generator type to be used in a wind energy system for residential and commercial applications.

TABLE 7.1
Electrical Criteria of a Wind Energy Generator

Electrical Criterion	PMSG	SEIG	SCIG	DFIG
Voltage regulation	Voltage boost procedures	Poor	Poor	Medium
Speed regulation	Total	Total $(10\% > n_s)$	Total	±30%
Frequency regulation	High	Low	Poor	Low
Number of poles	Typically, 24	4	4	4
Power converter	Full scale	Lower switching frequency	Lower switching frequency	Small PWM
Losses in power converter	Very high stress levels	Small	Small	High
Field losses	None	Medium	Medium	Medium
Brushes and slip rings	None	None	None	Necessary
Ride-through capability	Not so complex	Complex	Complex	Complex
Connectivity	Medium	Easy	Easy	Medium
Efficiency	91%	75%	89%	88%
Flexibility	High	High	High	High

Abbreviations: PMSG, permanent magnet SG; SEIG, self-excited IG; SCIG, squirrel cage IG; DFIG, doubly fed IG.

TABLE 7.2
Mechanical Criteria of a Wind Energy Generator

Criterion	PMSG	SEIG	SCIG	DFIG
Thermal dissipation	Minimum losses	Medium	Medium	Medium
Manufacturing requirements	Moderate	Easy	Easy	Involving
Adaptability	High	High	High	High
Expertise for using the generator	Medium	Low	Low	High
Pollution levels, environment impacts	Medium	Low	Minimum	Low
Portability	Medium	High	High	Medium
Reliability	High	High	High	Top medium
Robustness	High	High	High	Medium
Service interruption	Low	Low	Low	Low
Knowledge and technology needs	New	Easy	Easy	Involving
Turbine-generator coupling	Possibly direct	Direct or with gearbox	Direct or with gearbox	Direct coupling
Power per weight and per volume	Toward light	Low	Low	Medium

TABLE 7.3

Construction Criteria of a Wind Energy Generator

Constructive Criterion	PMSG	SEIG	SCIG	DFIG
Technology	Fast evolution	Proven	Proven	In evolution
Manufacture	Moderate	Easy	Easy	Involving
Acquisition costs	Still high	Low	Low	Medium
Adaptability	High	High	High	High
Capital return	Medium	Fast	Fast	Medium
Commercial availability	Low	Highest	Highest	High
Commissioning	Medium	Minimum	Minimum	Medium
Specific costs	Permanent magnets	Overall low	Overall low	Power converter
Expertise to run	Medium	Low	Low	High
OM	Low	Low	Low	High
Portability	Medium	High	High	Medium
Reliability	High	High	High	Top medium
Robustness	High	High	High	Medium
Today standardization	Low	High	High	Medium
Useful life	20 years	30 years	30 years	15 years
Specific problems	PM high cost and easy demagnetization at high temp.	Demands high external reactive power	Demands high reactive power from grid	Requires precise slip measurement

TABLE 7.4

Control Criteria of a Wind Energy Generator

Control Criterion	PMSG	SEIG	SCIG	DFIG
Active and reactive power control	Full	Needs compensation	Needs compensation	Full
Torque oscillations	Flat	Wavy	Flat	Wavy
Controllability	Needs voltage boost controller	Difficult for constant speed	Constant speed depends on power electronics control	Wide range around synchronous speed
Power factor control	Possibly external	Not really, variable frequency	Not really, variable frequency	Allowed
Speed regulation	Total	Loose $(10\% > n_s)$	Total	±30%
Speed response against surges	Fast	Slow	Slow	Fast
Connectivity	Medium	Easy	Easy	Medium
Synchronization for grid connection	Tight	Loose	Loose	Loose
Speed regulation	Tight	Difficult	Difficult	Loose

7.3 TURBINE SELECTION FOR WIND ENERGY

Figure 7.3 shows that a wind turbine system has four ranges, or wind speed bands. One is below cut-in; another is above maximum wind speed, and therefore, in practice there are only two really different operational ranges. The first band goes from zero to the minimum speed of generation (cut-in). Below the cut-in speed, the generated power just only overcomes barely the friction losses and turbine wearing outs. The second band (optimized power coefficient C_p, constant) is the normal operation maintained by a system, where the turbine blade position faces the direction of the wind attack (pitch control) for maximum power extraction. The third band (high-speed operation) has a speed control that maintains the maximum output constant limited by the generator rated capacity, because the whole turbine structure has mechanical safety constraints. Above this band (at wind speeds around 25 m/s), the nacelle or the rotor blades are aligned in the wind direction and prevent mechanical failure to the turbine, electrical generator and driving system and shutting off. IGs with self-excitation have typically been used for low-power wind turbines [4].

The energy captured by the rotor of a wind turbine must be considered with certain historical wind power intensity data (in W/m²) in order to access the economical viability of the site. Obviously, seasonal variation as well as year-to-year variations in the local climate should also be considered. The effective power extracted from the wind is derived from the airflow speed just reaching the turbine, v_1 and the velocity just leaving it, v_2. Considering the average speed $(v_1 + v_2)/2$ passing through the blade-swept area A, the kinetic energy imposes the net wind mechanical power of the turbine as given in Equation 7.1. Since the derivative of the air mass in terms of air density expresses the instantaneous power for the turbine, the following two equations can be calculated:

$$P_{turbine} = \frac{dK_e}{dt} = \frac{1}{2}\left(v_1^2 - v_2^2\right)\frac{dm}{dt} \text{ in W/m}^2 \tag{7.1}$$

$$P_{turbine} = \frac{1}{2}\rho C_p A v_1^3 \tag{7.2}$$

where

$C_p = \left(1 - \dfrac{v_2^2}{v_1^2}\right)\left(1 + \dfrac{v_2}{v_1}\right)\Big/2$ is the power coefficient to establish the rotor efficiency

ρ is the air density
v_1 is the wind velocity
A is the net blade area

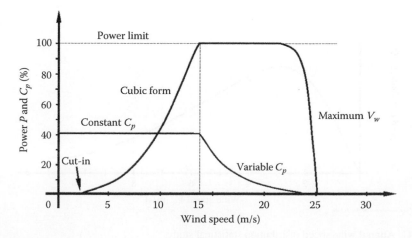

FIGURE 7.3 Speed control range for wind turbines.

The coefficient C_p has a theoretical maximum value of $C_p = 16/27 = 0.5926$ (Betz limit). In practice, the efficiency of a rotor is not as high as the ideal value, but more typical efficiencies range from 35% to 45%. Since (7.2) defines the instantaneous power and the characterization of a wind turbine, the associated generator must be sized in accordance with a random nature; it is very appropriate to define the local wind power as proportional to the statistical distribution of speed instead of instantaneous values [4].

The statistical-based analysis allows different sites, with the same annual average speed to find out distinct features of wind power availability. Figure 7.4 displays a typical occurrence curve of wind speed distribution in percent for a given site. That distribution can be evaluated monthly or annually. It is determined through bars of occurrence numbers, or percentage of occurrence, for each range of wind speed during a long period. It is usually observed as a variation of wind related to climate changes in that particular area. If the wind speed is lower than 3 m/s (denominated by "calm periods"), the wind power becomes very low for extraction and the system should stop. Therefore, studying the calm periods helps to determine the necessary timing for long-term energy storage or alternative power generation.

The power distribution varies according to the intensity of the wind multiplied by the power coefficient C_p of the turbine. Then, a typical distribution curve of power is depicted in Figure 7.4 as a function of the average wind speed. One year has 8760 h, so the vertical axis of Figure 7.4 represents a percentage of hours per year per meter per second. Sites with high average wind speeds do not have calm periods, and there is not much need of energy storage. Although higher speeds have more energy, there are limits, because there are structural problems in the system or in the turbine for high wind speed.

The Weibull probability density function is a general formula that describes the features of wind resources. However, it is more convenient to use the Rayleigh distribution (in practice), which is given by the following equation:

$$h(v) = \left(\frac{2}{c}\right)\left(\frac{v}{c}\right)e^{-(v/c)^2} \tag{7.3}$$

where the "factor c" is defined as the "scale factor"; it is related to the average wind velocity. Such function is represented in Figure 7.5; "factor c" is directly related to the number of days with high wind speeds, that is, the higher the "c" is, the higher the number of windy days is.

Such probability density function is capable of representing the statistical nature of the wind speed for most practical cases. Considering that weather may have the same cyclic seasons from

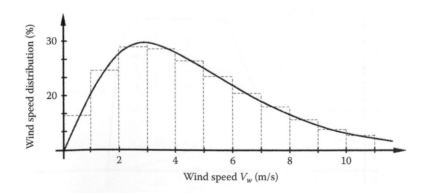

FIGURE 7.4 Annual wind speed distribution statistical study.

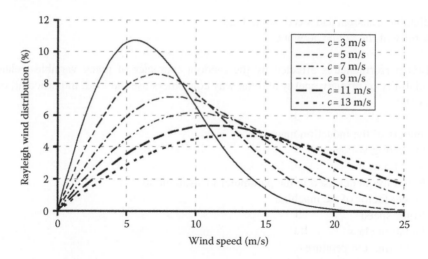

FIGURE 7.5 Annual wind speed study fit with a Rayleigh probability density function.

1 year to the next, it is natural that a cycle of 1 year should be the minimum when used for evaluating the wind resources.

For optimally designing an electrical generator, the random nature of the wind distribution at a particular site is considered, defining the machine frequency and voltage ratings. The majority of the losses are due to the necessity of a gearbox to match the generator speed with the turbine speed. Therefore, it is interesting to consider eliminating the gearbox in order to design an overall optimal system.

7.4 SELF-EXCITED INDUCTION GENERATORS FOR SMALL WIND ENERGY APPLICATIONS

A small wind power plant may use an SCIG as an IG, and it is common to have capacitors for providing the generator the required magnetizing current for self-starting conditions from black start. An SEIG usually has simplified controllers for speed, voltage, and frequency. It is also common to use IGs for small wind energy systems, because they are cheaper, more rugged, and robust than other electrical machines. However, they have larger losses and also need a gearbox. Nevertheless, IGs do not need sophisticated synchronization devices and have natural, or intrinsic, protection against short circuits. In stand-alone operation, an SCIG can be used; sometimes, a small PMSG is preferred together with a power converter. DFIGs are used for larger power applications, with a power converter connected on the rotor of the machine. SCIGs are used for many industrial applications, so their widespread use facilitates the application as a generator in developing countries. An IG requires either reactive power from the utility grid to operate or will need an inverter with a battery or capacitors for self-excitation.

A correct design of an SEIG involves many factors, such as

1. Instantaneous power and amount of energy in the primary source
2. Machine construction characteristics
3. Self-excitation process
4. Load starting procedures
5. Speed ratio of a multiple-stage gearbox

6. Predictable transient and steady-state loads
7. Proximity of the public network

The SEIG performance is affected by the random character of many variables related to the instant availability of primary energy and the way consumers use the load, in particular, on the following aspects:

- Parameters of the induction machine
  - Operating voltage
  - Rated power
  - Rated frequency used in the parameter measurements
  - Power factor of the machine
  - Rotor speed
  - Isolation class
  - Operating temperature
  - Carcass (cylindrical yoke) type
  - Ventilation system
  - Service factor
  - Noise level
  - Resonant conditions
- IG self-exciting process
  - Degree of iron saturation of the generator caused by capacitors
  - Fixed or controlled self-excitation process
  - Speed control
- Load parameters
  - Rated voltage
  - Starting torque and current
  - Maximum torque and current
  - Power factor
  - Generated harmonics
  - Load connected directly to the distribution network or through converters
  - Load type: passive, active, linear, and nonlinear
  - Load evolution over time
- Type of primary source (in order to define the shaft power)
  - Wind or hydro for small-scale applications
  - Primary machine conditions for acceleration or deceleration
  - Needs for energy storage

The induction machine parameters are related to their iron magnetization characteristics and consequently to their degree of iron saturation and operating rotor speed. Therefore, experimental tests are required to obtain their magnetization curve. The designer of a SEIG needs a setup capable of keeping the shaft speed constant for several conditions of the applied voltage. By experiments, it is easy to find out the correct excitation capacitor for the measured magnetization curve.

In the case of stand-alone operation (in small power plants), the connection of a capacitor bank across the IG terminals is necessary to supply their needs for reactive power, as illustrated in Figure 7.6. To keep the phase voltage balanced, it is advisable to connect each excitation capacitor C across each generator winding; one should keep the same capacitor connection of the winding machine wiring, either both in $\Delta-\Delta$ or in $Y-Y$ connection. Furthermore, every load can be individually compensated by a capacitor in such a way that whenever it is connected to the generator, it will not change the necessary excitation capacitance and the output voltage will remain nearly constant.

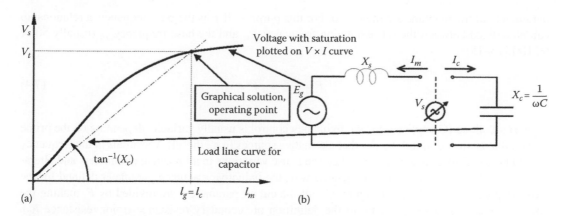

(a)

(b)

FIGURE 7.6 Scheme to obtain: (a) the magnetization curve and (b) the capacitor value for a self-excited induction generator (SEIG).

Electronic power converters when used as a load for IGs may cause harmonic distortion and losses. A variable power factor and some special control techniques should be adopted when using electronic power converters. The IEEE Std. 519 establishes the limits of 2% of harmonic content for single- and three-phase induction motors (except the category N, i.e., conventional) and 3% for high efficiency (H and D). Active filters provide signal injection to minimize such harmonics. The cost of passive filters can be relatively small, and the designed speed control of power plants is possible by electronic variation of frequency, using, for example, a droop control method. The self-excitation capacitor in stand-alone wind turbine systems, with electric or electronic control of the load, contributes favorably in these cases [4].

When an RLC load is connected across the SEIG terminals, the combination of the inductive reactance with the necessary self-excitation capacitance results in a new self-exciting reactance and a new output voltage condition. The equivalent circuit in per unit (p.u.) of an SEIG connected to an RL load is shown in Figure 7.7. This circuit represents a more generic per phase of the steady-state induction machine [2, 3]. It also shows that the frequency effects on the reactance should be considered if the generator is used at different frequencies from its base frequency f_b (in hertz) at which the

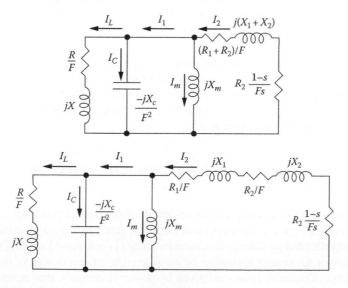

FIGURE 7.7 Per-phase equivalent circuit of a loaded IG.

parameters of the machine are measured. For this purpose, if F is the p.u. frequency, a relationship can be defined between the self-excitation frequency f_{exc} and the base frequency f_b (usually 50 or 60 Hz) [13–15]:

$$F = \frac{f_{exc}}{f_b} = \frac{\omega_{exc}}{\omega_b} \tag{7.4}$$

In stand-alone IG applications, the frequency control is usually variable, depending on the prime mover, that is, the wind turbine or other alternative sources (e.g., diesel). Variations in the frequency should be carefully considered since they can cause variations in all reactive parameters and alterations in the load voltage. In a more generic way, the inductive reactance parameter defined for the base frequency is $X = F\omega L$. In Figure 7.7, all the circuit parameters are divided by F, making the source voltage equal to V_{ph}/F. From the definition of secondary resistance (rotor resistance R_2), the following modification is used to correct R_2/s in order to take into account the variations in the stator and rotor p.u. frequencies:

$$\frac{R_2}{F \cdot s} = \frac{R_2}{F\left(1 - \dfrac{n_r}{n_s}\right)} = \frac{R_2}{F - v} \tag{7.5}$$

where v is the rotor speed in p.u. referred to the test speed used for the rotor.

The disadvantages of an SCIG are as follows:

1. Any wind speed fluctuations are directly translated into electromechanical torque variations, rather than rotational speed variations since the speed is not variable. This may cause high electromechanical stresses on the system (generator windings, turbine blades, and gearbox) and may result in resonance and oscillations between turbine and generator shaft. Fluctuations in the power output are not damped, and even small wind speed fluctuations impose an oscillating power [4]. Also, the periodical torque dips caused by the tower shadow (when the blades cross a line parallel to the tower) and shear effect are not damped by speed variations and result in higher flicker values. The turbine speed cannot be adjusted to the wind speed to optimize the aerodynamic efficiency, though many commercial wind turbines can switch the pole-pair numbers by a rearrangement of the stator windings connection to optimize discretely under lower or higher wind speeds.
2. As discussed earlier, a gearbox is necessary for small power wind turbines.
3. The IG needs reactive power from a permanent external reactive source connected to the stator windings to supply the stator excitation current terminals. Such reactive power must be supplied by the grid connection, by a capacitor bank, or from an electronic converter operating as a static VAR compensator.

7.5 PERMANENT MAGNET SYNCHRONOUS GENERATORS FOR SMALL WIND POWER APPLICATIONS

This section shows how a PM machine can be designed for small wind power systems. As discussed in the previous sections, Figure 7.2 portrays that electrical generators used for wind turbine systems have their efficiency dictated by three main characteristics: (1) generator losses, (2) converter losses, and (3) gearbox losses. Generator losses can be considered by proper design of the machine for the right operating range. Converter losses are given by power electronics, that is, on-state conduction losses of transistors and diodes as well as switching losses. Mechanical losses in a gearbox are proportional to their operating speed. In accordance with Figure 7.2, the gearbox efficiency can be calculated as given by the following equation:

$$P_{gear} = P_{gear,rated} \frac{\eta}{\eta_{rated}}$$ (7.6)

where

$P_{gear,rated}$ is the loss in the gearbox at rated speed (in the order of 3% at rated power)
η is the rotor speed (r/min)
η_{rated} is the rated rotor speed (r/min)

The losses in the gearbox dominate the efficiency on most wind turbines, and a simple calculation shows that small wind turbine systems, with low wind velocity range, have roughly 70% of their annual energy dissipation because of the mechanical various losses plus the gearbox efficiency [13].

The design of an electrical machine is often considered from the point of view of obtaining the maximum torque. Usually, the external volume (or the weight) of the machine is sized by their maximum torque for the considered application, but for a generator application, the designer should consider the shaft power production profile. The wind turbine power has a performance coefficient C_p, which considers the turbine mechanical design together with their aerodynamic efficiency. The tip-speed ratio (TSR) λ is a function of ω, the generator rotational speed, the radius of the blade, and the linear wind velocity as given by Equation 7.7. One can design a generator by maximizing the C_p coefficient at the minimum speed (v_{min}) in order to obtain a given turbine power, for example, for a very small system such as $P_{turbine} = 5$ kW.

$$\lambda = \frac{\omega \cdot r}{v}$$ (7.7)

Then, for such a 5 kW generator, the maximum torque (T_{max}) is calculated. Figure 7.3 shows that for a minimum wind speed up to the maximum wind speed (v_{max}), the power is limited to a maximum power (P_{max}), and also when the wind velocity is less than a certain limit speed (v_{lim}), the wind generator will stop.

Three wind power generator designs have been compared—one with ferrite, another with bounded NdFeB, and a third one with sintered NdFeB as discussed in [13]. The active length is calculated from their required torque for a given operating range and simulated by 2D finite element analysis. Both copper (P_c) and iron (P_i) losses are calculated by

$$P_c(v_{min}) = \sigma_c J^2 V_v(v_{min})$$ (7.8)

where

σ_c is the copper resistivity
J is the current density (in this case, equal to 5 A/mm^2)
V_c is the copper volume, which depends on the minimum wind velocity (v_{min})

$$P_i(B_s, f_s, v_{min}, v) = k(B_s, v) M_i(v_{min}, v)$$ (7.9)

where

k is an iron loss coefficient determined by the iron flux density (B_s) and the frequency (in this case, which is determined by the wind velocity)
M_i is the iron mass, which depends on the minimum and actual wind velocity

Copper and iron losses are represented in Figure 7.8 and compared for the increasing power level. The generator has a certain operating region around the power and velocity curves; the integral of

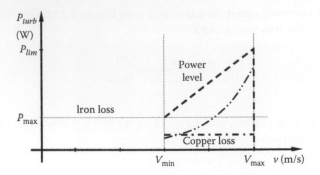

FIGURE 7.8   Copper and iron losses with power level versus wind velocity.

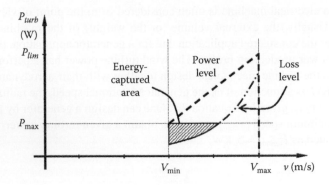

FIGURE 7.9   Energy-captured area for a large operating range.

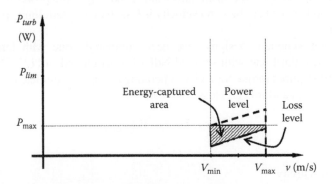

FIGURE 7.10   Energy-captured area for a small operating range.

this curve gives the captured energy, as represented in Figures 7.9 and 7.10. Such figures show the optimal range for maximizing the energy captured by the procedure used for a small wind turbine design. If a large operating range is used in the design, the machine will not capture enough energy for low wind speed variation (Figure 7.9), whereas a machine designed for a more constrained range will have a larger energy capture (Figure 7.10). However, the smallest shaft angular speed has to be accounted in having more turns in the generator stator for a larger voltage, and this angular velocity defines how many poles the machine should have in order to minimize the gear ratio. In addition, depending on the choice of the PMs and on the typical wind distribution, the optimal range may change.

A generator with ferrite magnets must be overdesigned to comply with the specifications of lower flux level of the ferrites. Therefore, the losses are more important for ferrite design than for NdFedB magnet design. This is the reason why most of the PMSGs in the market use rare-earth magnets. Actually, the optimal range for ferrite magnets is smaller and the captured wind energy decreases. If only technical considerations are used, it is easy to disregard completely ferrite-based magnets and just use rare-earth magnets instead. Only future high price of rare-earth magnets and their low availability make ferrite-based PM machines to be implemented. Ferrites have an advantage: they can be made anywhere as long as iron and ceramics are available, and the knowledge of making PM ferrites is accessible to anyone in the world, making them a "sustainable generator" option forever.

The energy-captured area can be plotted to show the power that can be extracted from the turbine shaft. This area is bounded by the maximum power from the turbine (P_{max}) and by the losses, which can be calculated by

$$E_{ca}\left(v_{min},v_{max}\right) = \int\limits_{v_{min}}^{v_{design}} \left[P(v)-P_c(v)-P_i(v)\right]dv \tag{7.10}$$

If the wind generator is overdesigned considering a high loss level, despite a large power range, the energy-captured area is smaller. Consequently, maximizing the power range area is not recommendable; several scenarios might support a study in order to have a compromise between an acceptable loss level and the power range. The overall efficiency and cost should be considered too. Table 7.5 shows the optimization results for the three machines designed for the same magnetic flux levels [13].

The flux density is lower for a ferrite magnet generator, so the total mass should be increased in order to obtain the same performance. In this case, the captured energy is less than that in the other cases. The best set of characteristics in terms of mass, energy stored, and consequently the mass–energy ratios are reached for the sintered NdFeB (1.2 T) magnet due to the high efficiency of such PM. The cost–energy ratio is computed using price data for all components (iron, copper, and PM). For the cost–energy ratio criterion, the ferrite magnet configuration has the smallest ratio. Although the mass is the highest one, the price of a ferrite magnet is about twenty times less than that of an NdFeB magnet. Thus, the total cost is smaller than the NdFeB magnet. Despite the ratio of mass to energy being the smallest, the ferrite magnet is a good alternative when compared to the rare-earth PM design option. Table 7.6 shows a comparison of the machine with a bounded NdFeB magnet designed and a high-torque off-the-shelf motor [13].

The PMSG offers many advantages as it is the most efficient of all electric machines since it has a movable magnetic source inside itself. The use of PMs for the excitation consumes no extra

TABLE 7.5

Comparison of Three Designed Permanent Magnet Machines

Magnet Type	Ferrite	Bounded NdFeB	Sintered NdFeB
Iron mass	67 kg	37 kg	32 kg
Copper mass	38 kg	22 kg	20 kg
Magnet mass	27 kg	23 kg	21 kg
Total mass	132 kg	82 kg	73 kg
Energy	1.7 kWh	2.1 kWh	2.3 kWh
Mass/energy	80 kg/kWh	38 kg/kWh	32 kg/kWh
Cost/energy	0.6 k€/kWh	2.2 k€/kWh	1.9 k€/kWh

Source: Ojeda, J. et al., *IEEE Trans. Ind. Appl.*, 48(6), 1808–1816, 2012.

TABLE 7.6

Comparison of a Commercial Permanent Magnet Generator and the Ferrite-Based Design

Machine Characteristics	PM Off Shelf	Ferrite
Turbine power	6.8 kW	5 kW
Maximum torque	1049 N m	1000 N m
Efficiency	79%	77%
Total mass	82 kg	132 kg
External radius	795 mm	240 mm
Air-gap radius	689 mm	208 mm
Active length	55 mm	476 mm

Source: Ojeda, J. et al., *IEEE Trans. Ind. Appl.*, 48(6), 1808–1816, 2012.

electrical power. Therefore, the copper losses of the exciter do not exist and the absence of the mechanical commutator and brushes as well as slip rings means low mechanical friction losses. An additional advantage is its compactness.

The recent introduction of high-energy density magnets (rare-earth magnets) has allowed the achievement of extremely high flux densities in the PMSG. Therefore, a rotor winding is not required. These in turn allow the generator to have a small, light, and rugged structure. As there is no current circulation in the rotor to create a magnetic field, the rotor of a PMSG does not heat up. The only heat production is on the stator, which is easier to cool down than the rotor because it is on the periphery of the generator and the stator.

The absence of brushes, mechanical commutators, and slip rings suppresses the need for regular maintenance and also the risk of failure in these elements. They have very long-lasting winding insulation, bearing, and magnet life length. Since no noise is associated with the mechanical contacts and also the drive converter switching frequency can be above 20 kHz, the system might produce only inaudible noise for human beings.

When compared with the conventional ones for low wind speed, PMSGs have the following advantages: (1) no speed multiplier or gears since there may be multiple permanent or electromagnets in the rotor for more current production; (2) few maintenance services because of its simplified mechanical design; (3) easy mechanical interface; (4) cost optimization; (5) highest power-to-weight ratio in a direct-drive system; (6) location of a moving magnetic field being generated in the center of the field; (7) more precise operations since a microprocessor controls the generator/motor electrical output and current instead of mechanical brushes; (8) higher efficiency for the brushless generation of electrical current and digitally controllable flexible adjustment of the generator speed with less friction, fewer moving components, less heat, and reduced electrical noise; and (9) can be kept cooler and thus have a longer life since the PMs are located on the rotor.

The PMSG has some disadvantages related to: (1) their high cost of PMs and (2) their commercial availability. The cost of higher energy density magnets prohibits their use in applications where initial cost is the major concern. Another disadvantage is the field-weakening operation for the PMSG machine, which is somewhat difficult due to the use of PMs, but usually the field-weakening mode is not a concern for wind power generators. Any accidental speed increase might damage the power electronic components above the converter rating, especially for generator-drive applications. In addition, surface-mounted PM generators cannot reach high speeds because of their limited mechanical strength of the assembly between the rotor yoke and the PMs. Finally, the demagnetization of the PM is possible by a large opposing magnetomotive force and high temperatures.

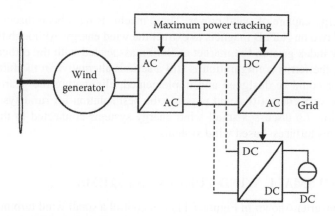

FIGURE 7.11   Power electronic converter topology for small wind turbines with maximum power tracking.

Very good ventilation should be implemented in order to cool down the generator, particularly for extremely compact applications.

The electrical generator is connected to a small-scale wind turbine, as indicated in Figure 7.11. The wind turbine typically has fixed attack angles of the blades. Either the wind energy system is connected to the distribution grid or supplies a DC load with power electronic interfaces. The control must be based on the wind acting on the turbine rotation and balance of the load power. The DC/DC converter connected to the DC link can serve as a storage, or it is sometimes implemented as an electrical braking for wind turbines, particularly for wind gusts or any runoff of the turbine that might cause power imbalance with the load or the grid. As the rotor speed changes, according to the wind speed, the speed control of the turbine should command low speed at low winds and high speed at high winds, in order to follow the maximum power operating point as indicated in Figure 7.12. A maximum power tracking may use a hill-climbing type controller or a fuzzy logic–based controller for commanding the speed of the generator [4].

This section discusses how statistics of wind resources can be taken into consideration to define a power envelope for the turbine. The electromechanical design can consider copper plus iron losses

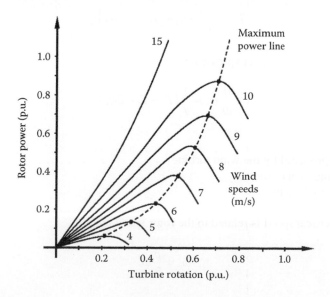

FIGURE 7.12   Turbine rotation versus power characteristics with wind speed.

and a possible energy capture of wind power. Three machines have been discussed, one based on ferrite magnet and two on NdFeB magnets to extract the wind energy. Although the NdFeB magnet has a mass–energy index greater, the market evolution associated with the difficulties in obtaining rare-earth magnet, the ferrite configuration is still a serious alternative in the future. A direct-drive electrical generator could be designed and implemented, eliminating the gearbox and making a small-scale turbine very competitive and possibly the best solution for rural systems, small farms, and villages. Section 7.6 discusses small wind energy systems connected to the utility grid plus principles of Magnus turbines–based wind systems.

7.6   GRID-TIED SMALL WIND TURBINE SYSTEMS

A back-to-back converter (shown in Figure 7.11) can control a small wind turbine with either an IG or a PMSG. Figure 7.13 shows the block diagram of a full-fledged PMSG back-to-back converter solution. Such a converter topology provides speed control for the machine-side converter to track the optimum power point, and the grid-side converter aims to control the power and keep DC-link voltage constant. The equivalent circuit of a PMSG is used in the rotating reference frame. The equations of the d-axis and q-axis in the rotating reference frame (not taken into account the core losses) are

$$v_{ds}^r = -R_s i_d - L_d \frac{d}{dt} i_d - \omega_e L_q i_q \tag{7.11}$$

$$v_{qs}^r = -R_s i_q - L_q \frac{d}{dt} i_q + \omega_e L_d i_d + \omega_e \psi_m \tag{7.12}$$

where the instantaneous power is given by

$$P_i = \frac{3}{2} \left[v_{ds}^r i_d + v_{qs}^r i_q \right] \tag{7.13}$$

and the electromagnetic torque is given by

$$T_e = \frac{3p}{4} \left[\psi_m + \left(L_d - L_q \right) i_d \right] i_q \tag{7.14}$$

The rotational angular speed is expressed as

$$\frac{d}{dt} \omega_r = \frac{1}{J} \left(T_w - T_e - B\omega_r \right) \tag{7.15}$$

where
     T_w is the torque produced by the wind turbine
     J is the turbine moment of inertia
     B is the friction coefficient

The angular electrical speed is related to the rotor speed by

$$\omega_e = \frac{p}{2} \omega_r \tag{7.16}$$

where p is the number of poles of the machine.

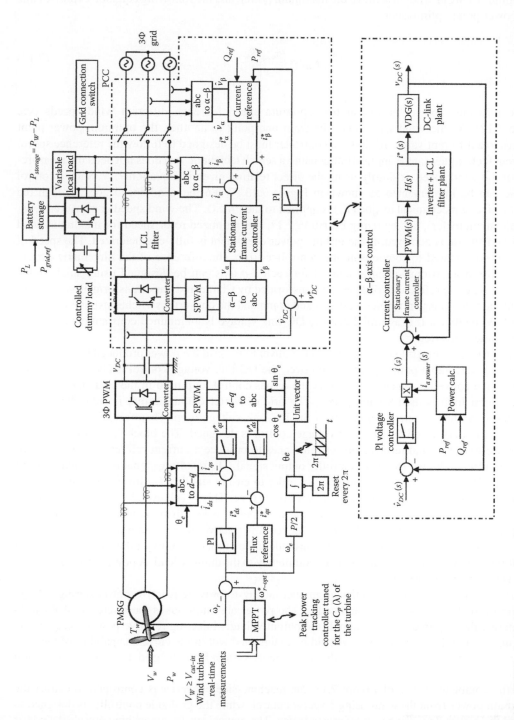

FIGURE 7.13 PMSG wind turbine controller with maximum power optimization and a back-to-back pulse-width modulation (PWM) grid-connected inverter with load management.

The relationship between the TSR and the power coefficient of the wind turbine for different rotor speeds has been discussed. For a given wind speed, there is an optimum rotor speed that gives the optimum TSR in order to achieve the maximum power, and the following equation expresses this peak power point optimization:

$$\lambda_{opt} = \frac{\omega_{w,opt} R_w}{v_w} \tag{7.17}$$

The maximum power occurs at this optimum speed for each different wind speeds (see Figure 7.12). As the wind speed changes from one point to another, the optimum power point changes to a different value. To do so, a controller must be designed to follow the reference speed. Different techniques can be applied either using a search technique that does not need the measurement of the wind speed or simply using the direct relation given in Equation 7.17. Such a control scheme can be implemented as shown in Figure 7.13. When the wind speed goes through the wind turbine, a mechanical torque (T_w) is applied to the PMSG. The wind speed is measured (e.g., with an anemometer), and by using Equation 7.17, the optimized rotor speed for maximum wind power conversion is achieved, if the turbine power coefficient is fully characterized. The speed of the rotor is measured to compensate the controller error. The reference of the direct axis current is zero. The proportional–integral control for each d-q axis can be designed using small signal modeling. The d-q reference voltages are transformed by Clarke and Park transformations in order to form the three-phase reference voltages and command the converter switches using PWM (e.g., the SPWM method). It is assumed that the DC-link voltage is maintained constant controlled by the grid-side converter.

The purpose of the grid-side converter is to deliver to the grid the power produced by the generator with an acceptable power quality. Moreover, the DC-link voltage control is also controlled by the grid-side converter. Figure 7.13 shows the α-axis control loop used in the grid-side converter and a similar loop is used for the β-axis. The grid converter output current is controlled by the inner loop and has a faster response than the DC-link voltage loop. Therefore, the inner loop is considered unitary when the DC-link voltage control is designed. The block "power calculation" makes use of active and reactive power references to produce a current reference, which in turn will be multiplied by the DC-link controller output signal. The resulting signal is the α-axis current reference. The power transfer is designed in order to ensure stability for different scenarios. The dynamic load should be provided with power all the time. This is achieved by the power from the wind generator or from the grid only when the wind power is less than the load demand. The grid could send a reference signal to the controller requiring active power, and a dummy load is connected in the system through a converter to provide full control capability of the whole system. The excess wind power can be sent partially or completely to the dummy load. A power transfer strategy chart is shown in Figure 7.14.

The grid-side converter is controlled in such a way to provide reactive power during voltage sags according to grid standards and codes. The power algorithm takes the reference of P and Q to generate the required VAR to support the output voltage. Injecting reactive power increases the line current, and the reactive power should be maintained within the current capability of the converter in order to avoid disconnecting the wind energy system from the grid due to some protection operation.

In the complete system of Figure 7.13, the machine-side converter is controlled to extract the maximum power from the wind using a vector control, while the grid-side controller is designed to control the power flow using α-β reference frame. The power transfer algorithm controls the power flow between the wind turbine and the grid. The turbine supplies a primary load that is variable in nature and a control dummy load to manage the power flow in the system considering the state of the grid and the load demand. The excess power is injected in the grid.

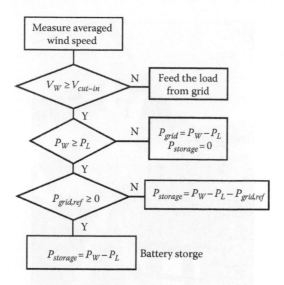

FIGURE 7.14   Wind power management including battery storage.

Such system can be used for smart-grid applications where the grid-side converter can support the grid voltage by injecting reactive power. The purpose of the machine-side converter is to track the optimum point of the rotor to extract the maximum power existing in the turbine. For a given wind turbine, the maximum power occurs at the maximum power coefficient of the turbine. It should be noted that a similar system could be implemented for an SCIG; using indirect vector control for the generator side and a modified direct vector control for the grid side, it is expected to be an easy retrofit for the system discussed in this chapter; the details of such implementation can be found in [4].

7.7   MAGNUS TURBINE–BASED WIND ENERGY SYSTEM

An example of a low-power wind turbine system is the Magnus turbine. Such a turbine has a distinctive feature with respect to conventional wind turbines: there are rotating cylinders around the turbine shaft instead of blades as depicted in Figure 7.15. The torque around the turbine rotor considers some characteristics of the rotating cylinders related to their geometry, kinematics, and energy capture parameters. The improvement in performance depends intrinsically on the type, number, surface pattern, and sizing of the cylinders and on the rotor shaft load as well as individual optimal rotational control of the turbine and cylinders.

In order to understand the torque produced by the Magnus effect, one can consider a rotating cylinder with a radius r_c and an angular velocity ω_c moving in the air with a speed $\omega_c r_c$. The air velocity nearer the cylinder is higher than $\omega_c r_c$ on one side and lower than that on the other side. This phenomenon is because the velocity of the air boundary layer surrounding the rotating cylinder is added in one side and subtracted in the other side [16, 18]. Therefore, there is a resulting pressure gradient impressing a net force on the cylinder in the perpendicular direction to the body velocity vector with respect to the fluid flow, as depicted in Figure 7.16. As the cylinders move around the rotor axis, some air layer tend to flow centrifugally out of the area swept by the moving cylinders. This is a complex phenomenon, considering all the effects of the air movement around the cylinders. It is very difficult to model, because the Bernoulli equation must be modeled, considering the outward motion of the rotors shafts, plus the variable spin around the turbine rotor, and the force that the air will impress on the combined structure of the turbine [17, 18].

Using the fundamental physics, we can understand that air pressures on the cylinders will cause the Magnus lift and drag torques T_L and T_D, as illustrated in Figure 7.16. These torques can be

FIGURE 7.15   Magnus turbine with nonsmooth rotating cylinders. (Courtesy of CEESP-UFSM/IFSC, Brazil.)

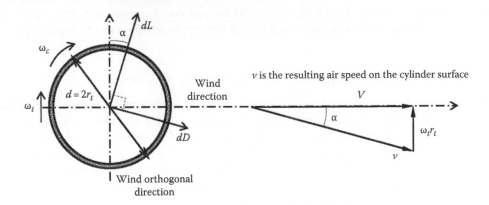

FIGURE 7.16   Lift and drag Magnus effects on their rotating cylinders.

expressed, respectively, by the mechanical drag and the lift actions on the wind turbine with a definition of power coefficient as given in Equation 7.18. The turbine mechanical power is similar to Equation 7.2 but it is used again, with the parameters defined by Equation 7.18, calculating the mechanical power in Equation 7.19:

$$C_p = \frac{2n_c\left(T_L - T_D\right)\cdot\omega_t}{\rho\cdot\pi\cdot r_t^2\cdot V^3} \tag{7.18}$$

$$P_{mec} = \frac{1}{2}\rho A V^3 C_p \tag{7.19}$$

where

n_c is the number of cylinders

ω_t is the angular speed of the turbine

$\rho = 1.225$ kg/m^3 is the air density at sea level at 15.5 °C, according to the International Standard Atmosphere

A is the area swept by the spinning cylinders (m^2)

V is the wind velocity (m/s)

As the terms T_L and T_D of the Magnus turbine torques are usually very complicated mathematical expressions, it is better to obtain them experimentally by a curve fitting procedure, that is, as a function of the rotor and the cylinders TSR, respectively, λ_t and λ_c. These coefficients are related to the radius of the area swept by the cylinder r_t, to the angular speed of the cylinders ω_c, to the cylinder radius r_c, and to the wind speed V as

$$\lambda_t = \frac{\omega_t r_t}{V} \tag{7.20}$$

$$\lambda_c = \frac{\omega_c r_c}{V} \tag{7.21}$$

Given the mechanical power, it is necessary to subtract the cylinder friction losses during its rotation in a laminar air flow and the motor losses in the cylinder drivers:

$$P_t = P_{mec} - P_{losses} \tag{7.22}$$

The electromechanical power losses due to air friction can be expressed by

$$P_{losses} = 1.328 \frac{n_c}{16\eta_{elect-mech}} \cdot \frac{\rho \pi d^4 \omega_c^3 (r_t - r_o)}{\sqrt{Re_d}} \tag{7.23}$$

where

r_o is the radius of the rotor hub

$Re_d = \rho \omega_c \pi (2r_c)^2 / 2\mu$ is the Reynolds number

μ is the air viscosity coefficient

$\eta_{elect-mech}$ is the electromechanical efficiency

These equations are quite challenging, and a practical engineer would prefer to consider experimental results or other forms of approximation to establish a polynomial equation whose form and coefficients are obtained by curve fitting. The optimal ratio dimensions of the cylinder have been established in the literature by $(r_t - r_o)/16$. Figure 7.17 illustrates the performance for a Magnus turbine specified in Table 7.7, where the different effects on the dimensionless power coefficients λ_c and λ_t are related, respectively, to the rotation of the cylinder around its shaft and the turbine rotor shaft. The reduced rotation of the Magnus cylinder–rotor is about two to three times lower when compared to the blades in the conventional turbines, ensuring much less air turbulence with higher operational safety and durability.

This is a major advantage compared to the low efficiency of other turbine types for most usual wind velocities (5–15 m/s) due to the small lift coefficient of an ordinary blade turbine under such conditions. It is clear that the Magnus wind turbine must be explored in a wider range of rotor and wind velocities, varying from 2 to 35 m/s compared to a traditional blade turbine, typically limited in a range from 3 to 25 m/s. This variable speed is well suited for the PMSG and IG characteristics and, certainly, when associated with a hill-climbing control (HCC) for the optimal speed of the turbine rotor and cylinders can be achieved for maximum generator power production. The power coefficient of the ordinary wind turbines drops rapidly to zero at about wind velocity on the order of

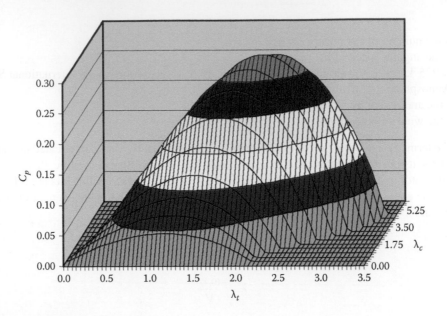

FIGURE 7.17   Variations of C_p with dimensionless coefficients λ_t and λ_c of a Magnus turbine.

TABLE 7.7

Magnus Turbine Parameters for the Example

Description	Parameter	Value	Units
Range of the wind speed	V	0.5–40	m/s
Air density at sea level and 15.5 °C	ρ	1.225	kg/m³
Air coefficient of viscosity	μ	$1.80e^{-05}$	N s/m²
Diameter of the turbine swept area	$2r_t$	11.50	m
Diameter of the cylinder	$2r_c$	0.35	m
Hub radius	r_o	0.50	m
Rollers' efficiency	$\eta_{elect-mech}$	0.50	—
Range of the turbine speed	ω_t	0.25–3.00	rad/s
Range of the cylinder speed	ω_c	0–150	rad/s
Number of cylinders	n_c	5	cyl

---

4 m/s, which is not the case for the Magnus turbines. Another advantage is when the wind velocities are higher than 35 m/s, there is a natural regulation to decrease the Magnus force, causing a self-braking. Therefore, the cylinder–rotor rotation aerodynamics with speed regulation prevents excessive spin-up and collapse due to excessive centrifugal forces. Therefore, a Magnus turbine has improved operation at low wind velocity as well as safe features toward very low wind velocities by its self-regulation nature.

A reasonable modeling of the Magnus turbine is quite complex because it is very difficult to take into accounts all variations and air turbulences through which the turbine may go through. A forward chaining speed controller for the cylinder and the Magnus wind turbine can be coupled to a PMSG or an IG. The 3D graph of Figure 7.17 clearly illustrates the opportunities for the PMSG and IG using an HCC in order to establish the point of maximum power for both the cylinders and the rotor. That means that it is possible to search for the maximum power point in accordance with the wind and variations of the IG load by continuously and sequentially adjusting the angular velocity of the cylinders and turbine rotor. It is also possible that a fuzzy logic–based controller will perform really well for a Magnus turbine coupled to a PMSG or an IG [20, 21].

7.8 SUMMARY

This chapter presents how a customized wind turbine has a typical operation for medium and low wind velocity sites since the power of any wind generator is directly affected by the wind speed. It is not possible to maintain a fixed speed of a generator always at a high efficiency level because commercially available generators for low wind speeds (5–15 m/s) are not currently designed to operate with their best efficiency over the whole speed range (typically from 3 to 25 m/s). In addition, high towers for wind turbines increase the overall costs and the turbine exposure to turbulences and wind gusts affects the generators performance.

Medium- or long-term statistics of wind resources must be taken into account to define the power envelope for turbines, considering copper and iron losses for three PMSG designs considered and the IG. In addition, a brief review of the IG and its advantages and disadvantages for small wind turbines has been discussed. A PMSG-based system has been described in details too, and it has been discussed how a low-speed electrical generator can be competitive and may be the best solution for small-scale wind turbines, typically used in rural systems, small farms, and villages.

Data, instrumentation, and measurements are necessary for small wind energy systems in order to approach the design and implementation for commercial and residential applications. In addition, it is necessary to manage the utility connection, assessing reactive power requirements, fault ride through, and power quality monitoring. Some schemes can be implemented for an SEIG using standard squirrel cage machines, and advanced control and signal processing system can be implemented for PMSG wind energy systems connected to the grid.

Although large-scale wind turbines may have optimized speed control, the market for small scale wind turbines does not allow expensive solutions. Therefore, an optimized electric generator must be designed to have the best efficiency for low wind velocities under cost constraints. It remains a challenge, even today, to design a generator capable of operating at wind velocities less than 3.5 m/s with a high efficiency and this is still an open-ended problem to be solved.

REFERENCES

1. Z. Zhang, A. Matveev, S. Øvrebø, R. Nilssen, and A. Nysveen, State of the art in generator technology for offshore wind energy conversion system. *Proceedings of IEMDC*, pp. 1131–1136, May 15–18, 2011.
2. L. Xu and W. Cheng, Torque and reactive power control of a doubly fed induction machine by position sensorless scheme. *IEEE Transactions on Industry Applications*, 31(2), 636–642, 1995.
3. V.D. Colli, F. Marignetti, and C. Attaianese, 2-D mechanical and magnetic analysis of a 10 MW doubly fed induction generator for direct-drive wind turbines. *Proceedings of IEEE IECON 09*, pp. 3863–3867, 2009.
4. M. Godoy Simões and F.A. Farret, *Modeling and Analysis with Induction Generators*, 3rd edn., Power Electronics and Applications Series, CRC Press, Boca Raton, FL, 2014.
5. V.D. Colli, F. Marignetti, and C. Attaianese, Feasibility of a 10 MW doubly fed induction generator for direct-drive wind turbines. *Proceedings of IEEE PES/IAS*, pp. 1–5, September 28–30, 2009.
6. V. Mostafa, A. Nysveen, R. Nilsen, R.D. Lorenz, and T. Rølvåg, Influence of pole and slot combinations on magnetic forces and vibration in low-speed PM wind generators. *IEEE Transactions on Magnetics*, 50(5).
7. H. Polinder, D. Bang, R.P.J.O.M. van Rooij, A.S. McDonald, and M.A. Mueller, 10 MW wind turbine direct-drive generator design with pitch or active speed stall control. *Proceedings of IEMDC*, pp. 1390–1395, May 3–5, 2007.
8. H. Kobayashi et al. Design of axial-flux permanent magnet coreless generator for the multi-megawatts wind turbines. *EWEC*, 2009.
9. H. Mellah and K.E. Hemsas, Simulations analysis with comparative study of PMSG performances for small WT application by FEM. *International Journal of Energy Engineering*, 3(2), 55–64, 2013.

10. R. O'Donnell, N. Schofield, A.C. Smith, and J. Cullen, Design concepts for high-voltage variable-capacitance DC generators. *IEEE Transactions on Industry Applications*, 45(5), 1778–1784, September–October 2009.
11. E. Spooner, P. Gordon, J.R. Bumby, and C.D. French, Lightweight ironless-stator PM generators for direct-drive wind turbines. *Proceedings of IEE Electrical Power Applications*, 152(1), 17–26, January 2005.
12. Small wind electric systems, http://energy.gov/energysaver/articles/small-wind-electric-systems, last accessed November 2014.
13. J. Ojeda, M. Godoy Simões, G. Li, and M. Gabsi, Design of a flux switching electrical generator for wind turbine systems. *IEEE Transactions on Industry Applications*, 48(6), 1808–1816, November–December 2012.
14. J. Soens, Impact of wind energy in a future, PhD dissertation, Katholieke Universiteit Leuven, Faculteit Wetenschappen, Leuven (Heverlee), Belgium, ISBN 90-5682-652-2, Wettelijk depot, UDC 621.548, December 2005.
15. H. Polinder, D.J. Bang, H. Li, Z. Chen, M. Mueller, and A. McDonald, Concept report on generator topologies, mechanical and electromagnetic optimization. Project UpWind, Part 1 (TUD, AAU), pp. 1–51; Part 2 (EDIN), pp. 52–79, Project UpWind, Contract No. 019945 (SES6), "Integrated Wind Turbine Design," project funded by the European Commission under the 6th EC Research and Technological Development, December 2007.
16. N.M. Bychkov, A.V. Dovgal, and V.V. Kozlov, Magnus wind turbines as an alternative to the blade ones. *Journal of Physics: Conference Series*, 75, 012004, 2007. doi:10.1088/1742-6596/75/1/012004.
17. H. Rouse, *Elementary Mechanics of Fluids*, Dover Publications Inc., New York, pp. 275–376, 1946.
18. A. Barbeiro, J.A. Garcia-Matos, A. Cantizano, and A. Arenas, Numerical tool for the optimization of wind turbines based on Magnus effect. *Ninth World Wind Energy Conference and Exhibition* (*WWEC 2010*), Istanbul, Turkey, 2010.
19. J. Maro, J.R. Cardoso, F.A. Farret, D.L. Hoss, and J.R. Dreher, Magnus wind turbine with DC servo-drive for the cylinders and boost converter. *Proceedings of the IEEE Industrial Electronics Society IECON*, Dallas, TX, 2014.
20. M. Godoy Simões, B.K. Bose, and R.J. Spiegel, Design and performance evaluation of a fuzzy-logic based variable-speed wind generation system. *IEEE Transactions on Industry Applications*, 33(2), 956–965, July/August 1997.
21. M. Godoy Simões, B.K. Bose, and R.J. Spiegel, Fuzzy logic based intelligent control of a variable speed cage machine wind generation system. *IEEE Transactions on Power Electronics*, 12(1), 87–95, January 1997.

FURTHER READING

N.M. Bychkov, A.V. Dovgal, and A.M. Sorokin, Parametric optimization of the Magnus wind turbine. *16th International Conference on Methods of Aerophysical Research* (*ICMAR*), Kazan, Russia, 2008.
L.C. Corrêa, J.M. Lenz, C.G. Trapp, and F.A. Farret, Maximum power point tracking for Magnus wind turbines. *Proceedings of the 39th Annual Conference of the IEEE Industrial Electronics Society*, pp. 1716–1720, Vienna, Austria, November 2013.
B.C. Doxey, Theory and application of the capacitor-excited induction generator. *The Engineer Magazine*, pp. 893–897, November 1963.
F.A. Farret, J.R. Gomes, and C.R. Rodrigues, Sensorless speed measurement associated to electronic control by the load for induction turbo generators. *Proceedings of SOBRAEP V Brazilian Power Electronics Conference*, pp. 88–93, Foz de Iguaçu, Brazil, September 1999.
R. Gono, S. Rusek, and M. Hrabcik, *Wind Turbine Cylinders with Spiral Fins*, Czech Science Foundation, Czech Republic, 2009.
D. Luo, D. Huang, and G. Wu, Analytical solution on Magnus wind turbine power performance based on the blade element momentum theory. *Journal of Renewable and Sustainable Energy*, 3(3), 033104, 2011.
S.S. Murthy, H.S. Nagaraj, and A. Kuriyan, Design-based computer procedures for performance prediction and analysis of self-excited induction generators using motor design packages. *Proceedings of IEEE*, 135(1), 8–16, January 1988.
T.K. Sengupta and S.B. Talla, Robins-magnus effect: The continuing saga. *Current Science*, 86(7), 2004.
J. Zhao, Q. Hou, H. Jin, Y. Zhu, and G. Li, CFD analysis of ducted-fan UAV based on Magnus effect. *Proceedings of IEEE International Conference on Mechatronics and Automation*, pp. 1722–1726, Chengdu, China, August 2012.

8 Power Electronics and Controls for Large Wind Turbines and Wind Farms

Ke Ma, Udai Shipurkar, Dan M. Ionel, and Frede Blaabjerg

CONTENTS

Abstract 177
8.1 Introduction 178
8.2 Wind Turbine Systems 178
8.3 Specific Issues for Wind Turbine Operation 181
    8.3.1 Complex Mission Profiles for Wind Turbines 181
    8.3.2 Strict Codes from Grid Side to Be Connected 183
    8.3.3 Growing Reliability Requirements 184
8.4 Power Electronics: Devices and Converters 185
    8.4.1 Power Semiconductor Devices 185
    8.4.2 Two-Level Converter Topologies 187
    8.4.3 Multilevel Converter Topologies 188
    8.4.4 Future Converter Topologies 189
        8.4.4.1 Cascaded H-Bridge Converter with Medium-Frequency Transformers 189
        8.4.4.2 Modular Multilevel Converter 189
8.5 Power Electronic Solutions for Wind Farm 189
    8.5.1 Solutions for Wind Power Transmission 189
    8.5.2 Solutions for Better Grid Support 192
8.6 MATLAB® Models for Thermal Analysis of Wind Turbines 193
    8.6.1 Overview of the Models and Relationship 194
    8.6.2 Mechanical Model of Turbine and Generator 194
    8.6.3 Electrical Model of Converter and Control 197
        8.6.3.1 Generator-Side Converter and Control 197
        8.6.3.2 Grid-Side Converter and Control 198
    8.6.4 Power Loss and Thermal Modeling 199
    8.6.5 Example of the Simulation Results 199
8.7 ANSYS® Models for Wind Turbine Systems 202
8.8 Summary 203
References 204

ABSTRACT

Wind power represents a major and growing source of renewable energy for electric power systems. This chapter provides an overview of state-of-the-art technologies and anticipated developments in the area of power electronic drives, controls, and electric generators for large multi-megawatt (MW) wind turbine (WT) systems. The principal components employed in a turbine for energy conversion from wind to electricity are described, and the main solutions that are commercially available are briefly reviewed. The specific issues of complex mission profiles, grid codes, and also reliability are discussed. The topics of power electronics, ranging from devices to circuit topologies, and similar matters for electric generators, together with the results of optimal design studies, are included. It is shown that the individual power rating of WTs has increased over the years and technologies required in order to reach and even exceed a power rating of 10 MW are discussed. The role of power electronics for improving the operation of WTs and ensuring compliance with power grid codes is analyzed with a view to produce fully controllable generation units suitable for a tight integration into the power grid and large-scale deployment in the future smart power systems.

8.1   INTRODUCTION

The cumulative installations of wind turbines (WTs) have grown at a fast pace over the last two decades. Even according to conservative estimations for continued developments, the installed wind power generation, which is currently greater than 360 GW, is expected to exceed 760 GW by 2020, making this form of renewable energy a significant component of modern power systems [1]. An example is set by Denmark, which has a very high penetration of wind generation that covered 42% of the electric energy consumption in 2015. As an important milestone, on November 3, 2013, Denmark set a record by having at national level wind power production in excess of power consumption [2], and now it happens regularly.

Not only that the installations of wind farms have grown significantly, but also the size and the power rating of WTs have increased dramatically. Just a few years ago, in 2011, the average rating of WTs was 1.7 megawatts (MW) for onshore and 3.6 MW for offshore installations. By the end of the decade, the number of high-power turbines, up to 10 MW rating, is expected to grow pushing the average rating further up [3–10]. Currently, the world's most powerful WT, the Vestas V-164, is rated for 8 MW and employs a 164 m rotor [9]. Most manufacturers are developing WTs larger than 4.5 MW, a trend that is aimed at significantly lowering the cost of wind energy delivered to the power grid.

All these continuous WT developments would not have been possible without significant technological advancements, including those for power electronic drives, controls, and electric generators that represent the scope of this chapter. Power electronics and related variable speed technologies enabled, among other things, the reduction in mechanical stress and the increase in energy production and made possible the operation of a wind turbine system (WTS) as a fully controllable generation unit suitable for a tight integration into the power grid.

This chapter first introduces the fundamentals, main components, and subsystems of WTS. The main solutions, which are commercially available, will also be briefly reviewed in Section 8.2. The specific issues of WTS, including complex mission profiles, power grid codes, and reliability, will be discussed in Section 8.3. Section 8.4 covers power electronics, starting from the device to the circuit level, including state-of-the-art technologies and future trends. The solutions and configurations of power electronics for wind farms will be introduced in Section 8.5. A MATLAB® simulation example is presented in Section 8.6, ANSYS® models are discussed in Section 8.7, and a final summary is provided.

8.2   WIND TURBINE SYSTEMS

The main components employed for energy conversion from wind to electricity in a state-of-the-art WTS include the rotor with turbine blades, possibly a gearbox (which is eliminated in direct drive solutions), an electric generator, a power electronics converter, and a transformer (see Figure 8.1) for interconnecting to the power grid.

Controlling a WTS involves both electrical and mechanical subsystems, as indicated in Figure 8.2, where a general control structure for a WTS including turbine, generator, and converter is illustrated at three different layers. The applied WT concept can either be the full-scale converter-based system

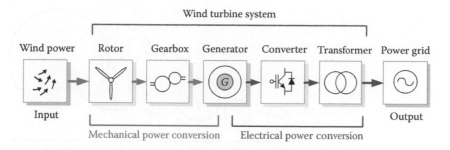

FIGURE 8.1 Energy conversion stage in a wind turbine system (WTS).

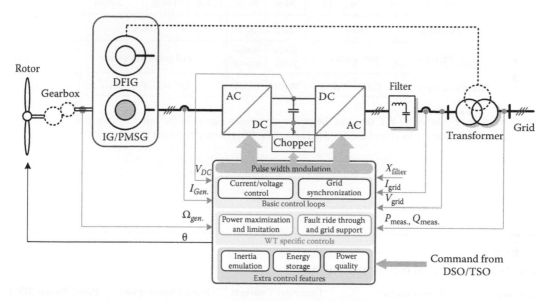

FIGURE 8.2 Generic power electronics converter control of wind power generators.

with a synchronous generator (SG) or a permanent magnet synchronous generator (PMSG) or the partial-scale converter-based system with a doubly fed induction generator (DFIG).

Generally, the power flow in and out of the system has to be managed carefully. The input mechanical power from the turbines should be limited by controlling the mechanical components such as pitch angle of rotors (θ) or direction of the yawing system. Meanwhile, the electrical power injected to the power grid should also be regulated according to the standards or commands given by the distribution system operator/transmission system operator (DSO/TSO). After the power flow in the system can be fully managed, more advanced features may be achieved by introducing extra control functions, such as the maximization of the generated power from turbines, ride-through operation of the grid faults, and supporting functions in both normal and abnormal operations of the power grid. In variable speed WTs, the current in the generator will typically be changed by controlling the generator-side converter, and thereby the rotational speed of the turbines can be adjusted to achieve maximum power production based on the available wind power. In respect to operation under grid fault, a coordinated control of several subsystems in the WT like the generator-/grid-side converters, braking chopper/crowbar, and pitch angle controller is necessary. Finally, the basic control functions of the electrical system like the current regulation, DC bus stabilization, and grid synchronization have to be quickly performed, where proportional–integral (PI) controllers or proportional-resonant controllers are typically used to track the reference.

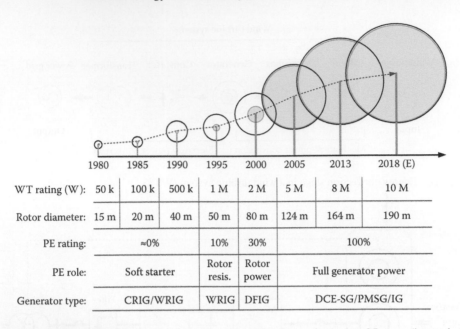

	1980	1985	1990	1995	2000	2005	2013	2018 (E)
WT rating (W):	50 k	100 k	500 k	1 M	2 M	5 M	8 M	10 M
Rotor diameter:	15 m	20 m	40 m	50 m	80 m	124 m	164 m	190 m
PE rating:	≈0%			10%	30%	100%		
PE role:	Soft starter			Rotor resis.	Rotor power	Full generator power		
Generator type:	CRIG/WRIG			WRIG	DFIG	DCE-SG/PMSG/IG		

FIGURE 8.3 Evolution of WT size and the power electronics impact in the last 35 years (inner circle indicates the power coverage by power electronics converters; PE, power electronics).

TABLE 8.1
Top Wind Turbine Manufacturers in 2014

Ranking	Manufacturer	Generator Concepts	Rotor Diameter (m)	Power Range (MW)
1	Vestas (Denmark)	3G-DFIG (Type III)	80–110	1.8–2
		1G-PMSG (Type IV)	105–164	3.3–8
2	Goldwind (China)	DD-PMSG (Type IV)	70–109	1.5–2.5
		3G-SCIG (Type IV)	110	3
3	Enercon (Germany)	DCE-SG (Type IV)	44–126	0.8–7.5
4	Siemens (Germany/Denmark)	3G-SCIG (Type IV)	82–120	2.3–3.6
		DD-PMSG (Type IV)	101–154	3–6
5	Suzlon (India)	3G-SCIG (Type IV)	52–88	0.6–2.1
		3G-DFIG (Type III)	95–97	2.1
6	General Electric (US)	3G-DFIG (Type III)	77–120	1.5–2.75
		DD-PMSG (Type IV)	113	4.1
7	Gamesa (Spain)	3G-DFIG (Type III)	52–114	0.85–2
		1G-PMSG (Type IV)	128	4.5
8	Guodian United Power (China)	3G-DFIG (Type III)	77–100	1.5–3
		DD-PMSG (Type IV)	136	6
9	Ming Yang (China)	3G-DFIG (Type III)	77–83	1.5–2
		1G-PMSG (Type IV)	92–108	2.5–3
10	Nordex (Germany)	3G-DFIG (Type III)	90–131	2.4–3.3

Note: 1G, one-stage gearbox; DD, direct drive; 3G, three-stage gearbox.

In order to achieve a better controllability and previously mentioned functions of the WTS, the power electronics control are becoming an essential part and it covers more and more power rating of WTs. The development of power electronics in the wind power application is illustrated in Figure 8.3. It can be seen that power electronics appeared in the WTS in the 1980s using a soft starter to initially interconnect the generator with the power grid. Then it started to process the electrical power in the system, first in the 1990s for the rotor resistance control and then in the 2000s for regulating the rotor power of the DFIG [28–32]. Since 2005, power electronics have been used to handle all of the generated power from the generator—leading to fully controlled WTs, which are suitable to be integrated into the power grid [33–38].

For WTSs, the most commonly used concepts can be categorized into four types: Type I, fixed speed WTSs with squirrel cage induction generator; Type II, partial variable speed WT with variable rotor resistance and wound rotor induction generator; Type III, variable speed WT with partial-scale frequency converter and DIFG; and Type IV, variable speed WT with full-scale power converter and PMSG or DC excited synchronous generator (DCE-SG) [6]. In Table 8.1, the type of generators, rotor sizes, and power ranges adopted by the major WT manufacturers for their MW product lines are listed and compared. It can be seen that the Type III and Type IV WT concepts with DFIG and PMSG/DCE-SG are currently dominant in the market of large WT due to the better power controllability, less mechanical stress due to advanced power electronics control, and also stronger grid support ability.

8.3 SPECIFIC ISSUES FOR WIND TURBINE OPERATION

Due to more significant impacts on the power grid and the relatively high cost of energy, there are many demands and challenges for WTs and thereby also for the wind power converter. In the past 30 years, these demands and challenges have always been the pushing force for the development of the power electronics technology used for WTs, and these demands can be, for example, addressed from the wind input, power delivered to the grid side, and inherent reliability performances.

8.3.1 COMPLEX MISSION PROFILES FOR WIND TURBINES

The power conversion stages for a typical WTS are as shown in Figure 8.1, in which the wind energy captured from the rotor blades is the input and electrical power injected into the power grid is the output. The total wind energy captured by the blades is closely related to the wind speed, and it is proportional to the area covered by the rotor blades and to a power coefficient C_p, which is a function of the pitch angle β, rotational speed ω_r, and wind speed v_w. The behavior of the wind speed will determine the loading of the power converter and behaviors of the output power into the grid, and thereby, it is one of the most important mission profiles to be considered during the design and selection of a WTS.

The wind speed behavior is normally complicated and can be grouped into wind classes that are mainly defined by three factors: the average annual wind speed, the speed of the extreme wind gust that could occur over 50 years, and how much turbulence exists at the wind site [9]. According to the IEC standard [11], there are three types of winds named Class I-high, Class II-medium, and Class III-low defined with annual average speeds of 10, 8.5, and 7.5 m/s, respectively. The distributions of the wind speed by different wind classes are shown in Figure 8.4, where the Weibull distribution is used to describe the characteristics. Also, a 1-year wind speed profile is shown in Figure 8.5 with a 3 h average at 80 m hub height, which was collected from the wind farm in Thyborøn, Denmark [12]. The shown wind speed belongs to the IEC wind Class I-high with the average wind speed of 8.5–10 m/s, and significant wind speed variations can be identified.

In the large WTS as shown in Figure 8.1, the generator should be regulated by the converter to control the electromagnetic torque, not only for maximizing the extracted power from the blades but also for the energy balancing in case of dynamics due to inertia mismatch between the mechanical

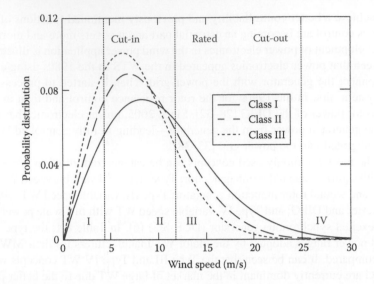

FIGURE 8.4 Distribution of the wind speed by different wind classes (region I, no power generation; region II, maximum power point tracking generation; region III, constant power generation; region IV, no power generation).

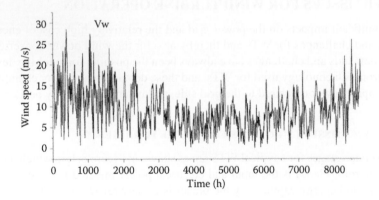

FIGURE 8.5 One-year wind speed variations at a wind farm of Thyborøn, Denmark (80 m height, 3 h averaged).

and electrical power. As a result, the complicated wind speed behaviors will be reflected by the flowing power in the converter and the loading/stress in power electronics components. The loading conditions will impose great challenges for the selection of converter topologies and devices and the design of the controls and the cooling system for the converter.

Besides the complicated loadings from the input power, there are some other challenges related to the mission profiles (i.e., operating conditions) of large WTs: Because of the large power capacity, the voltage level of the electrical power conversion may need to be boosted up to facilitate the power transmission; thus, a bulky transformer is normally required. The voltage is typically boosted up to 30 kV but recently seen to be raised to 60 kV. Because the space is limited in the nacelle or tower of the WT, the power density and strong cooling capability are crucial performances for the converter to be designed. Finally, because of the mismatch inertia between the mechanical power generated from turbine and the electrical power injected into the grid, energy storage and balancing control schemes are important issues and may result in extra cost of the converter system.

8.3.2 Strict Codes from Grid Side to Be Connected

The fluctuation and unpredictable features of wind energy are not preferred for the grid operation. Most countries have strict requirements for the behavior of WTs, also known as "grid codes," which are updated regularly [13–17]. Basically, grid codes are always trying to make the WTS to act as a conventional power plant seen from the electrical utility point of view. That means the WTS should not only be a passive power source simply injecting available power from the wind but also behave like an active generation unit, which can wisely manage the delivered active/reactive power according to the demands and provide frequency/voltage support functions for the power grid. Examples of the state-of-the-art grid supporting requirements are given in the following. They are specified either for the individual wind turbine or for whole wind farm connected, for example, the transmission system.

According to most grid codes, individual WTs must be able to control the active power at the point of common coupling. Normally, the active power has to be regulated based on the grid frequency, for example, in Denmark, Ireland, and Germany, so that the overall grid frequency can be maintained [15, 17]. Similarly, the reactive power delivered by the WTS has also to be regulated in a certain range. This leads to larger MVA capacity of the WTS when designing the whole converter system. The TSO normally specifies the reactive power range of the WTS according to the grid voltage levels.

Besides the normal operation, TSOs in different countries have issued strict grid supporting requirements for the WTS under grid faults. Figure 8.6 shows that the boundaries with various grid voltage dip amplitudes as well as the allowable disturbing time are defined for a wind farm. It has become a need that the WTS should provide reactive power (up to 100% current capacity) to contribute to the voltage recovery, when the grid voltage sag is present, as shown in Figure 8.7. Future grid codes have even requirements for large WTs to inject underexcited reactive power when the grid voltage is 20% higher than normal condition, and the reoccurred or multiply voltage dips could also become demands for large WTs [18].

The requirements for more grid support functions by WTs have on one hand increased the cost per produced kWh but on the other hand have enabled wind energy to be better utilized and smoothly integrated into the power grid. It can be predicted that stricter grid codes in the future will keep challenging the WTS and also continuously pushing forward the demands for the power electronics technology.

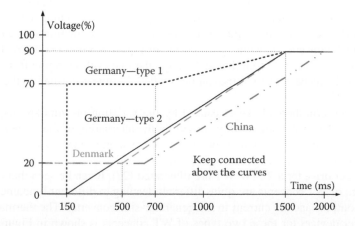

FIGURE 8.6   Grid voltage profile for low-voltage ride-through capability of WTs by different countries.

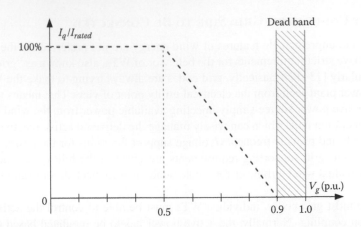

FIGURE 8.7 Reactive current requirements for a wind farm during grid voltage sags by the German-onshore and Danish grid codes.

8.3.3 GROWING RELIABILITY REQUIREMENTS

Due to the relative larger power capacity, the failures of the wind power conversion system may impose a stronger impact on the grid stability; the reliability performance is especially emphasized in the view of high cost to repair as well as loss of production. It is generally required that the power electronics used for WTs should have a lifetime of 20 years, which is at the same level of requirements for aircraft application in terms of running hours. However, according to some studies, it has been discovered that the reliability performance of WTs could be improved, especially for larger WTs ranging at multi-MW [19–22], as it is a complicated system with many components, and it has been found that power electronics is one of the most sensitive parts in the whole WTS [23–26].

It has also been pointed out that the thermal cycling or temperature fluctuations of power semiconductor devices could be one of the main causes of failures for the power electronics components [23–26]; the relationship between the characteristics of thermal cycling and the corresponding lifetime of components has been tested in the last decades [62, 63]. It has been found that the lifetime of the device is generally shorter under the thermal cycling with larger fluctuation amplitude and mean value. As mentioned before, the converted power by power electronics in the wind power application is closely related to the wind speed, which can indicate more adverse loading conditions in respect to reliability performances. An example as demonstrated by [12] is shown in Figure 8.8, which converts the wind speed profile of Figure 8.5 into the thermal stress of power semiconductor devices. It can be seen that many large thermal cycles ranging from 15 to 90 Kelvin are identified, which can be converted into an unsatisfied lifetime according to the lifetime models of power devices [23].

Besides the long-term thermal cycles caused by the variation of wind speeds, there are other types of thermal cycles, which are mainly caused by the alternating of the current and the control schemes of the converter, which are distributed in a much shorter time scale. Examples are shown in Figures 8.9 and 8.10, where the rotational speed for a DFIG with a two stage gearbox and for a PMSG with a direct drive for a 2 MW WT are illustrated [27]. It can be seen that the speed ranges of these two types of WT concepts are quite different, leading to different fundamental frequencies of the generator outputs and the current in the generator-side converter. The thermal cycling within 0.2 s of the power devices for these two types of WT concepts is shown in Figure 8.10. It can be seen that in the DFIG system, the converter could suffer from high thermal cycling compared to the PMSG system, resulting in worse loading conditions for the device in respect to reliability [27].

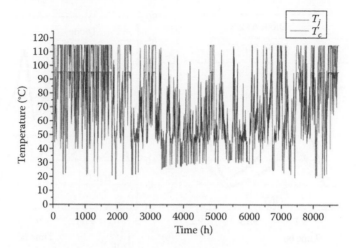

FIGURE 8.8 One-year thermal profile under the given mission profile in Figure 8.5 (junction temperature T_j and case temperature T_c of the IGBT, time step of 3 h). (From Ma, K. et al., *IEEE Trans. Power Electron.*, 30(2), 590, 2015.)

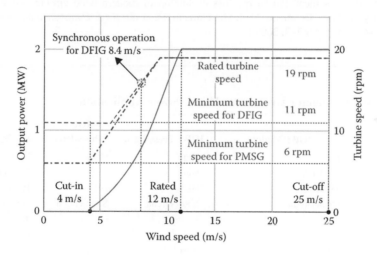

FIGURE 8.9 Speed and power curve of a 2 MW WT with different generators (DFIG, PMSG). (From Zhou, D., Reliability assessment and energy loss evaluation of modern wind turbine systems, PhD thesis, Aalborg University, Aalborg, Denmark, 2014.)

8.4 POWER ELECTRONICS: DEVICES AND CONVERTERS

8.4.1 POWER SEMICONDUCTOR DEVICES

Power semiconductor devices are the backbone in the wind power converter and are related to many critical performances of the WTS such as cost, efficiency, reliability, and modularity. Potential high-power silicon-based semiconductor technologies in the wind power application are among the module packaging *insulated gate bipolar transistor* (IGBT), press-pack packaging IGBT, and the press-pack packaging *integrated gate commutated thyristor* (IGCT) [59–61]. Recently, there has also been a booming development of silicon carbide (SiC)-based devices, which are majorly in the form of *metal-oxide-semiconductor field-effect transistor* (MOSFET) and diodes.

The four types of power semiconductor devices have quite different characteristics, and they are generally compared in Table 8.2. The module packaging technology of IGBT has a longer track

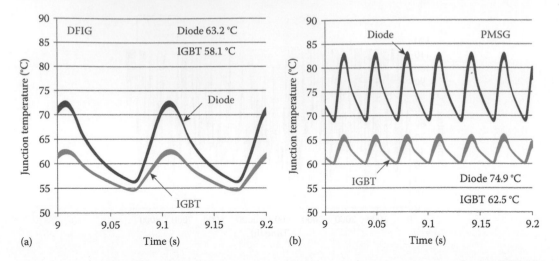

(a)   Time (s)   (b)   Time (s)

FIGURE 8.10   Junction temperature of power devices of a 2 MW wind power converter with different genera-tors using the operation profiles shown in Figure 8.9 (DFIG-based converter, PMSG-based converter). (From Zhou, D., Reliability assessment and energy loss evaluation of modern wind turbine systems, PhD thesis, Aalborg University, Aalborg, Denmark, 2014; Hirschmann, D. et al., *IEEE Conference on Vehicle Power and Propulsion*, pp. 1–6, September 7–9, 2005.)

TABLE 8.2

Silicon Power Semiconductor Devices for Wind Power Application

	IGBT Module	IGBT Press-Pack	IGCT Press-Pack	SiC-MOSFET Module
Power density	Low	High	High	Low
Reliability	Moderate	High	High	Unknown
Cost	Moderate	High	High	High
Failure mode	Open circuit	Short circuit	Short circuit	Open circuit
Insulation of heat sink	+	−	−	+
Snubber requirement	−	−	+	−
Thermal resistance	Large	Small	Small	Moderate
Cost	Moderate	High	High	High
Gate driver	Moderate	Moderate	Large	Small
Major manufacturers	Infineon, Semikron, Mitsubishi, ABB, Danfoss	Westcode, ABB	ABB	Cree, Rohm, Mitsubishi
Voltage ratings	1.7 kV–6.5	2.5/4.5 kV	4.5 kV/6.5 kV	1.2 kV/10 kV
Max. current ratings	1.5 kV–750 A	2.3 kA/2.4 kA	3.6 kA/3.8 kA	180 A/10 A

record of applications and fewer mounting restrictions. However, module packaging devices may suf-fer from larger thermal resistance and lower power density and might have higher failure rates [62]. The main trends to improve the packaging technology of the IGBT module are to introduce pres-sure contact to eliminate the base plate and thus base plate soldering, sinter technology to avoid the chip soldering, and replace bond wire material to reduce the coefficient of thermal expansion—all leading to increased lifetime of module packaging IGBTs as reported in [63]. However, the cost is always a critical factor in these new technologies.

The press-pack packaging technology improves the connection of chips by direct press-pack contacting, which leads to improved reliability, higher power density (easier stacking for series connection), and better cooling capability. However, IGCTs have not yet been mass adopted in the WTS due to the relatively high cost. As the power capacity of WTs grows even up to 10 MW, it can be expected that the press-pack packaging technology may become more promising for the future WTS, where medium-voltage level could become a need.

Besides silicon power devices, SiC-based devices, which are claimed to have better switching characteristics and lower power losses, are a promising technology in future wind power systems. Although the existing power capacity of SiC devices is still not high enough for the wind power conversion, these new devices have shown great potential for some future wind converter structures, which consist of paralleled/cascaded converter units.

It is expected that the WT concepts Type III and Type IV will continue to be the dominant systems in the next decades. However, in respect to the converter topologies, there are more flexibilities, which will be discussed in the following.

8.4.2 Two-Level Converter Topologies

The pulse width modulation–voltage source converter with two-level output voltage (2L-PWM-VSC) is the most frequently used topology in wind power applications. Because of full power controllability (4-quadrant operation) with a relatively simple structure and less components, it is popular to configure two 2L-PWM-VSCs as a back-to-back structure (2L-BTB) as shown in Figure 8.11. The 2L-BTB topology is the state-of-the-art solution in the DFIG-based WT concept. Because there is no reactive power required in PMSG- or SG-based solutions, a simple diode rectifier is an alternative to be applied on the generator side to achieve a cost-efficient variation of the 2L-BTB. In order to achieve a variable speed operation and a stable DC bus voltage, a boost DC–DC converter can be inserted into the DC link. But the diode rectifier may introduce low-frequency torque pulsations in the drive train.

However, the 2L-PWM-VSC topology may suffer from larger switching losses and lower efficiency at the MW power level. Also due to the relatively higher dv/dt stresses to the generator and transformer windings, bulky passive filters may be needed, especially for large WTSs.

In order to extend the power capability of 2L-BTB converters, Figure 8.12 shows a solution that has several 2L-BTB converters connected in parallel both on the generator side and on the grid side. This configuration is the state-of-the-art solution in the industry for the WTs with power levels higher than 3 MW [39, 40]. The standard and proven low-voltage 2L-BTB converter cells together with redundant and modular characteristics are the main advantages. However, if the DC link of the paralleled converter is connected, existing circulating current may need to be damped by extra filters or by using special PWM methods [41], which will add more cost and losses to the power conversion system.

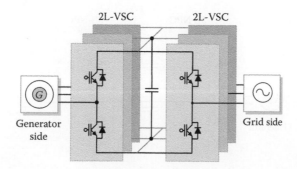

FIGURE 8.11 Two-level voltage source power converter in back-to-back (2L-PWM-VSC BTB).

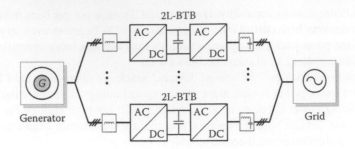

FIGURE 8.12   Topology with paralleled voltage source converters using a winding generator with regular windings.

8.4.3   Multilevel Converter Topologies

With the abilities of higher voltage amplitude and larger power capability, multilevel converter topologies are becoming promising candidates for large WTs [42–44]. In order to achieve a cost-effective solution, multilevel converters at present are normally used in 3 MW and upto 8 MW WTs with the full-scale power converter.

The three-level neutral point diode clamped (3L-NPC) topology is one of the most commercialized multilevel topologies on the market [45–47]. It is usually configured as a back-to-back (BTB) structure in WTs, as shown in Figure 8.13.

Because the 3L-NPC BTB achieves one more output voltage level compared to the 2L-BTB solution, the output filter size can be smaller. More importantly, the 3L-NPC BTB can double the output voltage with the same switching devices compared to 2L-BTB—which means an extended power capability can be obtained. However, it is found that the loss distribution is unequal between the outer and inner switching devices in a switching arm, and this problem might lead to derated power capacity when it is practically designed. In order to further extend the power handling ability, it is also possible to configure several 3L-NPC BTB converters in parallel, which is similar as the case shown in Figure 8.12.

The three-level H-bridge back-to-back (3L-HB BTB) converter is another interesting solution, which is composed of two three-phase H-bridge converters configured in a BTB structure. It achieves a similar output performance like the 3L-NPC BTB solution, but the unequal loss distribution and clamped diodes can be avoided. Thereby, a more efficient and equal loading of power switching devices and higher designed power capacity might be obtained [48–52]. A similar topology can also be configured to have five-voltage-level output per phase, with the same half bridge of 3L-NPC BTB [51]. However, either the 3L-HB BTB or the 5L-HB BTB solutions need an open-winding configuration both for the generator and for the transformer in order to achieve isolation among each phase. This feature has both advantages and disadvantages: On one hand, a potential fault-tolerant ability is

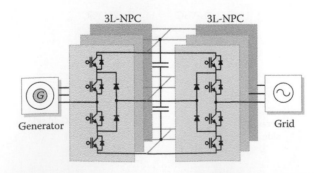

FIGURE 8.13   Three-level neutral point clamped back-to-back converter for wind turbines (3L-NPC BTB).

obtained if a part of the generator phases is out of operation [52]. On the other hand, doubled cable length is needed; extra cost, weight, loss, and inductance can be the major drawbacks. Moreover, paralleling the converter cells will be difficult in order to further extend the power capability.

Some benchmarking studies of potential converter topologies for wind power applications under either normal or abnormal operating conditions have been conducted in [48, 53], which are not included in this chapter.

8.4.4　Future Converter Topologies

8.4.4.1　Cascaded H-Bridge Converter with Medium-Frequency Transformers

A configuration that shares the similar idea with next-generation traction converters [54, 55] and is also proposed in the European UNIFLEX-PM Project [56] could be an interesting solution for the future WTS. It is based on a structure of the BTB cascaded H-bridge converter, with galvanic isolated DC/DC converters as interface. The transformer size can be significantly reduced in both weight and volume due to high-frequency operation. Moreover, it can be directly connected to the distribution power grid (10–20 kV) with high output voltage quality using a filterless design and having redundant ability. This solution would become attractive for future large WTs if it can be placed in the nacelle, where the bulky line frequency transformer can be replaced by the more compact and flexibly configured power semiconductor devices—leading to a promising increase of the power density.

8.4.4.2　Modular Multilevel Converter

Another potential configuration for the future wind turbines shares the similar idea with some of the new and emerging converters used for high-voltage direct current (HVDC) transmission [57, 58]. One advantage of this configuration is easily scalable voltage/power capability; therefore, it can achieve very high power conversion at dozens of kV with good modularity and redundant performance. The output filter can also be eliminated because of the increased voltage levels.

It can be seen that the topologies with multiconverter cells have modular and fault-tolerant abilities, which may contribute to achieving higher reliability and power capability. But, on the other hand, these configurations have significantly increased component count, which could compromise the system reliability and significantly increase the cost. The overall merits and defects of these multicell converters used in the wind power application still need to be further evaluated—also because the technologies for power semiconductor devices are developing rapidly.

8.5　POWER ELECTRONIC SOLUTIONS FOR WIND FARM

As the WT capacity is getting larger and larger, on one hand, the high cost of energy will require the transmission of wind power as efficient as possible, and on the other hand, the more significant impacts to the power grid would require the WTs to play more active role in the power grid according to grid codes. As a result, the design and configuration of wind farms, which normally involve very larger-scale wind power integration, are becoming critical to achieve both highly efficient wind power delivery and grid code compatibility.

8.5.1　Solutions for Wind Power Transmission

Wind farms may have significant impacts on the grids, and therefore, they play an important role in the power quality and the control of the grid systems. The power electronics technology is again an important part of the system configurations and control of wind farms in order to fulfill the growing demands. Some existing and potential configurations of wind farms are shown in Figure 8.14.

A wind farm equipped with DFIG-based WTSs is shown in Figure 8.14a. Such a wind farm system is, for example, in operation in Denmark as a 160 MW offshore wind power station. It is noted that due to the limitation of the reactive power capability, a centralized reactive power compensator like a static synchronous compensator (STATCOM) may be needed in order to fully satisfy the grid requirements.

Figure 8.14b shows another wind farm configuration equipped with a WTS based on a full-scale power converter. Because the reactive power controllability is significantly extended, the grid-side converter in each of the generation unit can be used to provide the required reactive power individually, leading to reactive power compensatorless solutions.

For long-distance power transmission from an offshore wind farm, HVDC is an interesting option because the efficiency is improved and no reactive power compensators are needed [64, 65].

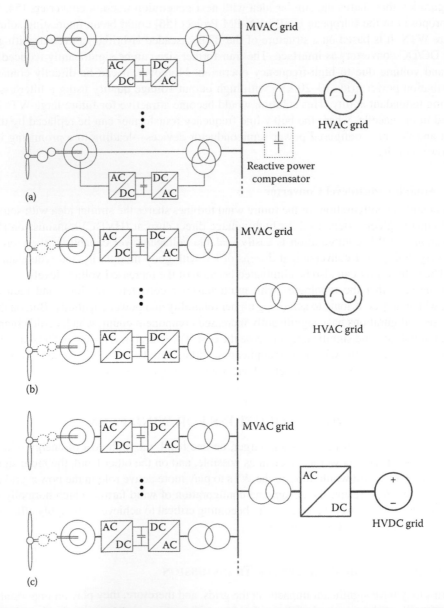

FIGURE 8.14 Potential wind farm configurations with AC and DC power transmission. (a) DFIG system with AC grid. (b) Full-scale converter system with AC grid. (c) Full-scale converter system with VSC rectifier and transmission DC grid.   *(Continued)*

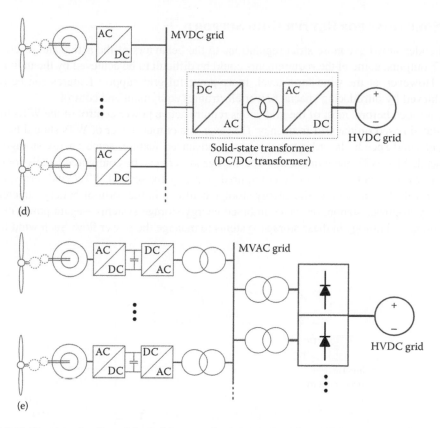

FIGURE 8.14 (*Continued*) Potential wind farm configurations with AC and DC power transmission. (d) Full-scale converter system with both distribution and transmission DC grid. (e) Full-scale converter system with multiple diode rectifiers and transmission DC grid. *Abbreviations*: MVAC, medium-voltage alternating current; MVDC, medium-voltage direct current; HVAC, high-voltage alternating current; HVDC, high-voltage direct current.

A typical HVDC transmission solution for wind power is shown in Figure 8.14c, in which the medium AC voltage of the wind farm output is converted into a high-voltage DC by a boost transformer and high-voltage source rectifier.

Another possible wind farm configuration with HVDC transmission is shown in Figure 8.14d where a solid-state transformer (or DC–DC transformer) [66] is used to convert the low/medium DC voltage of each WT output to medium/high DC voltage for transmission; thus, a full DC power delivery in both the distribution and transmission line can be realized. It is claimed in [67] that the overall efficiency of the power delivery can be significantly improved compared to the configuration in Figure 8.14c—because of less converters and transformers in this system, and it can be a future solution for large wind farms to increase the overall efficiency of power delivery. Moreover, the 4-quadrant operation of the "DC transformer," thanks to the use of power electronics, could bring some interesting features like power flow management for the future "smarter" grid.

In order to achieve more robust HVDC conversion and save the space/weight of offshore platform, a HVDC concept for an offshore wind farm was proposed in [68, 69]. In this configuration, the power control and power quality regulations are mainly performed by the distributed low-voltage wind power converter, while the rectifier is simply composed of diodes and has no control complexity. It is claimed that this solution will save 20% loss and 65% weight compared to the conventional VSC-based HVDC system at 200 MW rated power. Moreover, the reduced number of components, easily scalable and redundant rectifier connection, and reliable power semiconductor packaging all make this solution attractive for the reliable and cost-effective HVDC transmission of offshore wind power.

8.5.2 Solutions for Better Grid Support

The grid codes now have more strict regulations to the behaviors of the active and reactive power of the WT outputs; some of the requirements could be difficult to be achieved by the individual WT concept. However, at the wind farm level, more powerful grid support features can be relatively easily achieved by introducing special power electronics equipment and control.

As mentioned before, most of the grid codes require active power control of the WTS to be flexibly regulated based on the grid frequency. However, the output power of WTs should be based on the current wind speeds. In order to achieve this advanced feature, some energy storage systems may be needed for WTs and wind farms. The storage system can be configured locally for each WT unit, as shown in Figure 8.15a, or be configured centrally for several WTs/wind farms, as shown in Figure 8.15b. It is noted that the energy storage could be in the form of battery, supercapacitor, flywheel, hydropower station, or even combined energy storage systems—again power electronics is an enabling technology in these storage systems to manage the power flow. Such wind farm with

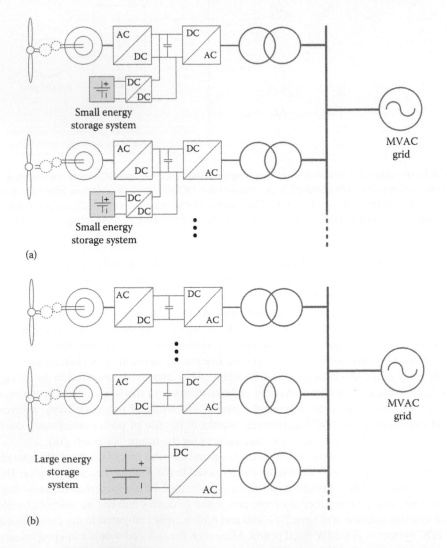

FIGURE 8.15 Potential configurations of energy storage for wind power plants to achieve more controllable active power output. (a) Distributed energy storage. (b) Centralized energy storage.

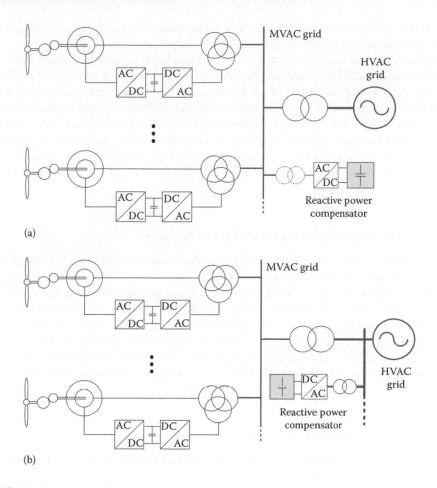

FIGURE 8.16 Potential configurations of reactive power compensation for wind power plants to achieve more controllable reactive power output. (a) Compensator connected to MVAC grid. (b) Compensator connected to HVAC grid.

energy storage will also be ready to operate as primary controller in the case that enough energy is prestored and the wind power plant is approved to support this power system control feature.

Similarly, in order to satisfy the grid requirements to provide the reactive power during grid voltage changes, some reactive power compensator can be introduced in the wind farm level to lower the burden and cost of individual WTS. This is especially beneficial for the DFIG-based WT concept, which has limited reactive power capability [70]. As shown in Figure 8.16a and b, this reactive power compensator, which could be in the form of STATCOM or static VAR compensator, can be connected either to the medium-voltage distribution grid or even directly to the high-voltage transmission grid with transformer, depending on grid codes and cost. Power electronics and the control are again an essential part in these reactive power compensator systems.

8.6 MATLAB® MODELS FOR THERMAL ANALYSIS OF WIND TURBINES

This section shows detail models used for the attached MATLAB® and Simulink® files, which are able to simulate a complete WTS for better understanding the operating principle. The models described in this section utilize a wind speed profile as input to simulate the mechanical behaviors of blades and generator, along with the electrical behaviors of a power electronics converter. As a further step ahead, the thermal models of power semiconductor are also included, which enable a deeper study on the reliability-related performance of the power electronics system.

The section starts with an overview of the complete WTS models in Section 8.6.1. Section 8.6.2 describes the model for the mechanical parts where the input is a wind speed profile and the output is the shaft torque and speed of the blades, and the generator models are also described. In the attached simulation files, two generator technologies for wind power application have been considered—the PMSG and the DFIG. Section 8.6.3 describes the electrical parts with the converter models and the control techniques used. To keep the descriptions simple, the thermal parts of the power semiconductors in the simulation file are not given in this chapter, but they are cited in several reference papers, which involve loss modeling and thermal impedance modeling of the power semiconductor devices.

8.6.1   OVERVIEW OF THE MODELS AND RELATIONSHIP

The structure of the system for the wind power generation is shown in Figure 8.17 [12], which is also implemented in the attached simulation files.

The input to the models is a time-domain wind speed profile. The wind speed is fed to the mechanical models, which generates the torque for the generator. The generator is controlled by power converter for maximum power extraction. Based on the generator and control algorithm, the electrical parameters of the converters are generated. These electrical parameters are fed into the power semiconductor loss model to calculate the losses in the transistors and diodes. The losses are finally converted into temperature profiles by the thermal impedance model of the power semiconductor devices.

8.6.2   MECHANICAL MODEL OF TURBINE AND GENERATOR

The mechanical model converts the input wind velocity to a load torque signal for the generator. The available shaft power of WTs can be expressed as

$$P = \frac{1}{2}\rho_{air}C_p\left(\lambda,\theta_p\right)\pi r^2 v_{wind}^3 \tag{8.1}$$

where
  ρ_{air} is the density of air
  r is the radius of the WT rotor
  v_{wind} is the velocity of wind
  $C_p(\lambda, \theta)$ is the power coefficient, which is a function of tip-speed ratio λ and pitch angle θ

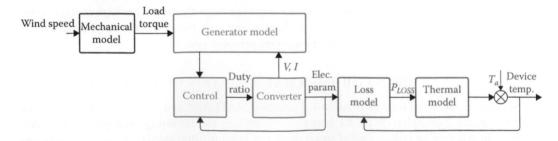

FIGURE 8.17   Overview of the model blocks in the attached simulation files.

The value of C_p can be numerically approximated [71] by Equations 8.2 and 8.3. The values C_1-C_6 are related to the design of the WT rotor and can be decided depending on the aerodynamic performance of selected WTs. These values can be found in Section 8.6.5.

$$C_p = C_1 \left(C_2 \frac{1}{\lambda_i} - C_3 \theta_p - C_4 \theta_p^{Cx} - C_5 \right) e^{C_6 \frac{1}{\lambda_i}} \tag{8.2}$$

$$\frac{1}{\lambda_i} = \frac{1}{\lambda + 0.08\theta_p} - \frac{0.035}{\theta_p^3 + 1} \tag{8.3}$$

Figure 8.18 gives the variation of C_p with λ and θ. It can be seen that one of the ways of controlling the power production from a WT is by controlling the pitch angle of the rotor blades. This property is utilized in cases where the rotor angular speed exceeds the rated rotational speed, and then the blades are pitched so as to reduce the shaft torque on the rotor, which in turn reduces the rotor speed to below rated speeds. The pitch control scheme used in the simulation files aims to pitch the blades of the turbine when the rotational speed exceeds the rated value.

The modeling of the DFIG and PMSG has been extensively covered in the literature [72–74]. Equations 8.4 through 8.8 describe the dynamic model of the DFIG:

$$\bar{u}_s^{dq} = R_s \bar{i}_s^{dq} + \omega_s \begin{bmatrix} 0 & -1 \\ 1 & 0 \end{bmatrix} \bar{\lambda}_s^{dq} + \frac{d\bar{\lambda}_s^{dq}}{dt} \tag{8.4}$$

$$\bar{u}_r^{dq} = R_r \bar{i}_r^{dq} + \left(\omega_s - p\omega_m \right) \begin{bmatrix} 0 & -1 \\ 1 & 0 \end{bmatrix} \bar{\lambda}_r^{dq} + \frac{d\bar{\lambda}_r^{dq}}{dt} \tag{8.5}$$

$$\bar{\lambda}_s^{dq} = L_s \bar{i}_s^{dq} + M_{sr} \bar{i}_r^{dq} \tag{8.6}$$

$$\bar{\lambda}_r^{dq} = L_r \bar{i}_r^{dq} + M_{sr} \bar{i}_s^{dq} \tag{8.7}$$

$$T = pM_{sr} \left(i_s^q i_r^d - i_s^d i_r^q \right) \tag{8.8}$$

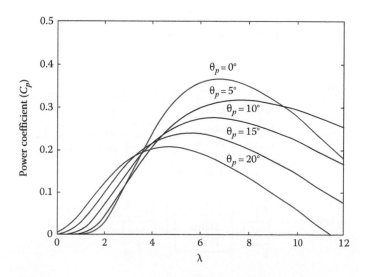

FIGURE 8.18 Power coefficient curve of WT in the attached simulation files.

Similarly, Equations 8.9 through 8.12 describe the dynamic model of the PMSG:

$$\bar{u}_s^{dq} = R_s \bar{i}_s^{dq} + p\omega_m \begin{bmatrix} 0 & -1 \\ 1 & 0 \end{bmatrix} \bar{\lambda}_s^{dq} + \frac{d\bar{\lambda}_s^{dq}}{dt} \tag{8.9}$$

$$\lambda_s^d = L_s i_s^d + \psi_r \tag{8.10}$$

$$\lambda_s^q = L_s i_s^q \tag{8.11}$$

$$Te = p\psi_r i_s^q \tag{8.12}$$

where
  u represents voltage
  R represents resistance
  i is the current
  λ represents flux linkage
  L and M are inductances
  p is the pole pairs
  Te is the electrical torque produced
  ψ_r is the flux linkage due to the permanent magnet

The subscripts s and r refer to stator and rotor quantities, respectively, while the superscript dq refers to the reference frame. As an example, the dynamic model of the permanent magnet synchronous machine (PMSM) and the DFIG in the dq reference frame are given in Figures 8.19 and 8.20, respectively, and they are based on the equations described earlier.

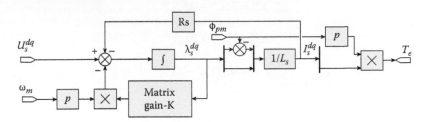

FIGURE 8.19 Block diagram of PMSM used in the simulation file.

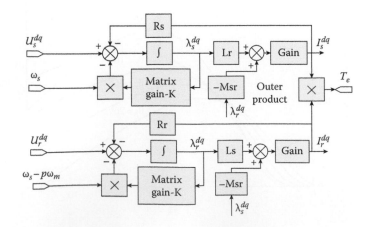

FIGURE 8.20 Block diagram of DFIG used in the simulation file.

8.6.3 ELECTRICAL MODEL OF CONVERTER AND CONTROL

The converter and electrical control models are divided into the generator side and the grid side. An overview of the control schemes for PMSG-based and DFIG-based WTS is shown in Figures 8.21 and 8.22, respectively.

8.6.3.1 Generator-Side Converter and Control

The control for the generator decides the duty ratio and the generator current. These two parameters along with the DC-link voltage can be used to model the generator-side converter according to the following equations:

$$U_s^{dq} = D^{dq} \times V_{DC} \tag{8.13}$$

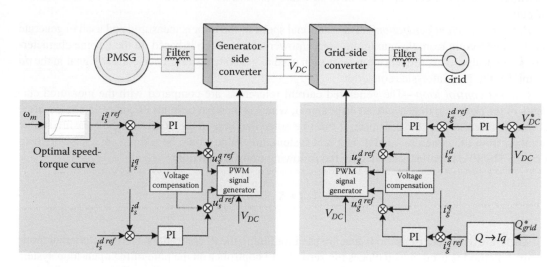

FIGURE 8.21 Schematic for PMSM control used in the simulation file.

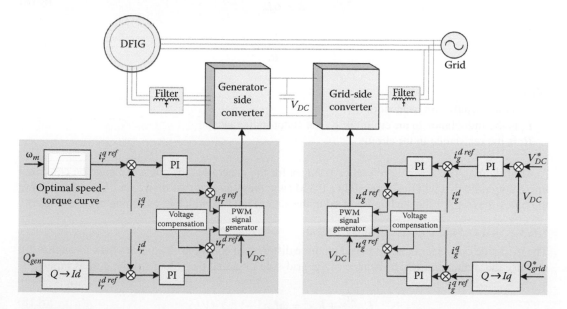

FIGURE 8.22 Schematic for DFIG control used in the simulation file.

$$I_{DC} = D^d \times I_s^d + D^q \times I_s^q \tag{8.14}$$

where

U_s is the stator voltage

V_{DC} is the DC-link voltage

I_{DC} is the current flow into DC link

I_s is the stator current

D is the duty ratio

The generator torque control for WTs is well established and can be found in [73–76]. The control algorithm implemented in this section is based on the vector control of the generators for maximum power extraction. The control structure for both the drive trains is made up of a number of parts.

Reference current generation—The rotational speed of the rotor is measured and used to generate a reference torque from the maximum torque/power curve, which is based on the turbine characteristics and design. This reference torque is then used to generate the reference current signal in the *dq* frame for the generator-side converter.

Current control loop—The generated current references are compared with the measured currents in the *dq* frame to generate an error signal, which is then fed through PI controllers to generate a voltage reference for the converter. It must be noted that here in the examples, the measured value is taken from the generator model. In a real system, this value would be taken from a measurement circuit. The PI controller is defined by the following transfer function:

$$G_{PI}(s) = K\left(1 + \frac{1}{\tau \cdot s}\right) \tag{8.15}$$

There are a number of methods that are used for tuning the PI controllers. Here, the internal mode control method is used, which places the zero of a PI controller on the pole of the open-loop system [77]. Therefore, the controller parameters are given by the following equations:

$$K = \alpha L_s \tag{8.16}$$

$$\tau = \frac{L_s}{R_s} \tag{8.17}$$

where

α is the closed-loop bandwidth

L_s is the inductance in the circuit (e.g., the stator inductance)

R_s is the resistance in the circuit

Modulation: The resulting reference voltages for the generator-side converter must be converted into a duty ratio that will finally result in a PWM switching signal for the converter. Figure 8.23 shows the block diagram of the modulation used in the simulation file, which is switching cycle averaged.

8.6.3.2   Grid-Side Converter and Control

The grid-side converter model is based on three differential equations (Equations 8.18 through 8.20) that use the inductance and resistance of the grid-side filter and the voltage of the grid as inputs:

$$L_f \frac{di_g^d}{dt} + R_f i_g^d = \omega L_f i_g^q + V_{conv}^d - V_{grid}^d \tag{8.18}$$

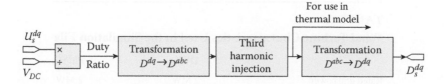

FIGURE 8.23 Modulation of generator-side converter used in the simulation file.

$$L_f \frac{di_g^q}{dt} + R_f i_g^q = -\omega L_f i_g^d + V_{conv}^q - V_{grid}^q \tag{8.19}$$

$$C_{DC} \frac{dV_{DC}}{dt} = i_{DC} - k\left(i_g^d D^d + i_g^q D^q\right) \tag{8.20}$$

where the value of k depends on the method of transformation used to convert the abc values to dq. If the normalized Clarke transform (or Concordia transform) is used, the value of k should be 1. However, if the nonnormalized transform is used, $k = 3/2$ will be used. Further, L_f is the filter inductance, R_f is the filter resistance, i_g is the grid current, V_{grid} is the grid voltage, C_{DC} is the DC-link capacitance, V_{DC} is the DC-link voltage, and D is the duty ratio.

The grid-side converter is controlled in a dq reference frame rotating with the grid voltage. The i_g^d current regulates the real power transferred to the grid by maintaining the DC-link voltage. Similarly, the i_g^q current controls the reactive power transferred to the grid. The grid-side control is thus based on two cascaded control loops, as shown in Figures 8.22 and 8.23. The modulation and limiting algorithms for the grid side are similar to those described for the generator-side controller, and they are not described here.

8.6.4 Power Loss and Thermal Modeling

The loss and thermal behaviors of power devices are closely related to the efficiency and reliability performances of the converter system, and they are also built in the attached simulation files. The thermal model takes the currents and duty ratio from the other blocks and calculates the temperatures in the power semiconductors (junction, case, and heat sink). This is done in two parts—first, the instantaneous loss model calculates the losses in the semiconductor, and the thermal model is then used to calculate the temperatures in the system based on the heat generated due to the losses. The modeling procedure is well detailed in [78–81] and it is not described in this chapter.

8.6.5 Example of the Simulation Results

The parameters of the WTs, generators, and converters used in the attached simulation files can be found in Tables 8.3 through 8.5:

Figure 8.24 shows the example wind speed profiles used in the attached simulation files. The rotational speed of the rotor in the PMSG system is shown in Figure 8.25. Based on the control described, the stator voltage responses of the generator can be seen in Figure 8.26.

The duty ratio and currents calculated are then used to develop the power loss profile for the semiconductors as shown in Figure 8.27. This is then used to generate temperature profiles, as shown in the junction temperature profile for the power semiconductors in Figure 8.28.

TABLE 8.3

Wind Turbine Characteristics Used in the Simulation File

	DFIG	PMSG
Rated grid power	2 MW	2 MW
Rotor diameter	82.6 m	82.6 m
Optimal tip-speed ratio	8.1	8.1
Max. power coefficient	0.37`	0.37

TABLE 8.4

Generator Parameters Used in the Simulation File

	DFIG	PMSG
Rated grid power	2 MW	2 MW
Pole pairs	2	102
Gear ratio	95	—
Rated shaft speed	1800 rpm	19 rpm
Stator leakage inductance	0.038 mH	0.276 mH
Magnetizing inductance	2.91 mH	
Rotor leakage inductance	0.034 mH	—
Stator/rotor turns ratio	0.369	—

TABLE 8.5

Back-to-Back Converter Data Used in the Simulation File

	DFIG	PMSG
Parallel converters	2	4
Rated active power	400 kW	500 kW
DC-link voltage	1150 V	1150 V
Switching frequency	2 kHz	2 kHz
Grid-side converter		
Rated output voltage	704 V	704 V
Filter inductance	0.5 mH	0.15 mH
Generator-side converter		
Rated output voltage	560 V	760 V

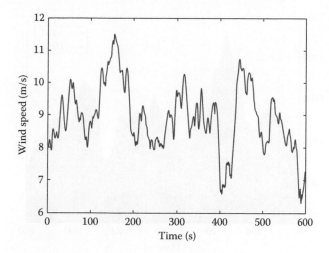

FIGURE 8.24 Example wind speed profile used in the simulation files.

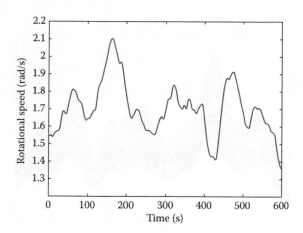

FIGURE 8.25 Rotational speed of PMSG generator.

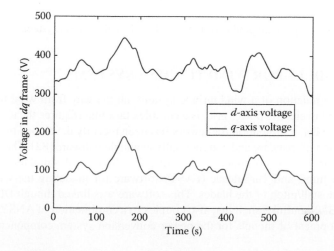

FIGURE 8.26 Stator voltages of PMSG in *dq* frame.

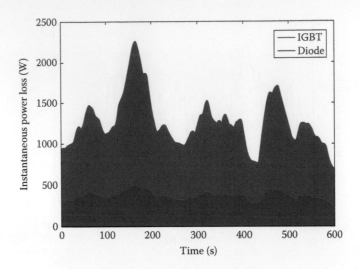

FIGURE 8.27   Power loss profile for IGBT and diode in the PMSG system.

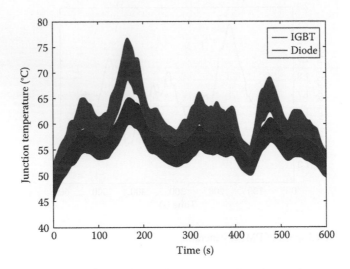

FIGURE 8.28   Junction temperature profile for IGBT and diode in the PMSG system.

8.7   ANSYS® MODELS FOR WIND TURBINE SYSTEMS

The comprehensive simulation of a wind turbine system, all the way from wind to rotor blades to energy conversion and to electric power grid, is a complex task that requires the use of multidomain techniques. In this respect, a simulation framework has been recently developed by Novakovic et al. based on a combination of freeware and commercially available software [82]. The method employs the National Renewable Energy Laboratory (NREL) TurbSim wind simulator and the Fatigue, Aerodynamics, Structures, and Turbulence (FAST) software for modeling the wind turbine rotor system, including the stall/pitch of the blades. This software was linked though DLL to a specially developed MATLAB®/Simulink® shell that also couples to a combination of ANSYS® software that served for the development of models for the power conversion system components, as illustrated in Figure 8.29.

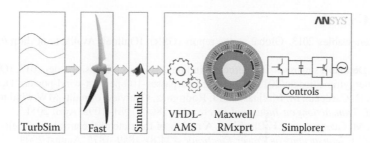

FIGURE 8.29 The comprehensive simulation of a wind turbine system using a framework of freeware and commercially available software. (Based on Novakovic et al., *IEEE Indus. Appl. Mag.*, 22(5), 73, 2016.)

The gearbox was modeled in the VHDL-AMS language within the ANSYS Simplorer® software. The back to back AC/DC and DC/AC power electronics converters, together with their associated generator side and grid side controllers, was also modeled using this software, but in an approach that considers the electric circuit schematics and details including those of the IGBTs used for practical implementation. As such, the model is able to simulate with very small steps time transients and can be successfully employed in order to estimate the converter power losses, and, based on this and on an additional simulator for the thermal field, can predict the temperature variation for the power electronic semiconductor devices. The permanent magnet (PM) generator can be modeled using the ANSYS Maxwell® software for electromagnetic finite element numerical analysis (FEA) or the analytical and equivalent circuit based ANSYS RMxprt® module. The former is superior in terms of detail and accuracy, while the latter is advantageous in terms of high computational speed and reduced effort.

Such a framework of tightly coupled software modules supports the complex design process at the component and system level, including the interactions involved in the implementation of maximum power point tracking (MPPT) controls that require the variable speed operation of generators below the rated wind speed and the pitch blade control at high speeds. The comprehensive model also supports high fidelity transient and stability studies under large power variations and grid fault conditions.

8.8 SUMMARY

The individual power rating and the installations of WTs have been significantly increasing over the decades, such that wind energy now plays an important and growing role in power systems. A main driving factor is represented by the continuous need for sustainable and renewable energy at competitive prices.

The review of state-of-the-art solutions indicates that power electronic technologies, including associated controls and variable speed generators, have significantly improved the operation and performance of the WTS. Through proper selection and configuration, controls, and grid regulations, it is now possible for the WTS and for wind farms to act similar to conventional power plants and actively contribute to the frequency and voltage control in the power grid.

These capabilities create opportunities for a continued large-scale deployment of WTs. Furthermore, driven by the need to lower the cost of energy and enhance the performance of energy conversion, there are yet many new possibilities for the advancement of power electronics and electric generator technologies, including the development and application of new topologies and power devices, the increase of reliability, and the incorporation of energy storage and power system control capabilities.

REFERENCES

1. REN21, Renewables 2013: Global status report (GSR), [Online]. Available at: http://www.ren21.net/, June 2013.
2. C. Morris, Denmark surpasses 100 percent wind power, Energy Transition.de, 2013, [Online]. Available at: http://energytransition.de/2013/11/denmark-surpasses-100-percent-wind-power/, December 2014.
3. M. Liserre, R. Cardenas, M. Molinas, and J. Rodriguez, Overview of multi-MW wind turbines and wind parks, *IEEE Transactions on Industrial Electronics*, 58(4), 1081–1095, April 2011.
4. Z. Chen, J. M. Guerrero, and F. Blaabjerg, A review of the state of the art of power electronics for wind turbines, *IEEE Transactions on Power Electronics*, 24(8), 1859–1875, August 2009.
5. F. Blaabjerg, Z. Chen, and S. B. Kjaer, Power electronics as efficient interface in dispersed power generation systems, *IEEE Transactions on Power Electronics*, 19(4), 1184–1194, 2004.
6. A. D. Hansen, F. Iov, F. Blaabjerg, and L. H. Hansen, Review of contemporary wind turbine concepts and their market penetration, *Journal of Wind Engineering*, 28(3), 247–263, 2004.
7. M. P. Kazmierkowski, R. Krishnan, and F. Blaabjerg, *Control in Power Electronics-Selected Problems*, Academic Press, Amsterdam, the Netherlands, 2002.
8. F. Blaabjerg, M. Liserre, and K. Ma, Power electronics converters for wind turbine systems, *IEEE Transactions on Industry Applications*, 48(2), 708–719, 2012.
9. Website of Vestas Wind Power, Wind turbines overview, Available: http://www.vestas.com/, April 2016.
10. UpWind Project, Design limits and solutions for very large wind turbines, March 2011.
11. Wind turbines: Part I: Design requirements, IEC 61400-1, 3rd edn.
12. K. Ma, M. Liserre, F. Blaabjerg, and T. Kerekes, Thermal loading and lifetime estimation for power device considering mission profiles in wind power converter, *IEEE Transactions on Power Electronics*, 30(2), 590–602, 2015.
13. M. Tsili, A review of grid code technical requirements for wind farms, *IET Journal of Renewable Power Generation*, 3(3), 308–332, 2009.
14. Energinet: Wind turbines connected to grids with voltages below 100 kV, January 2003.
15. Energinet: Technical regulation 3.2.5 for wind power plants with a power output greater than 11 kW, September 2010.
16. TenneT TSO GmbH, Grid code: High and extra high voltage, December 2012.
17. TenneT TSOGmbH, Requirements for offshore grid connections in the grid of TenneT TSO GmbH, Dec 2012.
18. W. Chen, F. Blaabjerg, M. Chen, and D. Xu, Capability of DFIG WTS to ride through recurring asymmetrical grid faults, in *Proceedings of ECCE 2014*, pp. 1827–1834, 2014.
19. S. Faulstich, P. Lyding, B. Hahn, and P. Tavner, Reliability of offshore turbines–identifying the risk by onshore experience, in *Proceedings of European Offshore Wind*, Stockholm, Sweden, 2009.
20. B. Hahn, M. Durstewitz, and K. Rohrig, Reliability of wind turbines: Experience of 15 years with 1500 WTs, *Wind Energy*, Spinger, Berlin, Germany, 2007.
21. E. Wolfgang, L. Amigues, N. Seliger, and G. Lugert, Building-in reliability into power electronics systems, *The World of Electronic Packaging and System Integration*, pp. 246–252, 2005.
22. D. Hirschmann, D. Tissen, S. Schroder, and R. W. De Doncker, Inverter design for hybrid electrical vehicles considering mission profiles, *IEEE Conference on Vehicle Power and Propulsion*, 7–9, pp. 1–6, September 2005.
23. C. Busca, R. Teodorescu, F. Blaabjerg, S. Munk-Nielsen, L. Helle, T. Abeyasekera, and P. Rodriguez, An overview of the reliability prediction related aspects of high power IGBTs in wind power applications, *Microelectronics Reliability*, 51(9–11), 1903–1907, 2011.
24. N. Kaminski and A. Kopta, Failure rates of HiPak modules due to cosmic rays, ABB application note 5SYA 2042-04, March 2011.
25. E. Wolfgang, Examples for failures in power electronics systems, Presented at *ECPE Tutorial on Reliability of Power Electronic Systems*, Nuremberg, Germany, April 2007.
26. S. Yang, A. T. Bryant, P. A. Mawby, D. Xiang, L. Ran, and P. Tavner, An industry-based survey of reliability in power electronic converters, *IEEE Transaction on Industry Applications*, 47(3), 1441–1451, May/June 2011.
27. D. Zhou, Reliability assessment and energy loss evaluation of modern wind turbine systems, PhD thesis, Department of Energy Technology, Aalborg University, Aalborg, Denmark, 2014.
28. S. Muller, M. Deicke, and R. W. De Doncker, Doubly fed induction generator systems for wind turbines, *IEEE Industry Applications Magazine*, 8(3), 26–33, May/June 2002.

29. D. Xiang, L. Ran, P. J. Tavner, and S. Yang, Control of a doubly fed induction generator in a wind turbine during grid fault ride-through, *IEEE Transactions on Energy Conversion*, 21(3), 652–662, September 2006.

30. F. K. A. Lima, A. Luna, P. Rodrigue, E. H. Watanabe, and F. Blaabjerg, Rotor voltage dynamics in the doubly fed induction generator during grid faults, *IEEE Transactions on Power Electronics*, 25(1), 118–130, January 2010.

31. D. Santos-Martin, J. L. Rodriguez-Amenedo, and S. Arnaltes, Providing ride-through capability to a doubly fed induction generator under unbalanced voltage dips, *IEEE Transactions on Power Electronics*, 24(7), 1747–1757, July 2009.

32. R. Pena, J. C. Clare, and G. M. Asher, Doubly fed induction generator using back-to-back PWM converters and its application to variable speed wind-energy generation, *Electric Power Application*, 143(3), 1996, 231–241.

33. J. Dai, D. D. Xu, and B. Wu, A novel control scheme for current-source-converter-based PMSG wind energy conversion systems, *IEEE Transactions on Power Electronics*, 24(4), 963–972, April 2009.

34. X. Yuan, F. Wang, D. Boroyevich, Y. Li, and R. Burgos, DC-link voltage control of a full power converter for wind generator operating in weak-grid systems, *IEEE Transactions on Power Electronics*, 24(9), 2178–2192, September 2009.

35. P. Rodriguez, A. Timbus, R. Teodorescu, M. Liserre, and F. Blaabjerg, Reactive power control for improving wind turbine system behavior under grid faults, *IEEE Transactions on Power Electronics*, 24(7), 1798–1801, July 2009.

36. A. Timbus, M. Liserre, R. Teodorescu, P. Rodriguez, and F. Blaabjerg, Evaluation of current controllers for distributed power generation systems, *IEEE Transactions on Power Electronics*, 24(3), 654–664, March 2009.

37. M. Liserre, F. Blaabjerg, and S. Hansen, Design and control of an LCL-filter-based three-phase active rectifier, *IEEE Transactions on Industry Applications*, 41(5), 1281–1291, September–October 2005.

38. P. Rodriguez, A. V. Timbus, R. Teodorescu, M. Liserre, and F. Blaabjerg, Flexible active power control of distributed power generation systems during grid faults, *IEEE Transactions on Industrial Electronics*, 54(5), 2583–2592, October 2007.

39. B. Andresen and J. Birk, A high power density converter system for the Gamesa G10x 4.5 MW Wind turbine, in *Proceedings of EPE'2007*, pp. 1–7, 2007.

40. R. Jones and P. Waite, Optimised power converter for multi-MW direct drive permanent magnet wind turbines, in *Proceedings of EPE'2011*, pp. 1–10, 2011.

41. G. Gohil, R. Maheshwari, L. Bede, T. Kerekes, R. Teodorescu, M. Liserre, and F. Blaabjerg, Modified discontinuous PWM for size reduction of the circulating current filter in parallel interleaved converters, *IEEE Transactions on Power Electronics*, 30(7), 3457–3470, 2015.

42. J. Rodriguez, S. Bernet, W. Bin, J. O. Pontt, and S. Kouro, Multilevel voltage-source-converter topologies for industrial medium-voltage drives, *IEEE Transactions on Industrial Electronics*, 54(6), 2930–2945, 2007.

43. S. Kouro, M. Malinowski, K. Gopakumar, J. Pou, L. G. Franquelo, B. Wu, J. Rodriguez, M. A. Perez, and J. I. Leon, Recent advances and industrial applications of multilevel converters, *IEEE Transactions on Power Electronics*, 57(8), 2553–2580, 2010.

44. A. Faulstich, J. K. Stinke, and F. Wittwer, Medium voltage converter for permanent magnet wind power generators up to 5 MW, in *Proceedings of EPE 2005*, pp. 1–9, 2005.

45. N. Celanovic and D. Boroyevich, A comprehensive study of neutral-point voltage balancing problem in three-level neutral-point-clamped voltage source PWM inverters, *IEEE Transactions on Power Electronics*, 15(2), 242–249, 2000.

46. S. Srikanthan and M. K. Mishra, DC capacitor voltage equalization in neutral clamped inverters for DSTATCOM application, *IEEE Transactions on Industrial Electronics*, 57(8), 2768–2775, August 2010.

47. J. Zaragoza, J. Pou, S. Ceballos, E. Robles, C. Jaen, and M. Corbalan, Voltage-balance compensator for a carrier-based modulation in the neutral-point-clamped converter, *IEEE Transactions on Industrial Electronics*, 56(2), 305–314, February 2009.

48. K. Ma, F. Blaabjerg, and D. Xu, Power devices loading in multilevel converters for 10 MW wind turbines, in *Proceedings of ISIE 2011*, pp. 340–346, June 2011.

49. K. Ma and F. Blaabjerg, Multilevel converters for 10 MW wind turbines, in *Proceedings of EPE'2011*, pp. 1–10, Birmingham, U.K., 2011.

50. J. Rodriguez, S. Bernet, P. K. Steimer, and I. E. Lizama, A survey on neutral-point-clamped inverters, *IEEE Transactions on Industrial Electronics*, 57(7), 2219–2230, 2010.

51. M. S. El-Moursi, B. Bak-Jensen, and M. H. Abdel-Rahman, Novel STATCOM controller for mitigating SSR and damping power system oscillations in a series compensated wind park, *IEEE Transactions on Power Electronics*, 25(2), 429–441, February 2010.

52. K. Ma, W. Chen, M. Liserre, and F. Blaabjerg, Power controllability of three-phase converter with unbalanced AC source, *IEEE Transactions on Power Electronics*, 30(3), 1591–1604, March 2014.

53. K. Ma, F. Blaabjerg, and M. Liserre, Operation and thermal loading of three-level neutral-point-clamped wind power converter under various grid faults, *IEEE Transactions on Industry Applications*, 50(1), 520–530, 2014.

54. B. Engel, M. Victor, G. Bachmann, and A. Falk, 15 kV/16.7 Hz energy supply system with medium frequency transformer and 6.5 kV IGBTs in resonant operation, in *Proceedings of EPE'2003*, pp. 1–10, Toulouse, France, September 2–4, 2003.

55. S. Inoue and H. Akagi, A bidirectional isolated DC–DC converter as a core circuit of the next-generation medium-voltage power conversion system, *IEEE Transactions on Power Electronics*, 22(2), 535–542, 2007.

56. F. Iov, F. Blaabjerg, J. Clare, O. Wheeler, A. Rufer, and A. Hyde, UNIFLEX-PM-A key-enabling technology for future european electricity networks, *EPE Journal*, 19(4), 6–16, 2009.

57. M. Davies, M. Dommaschk, and J. Dorn, J. Lang, D. Retzmann, and D. Soerangr, HVDC PLUS: Basics and principles of operation, Siemens Technical articles, 2008.

58. A. Lesnicar and R. Marquardt, An innovative modular multilevel converter topology suitable for a wide power range, in *Proceedings of IEEE Bologna PowerTech Conference*, pp. 1–6, 2003.

59. K. Ma and F. Blaabjerg, The impact of power switching devices on the thermal performance of a 10 MW wind power NPC converter, *Energies*, 5(7), 2559–2577.

60. R. Jakob, C. Keller, and B. Gollentz, 3-Level high power converter with press pack IGBT, in *Proceedings of EPE' 2007*, pp. 2–5, September 2007.

61. R. Alvarez, F. Filsecker, and S. Bernet, Comparison of press-pack IGBT at hard switching and clamp operation for medium voltage converters, in *Proceedings of EPE'2011*, pp. 1–10, 2011.

62. U. Scheuermann, Reliability challenges of automotive power electronics, *Microelectronics Reliability*, 49(9–11), 1319–1325, 2009.

63. U. Scheuermann and Ralf Schmidt, A new lifetime model for advanced power modules with sintered chips and optimized al wire bonds, in *Proceedings of PCIM' 2013*, pp. 810–813, 2013.

64. A. Prasai, Y. Jung-Sik, D. Divan, A. Bendre, and S.-K. Sul, A new architecture for offshore wind farms, *IEEE Transactions on Power Electronics*, 23(3), 1198–1204, May 2008.

65. F. Iov, P. Soerensen, A. Hansen, and F. Blaabjerg, Modelling, analysis and control of DC-connected wind farms to grid, *International Review of Electrical Engineering*, Praise Worthy Prize, 10, February 2006.

66. S. Engel, N. Soltau, H. Stagge, and R. W. De Doncker Dynamic and balanced control of three-phase high-power dual-active bridge DC-DC converters in DC-grid applications, *IEEE Transactions on Power Electronics*, 28(4), 1880–1889, 2012.

67. C. Meyer, M. Hoing, A. Peterson, and R. De Doncker, Control and design of dc grids for offshore wind farms, *IEEE Transactions of Industrial Applications*, 43(6), 1475–1482, November/December 2007.

68. R. B. Gimenez, S. A. Billalba, J. R. Derlee, F. Morant, and S. B. Perez, Distributed voltage and frequency control of offshore wind farms connected with a diode-based HVdc link, *IEEE Transactions on Power Electronics*, 25(12), 3095–3105, 2010.

69. P. Menke, R. Zurowski, T. Christ, S. Seman, G. Giering, T. H. Mer, W. Zink et al., Latest DC grid access using diode rectifier units (DRU), *Siemens AG, Wind Integration Workshop*, Brussel, Belgium, October 2015.

70. D. Zhou, F. Blaabjerg, T. Franke, M. Tonnes, and L. Mogens, Reduced cost of reactive power in doubly fed induction generator wind turbine system with optimized grid filter, in *Proceedings of IEEE ECCE 2014*, pp. 1490–1499, 2014.

71. J. G. Slootweg, H. Polinder, and W. L. Kling, Dynamic modelling of a wind turbine with doubly fed induction generator, *2001 Power Engineering Society Summer Meeting Conference Proceeding* (Cat. No. 01CH37262), vol. 1, pp. 644–649, 2001.

72. R. Pena, J. C. Clare, and G. M. Asher, Doubly fed induction generator using back-to-back PWM converters and its application to variable-speed wind-energy generation, *IEE Proceedings: Electric Power Applications*, 143(3), 231, 1996.

73. Y. Errami, M. Maaroufi, and M. Ouassaid, Modelling and control strategy of PMSG based variable speed wind energy conversion system, *2011 International Conference on Multimedia Computing and Systems (ICMCS)*, vol. 2, pp. 1–6, 2011.

74. P. Kundur, *Power Systems Stability and Control*, 1st edn., McGraw-Hill Education, New York, 1994.

75. O. Anaya-Lara, N. Jenkins, J. Ekanayake, P. Cartwright, and M. Hughes, *Wind Energy Generation: Modelling and Control*. Wiley, Hoboken, NJ, 2009.
76. A. Tapia, G. Tapia, J. X. Ostolaza, and J. R. Saenz, Modeling and control of a wind turbine driven doubly fed induction generator, *IEEE Transactions on Energy Conversion*, 18(2), pp. 194–204, June 2003.
77. L. Harnefors and H.-P. Nee, Model-based current control of AC machines using the internal model control method, *IEEE Transactions on Industry Applications*, 34(1), pp. 133–141, 1998.
78. A. Glumineau and J. de León Morales, *Sensorless AC Electric Motor Control*. Springer International Publishing, Cham, Switzerland, 2015.
79. K. Ma, A. S. Bahman, S. Beczkowski, and F. Blaabjerg, Complete loss and thermal model of power semiconductors including device rating information, *IEEE Transactions on Power Electronics*, 30(5), 2556–2569, 2015.
80. K. Ma, Y. Yang, and F. Blaabjerg, Transient modelling of loss and thermal dynamics in power semiconductor devices, in *Proceedings of IEEE ECCE 2014*, pp. 5495–5501, 2014.
81. K. Ma, F. Blaabjerg, and M. Liserre, Electro-thermal model of power semiconductors dedicated for both case and junction temperature estimation, in *Proceedings of PCIM Europe*, pp. 1042–1046, 2013.
82. B. Novakovic, Y. Duan, M. Solveson, A. Nasiri, and D. M. Ionel, Comprehensive modeling of turbine systems from wind to electric grid, *IEEE Industry Applications Magazine*, 22(5), 73–84, 2016.

9 Electric Generators and their Control for Large Wind Turbines

Ion G. Boldea, Lucian N. Tutelea, Vandana Rallabandi, Dan M. Ionel, and Frede Blaabjerg

CONTENTS

Abstract 209
9.1 Introduction 210
9.2 DFIG 213
    9.2.1 Introduction 213
    9.2.2 DFIG Topology and Circuit Model 215
    9.2.3 DFIG *dq* Model and Control 218
9.3 CRIG 221
    9.3.1 Introduction 221
    9.3.2 CRIG Circuit Model and Performance 223
    9.3.3 CRIG Control 225
9.4 PMSG 227
    9.4.1 Introduction 227
    9.4.2 Optimal Design of PMSGs 229
    9.4.3 Circuit Modeling and Control of PMSG 231
9.5 DCE-SG 236
    9.5.1 Optimal Design of DCE-SG 237
    9.5.2 Active Flux–Based Sensorless Vector Control of DCE-SG 242
9.6 Modeling of Electric Generators by Finite Element Analysis (FEA) 246
9.7 Summary 247
References 248

ABSTRACT

The electric generator and its power electronics interface for wind turbines (WTs) have evolved rapidly toward higher reliability and reduced cost of energy in the last 40 years. This chapter describes the up-to-date electric generators existing in the wind power industry, namely, the doubly fed induction generator, the cage rotor induction generator, and the synchronous generator with DC or permanent magnet excitation. The operating principle, performance, optimal design, and the modeling and control of the machine-side converter for each kind of generator are addressed and evaluated. In view of the fact that individual power rating of WTs has increased to around 10 MW, generator design and control technologies required to reach this power rating are discussed.

9.1 INTRODUCTION

Wind energy is penetrating electric power systems at a high pace, such that in 2016 it accounted for more than 360 GW of installed power [1–24]. The technology of wind energy harnessing has evolved rapidly in the last 40 years with decisive progress in the wind turbine (WT) design, tower construction, electric generator, and power electronics technology, including interfacing to local and national electric power grids [1–4]. The power rating of WTs has increased steadily since 1980 with the main motivation being the reduction of the energy cost [25], and large 8 MW units were first installed and commissioned in 2014. From an economic point of view, it is interesting to note that the levelized cost of electricity for onshore wind farms, which is expected to continue to decrease due to technological developments, is already, in regions such as Germany, comparable with that of coal power plants [5].

A WT system is complex, has thousands of components, and includes, as main subsystems, a tower, rotor blades and hub, a mechanical drive train with or without gearbox, and electric power conversion with a generator, as schematically shown in Figure 9.1. Examples of WTs are illustrated in Figures 9.2 and 9.3.

The typical cost breakdown for a state-of-the-art WT is listed in Table 9.1 based on industry information available [26]. Screws and cables contribute an additional 1% each. At first look, the elimination of the gearbox, which accounts for a significant cost proportion, may be advantageous, but it should be kept in mind that the increased cost of a direct drive (DD) generator, higher cost of full power electronics conversion, reliability, and serviceability all play an important role in the decision process for new WT designs.

The three main variable speed WT generators: doubly fed induction generators (DFIGs), cage rotor induction generators (CRIGs), and synchronous generators (SG) with permanent magnet (PM) or DC excitation are all schematically represented in Figure 9.4. AC–DC–AC PWM converters

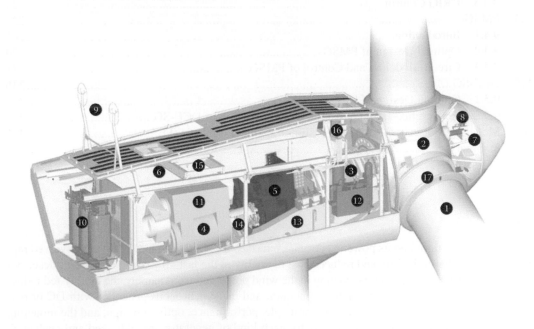

FIGURE 9.1 Blowing wind causes the WT blades (1), which are connected to a hub (2), to rotate the main shaft (3) and ultimately generate electricity using a generator (4) coupled at the end of a mechanical gearbox drive train (5) and power conversion system that is placed in the WT nacelle (6) on top. Other components shown include hub controller (7), pitch cylinders (8), ultrasonic wind sensors (9), HV transformer (10), air cooler for the generator (11), hydraulic unit (12), machine foundation (13), composite disk coupling (14), service crane (15), oil cooler (16), and blade bearing (17). (Courtesy of Vestas Wind Systems A/S.)

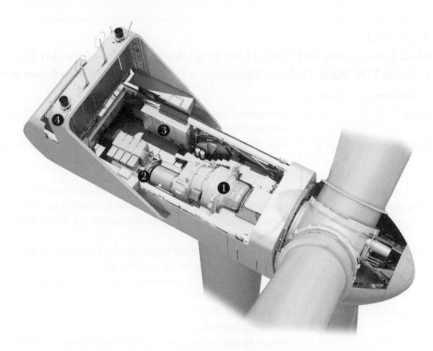

FIGURE 9.2 Vestas multi-MW variable speed WT system including mechanical gearbox (1), generator (2), power electronic controls (3), and cooler top (4). (Courtesy of Vestas Wind Systems A/S.)

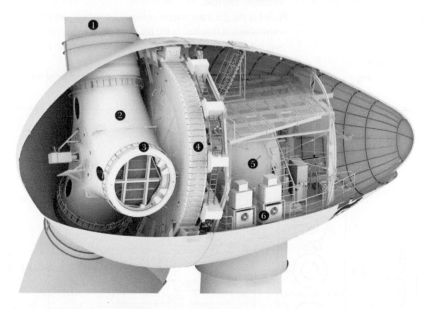

FIGURE 9.3 Enercon E-126 DD WT including rotor blades (1), rotor hub (2), hub adapter (3), annular generator (4), main carrier (5), and yaw drive (6). (Courtesy of Enercon GmbH.)

are employed as an interface by all of the generator types but in the rotor (at partial rating: 30%) for DFIG and in the stator (at full rating: 100%) for SG and CRIG. The AC–DC–AC bidirectional PWM voltage source converters may be designed at 400/690 Vrms (line voltage) in two-level configurations and up to 3.2–3.6 or even 6 kVrms (line voltage) in multilevel configurations. Multilevel inverters are generally used at powers above 3 MVA/unit, in order to provide, with filtering, a reasonable total harmonic distortion (THD) of output voltage [8]. For more details, see also Chapter 8.

TABLE 9.1

Main Components and Typical Cost Breakdown for State-of-the-Art Type C Multi-MW Wind Turbines Employing a Doubly Fed Induction Generator

Component	Cost %	Comments
Tower	26	Manufactured from sections of rolled steel or from concrete. Typical height is 100 m.
Rotor blades	22	Made of composite materials with carbon fiber added for mechanical strength. Blades are typically longer than 60 m each.
Rotor hub	1	Casted from iron in a single piece.
Rotor bearings	1	Designed to withstand significant and variable radial and axial loads.
Main shaft	2	Transfers the mechanical power from the WT rotor to the gearbox.
Main frame	3	Manufactured from steel, must be designed for low weight and high mechanical strength.
Gearbox	13	Converts the low speed and high torque from the wind side to high speed and low torque for the generator side.
Generator	3	Apart from DFIG, other commonly used generator types include PMSM, CRIG, and DCE-SG.
Yaw system	1	Rotates the nacelle to follow the wind direction in order to maximize the energy harvesting.
Pitch system	3	Adjusts the angle of the blades for best wind incidence in order to maximize the energy harvesting.
Power converter	5	Power electronics for AC–AC conversion, with or without (less common) DC link.
Transformer	4	Placed at the generator output; it increases voltage and reduces current and, if installed in the nacelle, results in smaller cable size and reduced losses from the nacelle to the base of the tower.
Brake system	1	Disk type used for safety and maintenance.
Nacelle housing	1	Typically made of fiber glass in order to achieve reduced weight and cost.

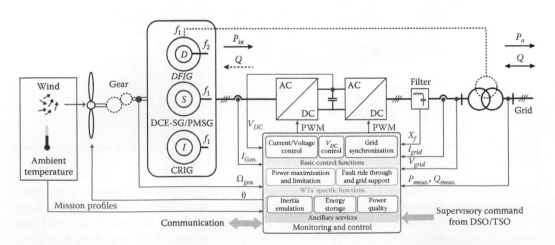

FIGURE 9.4 Generic wind generators' control system with AC–DC–AC bidirectional static power converters: DFIG, DCE-SG, PMSG, and CRIG. (From Ma, K. et al., *EPCS*, 43, 1406, 2015.)

TABLE 9.2

State-of-the-Art Multi-MW Wind Generators

Principle	DFIG	CRIG	PMSG	DCE-SG
By voltage sags	Medium/low	Acceptable	Good	Very good
By reactive power capability	Medium/low	High	High	High
By gear ratio	100/1 (3G)	100/1 (3G)	None (up to 5 MW) 50/1 (1/2G) (at 8 MVA)	None (direct drive)

In a DFIG system, the stator frequency f_1 is constant at the power line value, while the rotor frequency $f_2 = S * f_1$ ($S_{max} = \pm 0.3$) is variable. For SGs and CRIGs, the stator frequency f_1 is variable in order to control the speed. In general, as the WT power is proportional to ω_r^3, $a \pm 30\%$, speed variation range is considered sufficient to collect most of the wind power efficiently. With large wind-tower turbines, the speed decreases steadily with power, from 15 rpm at 3 MVA to 10(9) rpm at 8(10) MVA. As the size (and cost) of large wind electric generators decreases almost in proportion with the increase in speed, single-stage (up to 10[20]/1 ratio: 1G—one stage gear) or multistage (up to 250/1 ratio: 3/4G—3/4 stage gear) mechanical transmission is employed in order to reach optimal solutions [9, 10].

So far, existing large power (above 1 MVA/unit) wind generators may be characterized as shown in Table 9.2 in terms of handling voltage sags, reactive power capability, and mechanical transmission/gearing.

A comparative study between DD and 3G transmission wind generator systems is provided in [11]. To summarize, the generation cost per kWh does not vary sharply with the power rating of the WT but it is smaller for higher powers. According to the calculations, DD transmissions may have slightly lower generation cost, but, nevertheless, long-time reliable operation is yet to be fully demonstrated by a large number of field installations. Also, according to the study, the investment cost per kW is larger when the power increases to 10 MW. The optimal solution also heavily depends on the wind speed range, available wind power, and the selected sites, that is, onshore or offshore. Design studies should consider the multiple aspects involved in terms of modeling, optimization, control, and grid integration in the local, regional, national, and international power grids.

Scaling above 10 MW, the state-of-the-art concept of large WT with three bladed rotors may face major challenge as both the rotor and tower dimensions would need to significantly increase and stretch technological limitations. Alternative concepts are being considered, such as the INVELOX system, in which the wind is guided through mechanically flexible towers/tubes, and the vortex tube principle is employed in order to accelerate the wind toward the turbine and generator that are placed at the ground level [5].

The following sections are dedicated to the detailed characterization of DFIG, CRIG, permanent magnet synchronous generator (PMSG), and DC-excited synchronous generator (DCE-SG). In addition, very recently proposed topologies are only mentioned with principle, merits, and pertinent comparisons.

9.2   DFIG

9.2.1   INTRODUCTION

Today, the DFIG represents about 50% of installed wind power, though the power/unit is still below 5 MVA/unit using 3G transmission. Typically used are 6-pole machines at $1000(1200) \pm 300(400)$ rpm at 50(60) Hz [12].

The main advantages of DFIG are related to the lower AC–DC–AC PWM converter rating according to the speed range (maximum $\pm 30\%$). The main demerit resides in the brush–copper ring transmission of electric energy from/to the rotor of DFIG at variable frequency ω_2. Stator frequency ω_1, with the stator being grid connected, is rather constant. The speed ω_m and slip S are

$$\omega_m = \omega_1 - \omega_2; \quad S = \frac{\omega_2}{\omega_1} \tag{9.1}$$

For synchronous operation, $\omega_2 = 0$ and $\omega_m = \omega_1$, while for subsynchronous operation, $\omega_2 > 0$, $S > 0$, $(\omega_m < \omega_1)$. For supersynchronous operation, $\omega_2 < 0$, $S < 0$, $(\omega_m > \omega_1)$. Electric active power is produced by the stator $P_s > 0$, while the rotor electric power is $P_r < 0$ (absorbed) for $\omega_2 > 0$, $S > 0$ $(\omega_m < \omega_1)$ and is delivered $P_r > 0$ for $\omega_2 < 0$, $S < 0$, $(\omega_m > \omega_1)$:

$$P_m = \sum P_{losses} + P_s + P_r \tag{9.2}$$

where
P_m is the mechanical (shaft) power
ΣP_{losses} is the total loss in the DFIG and in the PWM converter connected to the rotor

The generic scheme and the circulation of active powers in the DFIG system are illustrated in Figure 9.5.

A negative ω_2 during supersynchronous mode implies that the sequence of phases a, b, and c in the rotor is changed to a, c, and b. The direction of rotation of the rotor magnetic field is changed so that it is synchronous with the stator field and thus ideally constant torque (power) is produced. The DFIG power should be controlled with the speed and thus the maximum power P_{max} is

$$P_{max} \approx P_s + P_{rmax} = P_s + |S_{max}| P_s \tag{9.3}$$

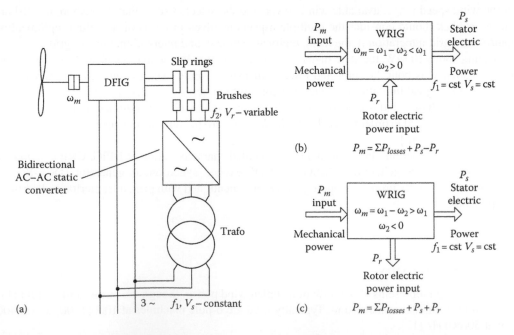

FIGURE 9.5 DFIG system: (a) general scheme with slip rings, (b) subsynchronous mode of operation, and (c) supersynchronous mode of operation.

with

$$\omega_{rmax} = \omega_1 \left(1 + \left| S_{max} \right| \right) \tag{9.4}$$

Before entering into some details of the DFIG model, performance, design, and control, it is worth mentioning that there have been recent technological attempts to perform the following:

- Design a DD DFIG at 10 MW and 10 rpm with a mechanically flexible frame (to keep the air gap lower than 2 mm at rotor diameters above 9 m—which is a challenge) [12].
- Investigate the use of a rotary transformer inverter on the stator. The rotor winding is connected to the secondary of the transformer. This system realizes control of torque, speed, current, and power factor by changing the amplitude, frequency, and phase of the voltage across the primary winding of the rotary transformer [27]. This leads to elimination of the brushes, thereby reducing maintenance requirements.
- Build a brushless dual AC stator winding with p_p, p_c pole pairs, and nested cage rotor with $(p_p + p_c)$ poles to the DFIG, where the control winding (p_c pole pairs) is fed through an AC–DC–AC PWM voltage source converter sized around $|S_{max}| P_s$ [13]. However, the weak coupling (50%) and high internal reactance of the stator windings through the rotor are the main disadvantages of the *brushless* DFIG.

9.2.2 DFIG Topology and Circuit Model

The DFIG is similar to a wound rotor induction motor. In the MW range, the laminated stator and rotor slotting are shown in Figure 9.6.

The distributed 1(2) layer winding of the stator and rotor have the same number of poles $2p$ but, in general, different (integer in general) numbers of slots/poles/phases $q_1 > q_2$. The ratio of phase turns $a_{rs} = w_r/w_s$ is chosen frequently in order to avoid the use of a voltage matching transformer to power grid:

$$a_{rs} = \frac{w_r}{w_s} = \frac{1}{S_{max}} \tag{9.5}$$

However, in this way, for accidental slip values higher than $|S_{max}|$, the rotor overvoltage will endanger the rotor windings and thus an overvoltage protection system becomes necessary in practice. The steady-state circuit equations per phase of a DFIG are

$$\begin{aligned}
\left(R_s + j\omega_1 L_{sl}\right)\underline{I}_s + \underline{V}_s &= \underline{E}_1; \quad \text{at} \quad \omega_1 \\
\left(R_r^r + jS\omega_1 L_{rl}^r\right)\underline{I}_r^r + \underline{V}_r^r &= \underline{E}_{2s}; \quad \text{at} \quad \omega_2; \quad \underline{E}_{2s} = S\underline{E}_1 a_{rs}
\end{aligned} \tag{9.6}$$

$R_s, R_r^r, L_{sl}, L_{rl}^r$ are stator and rotor resistances and leakage inductance per phase.

Reducing the rotor to the stator by $R_r = R_r^r/a_{rs}^2$; $L_{rl} = L_{rl}^r/a_{rs}^2$; $\underline{V}_r = \underline{V}_r^r/a_{rs}$; $\underline{I}_r = \underline{I}_r^r a_{rs}$ and dividing the rotor equations in (9.6) by S, one obtains

$$\left(\frac{R_r}{S} + j\omega_1 L_{rl}\right)\underline{I}_r + \frac{\underline{V}_r}{S} = \underline{E}_1 \quad \text{at} \quad \omega_1 \tag{9.7}$$

Equation 9.7 is now expressed at stator frequency ω_1, as in (9.6), and yields the equivalent circuit shown in Figure 9.7.

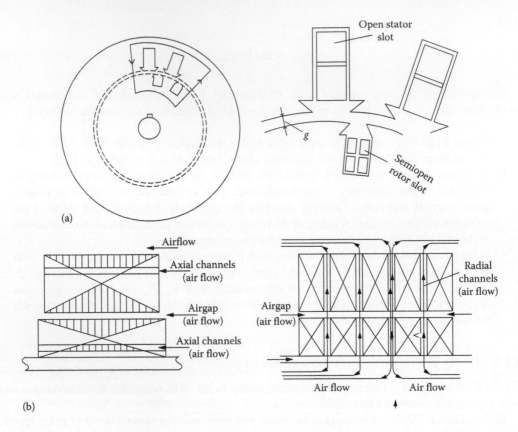

(a)

(b)

FIGURE 9.6 Layout of DFIG: (a) stator and rotor laminations, rotor and stator slotting; (b) stator and rotor stacks for axial and radial–axial cooling.

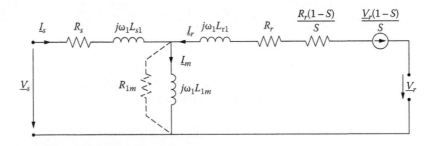

FIGURE 9.7 DFIG electrical equivalent circuit.

Adding the flux/current relationships for stator flux, ψ_s, and rotor flux, ψ_r,

$$\underline{\psi}_s = \underline{\psi}_r \frac{L_m}{L_r} + L_{sc}\underline{I}_s; \quad L_r = L_m + L_{rl}; \quad L_{sc} = L_s - \frac{L_m^2}{L_r} \approx L_{sl} + L_{rl} \tag{9.8}$$

Equations 9.6 and 9.7 then become

$$\underline{I}_r = \frac{\underline{\psi}_r - I_s L_m}{L_r} \tag{9.9}$$

$$\left(R_s + j\omega_1 L_{sc}\right)\underline{I}_s + j\omega_1 \frac{L_m}{L_r}\underline{\psi}_r = -\underline{V}_s$$

$$-R_r \frac{L_m}{L_r}\underline{I}_s + \left(\frac{R_r}{L_r} + jS\omega_1\right)\underline{\psi}_r = -\underline{V}_r \tag{9.10}$$

with V_s and R_s given, and stator power factor assigned a value $\cos \varphi_s$, the stator current I_s is

$$I_s = \frac{P_s}{3V_s \cos(\varphi_s)} \tag{9.11}$$

$$\underline{I}_s = I_s\left(\cos(\varphi_s) - j\sin(\varphi_s)\right) \tag{9.12}$$

Now from (9.10), φ_s is first calculated and then V_r is obtained for a given slip frequency $S\omega_1$ and stator frequency ω_1. It is feasible to absorb some reactive power through the stator from the power grid and magnetize the machine partially from the rotor via a PWM converter or to operate the stator at unity power factor and thus fully magnetize the machine from the rotor when $|\psi_r| > |\psi_s|$ and $|I_r| > |I_s|$.

An example follows.

Example 9.1

Consider a DFIG with the following data:

$P_{sn} = 3$ MW, $\cos \varphi_{sn} = 1$, $V_{snl} = 3.2$ kV/Y line voltage rms, at $S_{max} = -0.25$, turns ratio $a_{rs} = 4/1$, $r_s = r_r = 0.01$ pu, $r_{1m} = \infty$, $l_{sl} = l_{rl} = 0.06$ pu, $l_{1m} = 3$ pu, $f_{1n} = 50$ Hz, $2p = 6$ poles. Calculate

a. Parameters R_s, R_r, X_{sl}, X_{rl}, and X_{1m}, in Ω
b. For $S = -0.25$ and maximum power P_{max}, at $\cos \varphi_{sn} = 1$, calculate the rotor current, rotor voltage, and its angle δ_v with respect to V_s (in stator coordinates), rotor active and reactive power P_r, Q_r, and total electric power $P_e = P_s + P_r$.
c. Calculate the rotor-side converter kVA and the corresponding phasor diagram.

Solution

a. The stator current from (9.11) is simply

$$\left(I_s\right)_{s=-0.25} = \frac{P_s}{\sqrt{3}V_{sn}\cos(\varphi_n)} = \frac{3\times10^6}{\sqrt{3}\times3200\times1} = 541.9 \text{ A}$$

The nominal (base) reactance X_n is

$$X_n = \frac{V_{snl}/\sqrt{3}}{I_n} = \frac{3200}{\sqrt{3}\times541.9} = 3.413 \ \Omega$$

Thus,

$$R_1 = R_r = r_s X_n = 0.01\times3.4134 = 0.034134 \ \Omega$$

$$X_{rl} = X_{ls} = l_{sl}X_n = 0.06\times3.4134 = 0.2048 \ \Omega$$

$$X_m = l_m X_n = 3\times3.4134 = 10.24 \ \Omega$$

b. The first equation in (9.10) may be solved for the rotor flux ψ_r with V_s and I_s in phase as real numbers:

$$\underline{\psi}_r = \left[-\frac{3200}{\sqrt{3}} - (0.03413 + j0.4096)541.9 \right] \times \frac{-j}{314} \times \frac{3+0.06}{3} = -0.72 + j6.06 \text{ Wb}$$

Now, from the second equation of (9.10)

$$\underline{V}_r = 0.03413 \times \frac{3}{3.06} \times 541.9 - \left(\frac{0.03413 \times 314}{3.06 \times 3.413} - j0.25 \times 314 \right)$$

$$\times (-0.72 + j6.06) = -456.8 - j62.73$$

note that the real rotor voltage is $\underline{V}_r^r = a_{rs}\underline{V}_r = 4\underline{V}_r$.

The rotor current $\underline{I}_r$ is from (9.9):

$$\underline{I}_r = \frac{-0.72 + j6.06 - 3 \times \dfrac{3.41}{314} \times 541.9}{\dfrac{3.41 \times 3.06}{314}} = -552.94 + j182.3 \text{ A}$$

The rotor active power P_r is simply

$$P_r = 3\mathrm{Re}\left(V_r I_r^*\right) = 3\mathrm{Re}(-456.8 - j62.73) \times (-552.94 - j182.3) = 723.45 \text{ kW}$$

The rotor reactive power Q_r (at the slip frequency) is

$$Q_r^r = 3\mathrm{Im}\left(V_r I_r^*\right) = 353.88 \text{ kVAR}$$

The total power at max speed is $P_t = P_r + P_s = 3000 + 723.45 = 3723.45 \text{ kW} = 3.72 \text{ MW}$.

The rotor-side converter has to be designed for $S_r^r = \sqrt{P_r^2 + Q_r^2} = \sqrt{723.45^2 + 353.88^2} = 805.36 \text{ kVA} = 0.80 \text{ MVA}$.

9.2.3 DFIG *dq* MODEL AND CONTROL

The *dq* model of a DFIG for generator operation in synchronous coordinates is [4]

$$\frac{d\psi_d}{dt} = -V_d - R_s i_d + \omega_1 \psi_q; \quad \frac{d\psi_q}{dt} = -V_q - R_s i_q - \omega_1 \psi_d$$

$$\frac{d\psi_{dr}}{dt} = -V_{dr} - R_r i_{dr} + (\omega_1 - \omega_r)\psi_{qr}; \quad \frac{d\psi_{qr}}{dt} = -V_{qr} - R_r i_{qr} - (\omega_1 - \omega_r)\psi_{dr}$$

$$\psi_{d,q} = L_s i_{d,q} + L_m i_{dr,qr}; \quad \psi_{dr,qr} = L_r i_{dr,qr} + L_m i_{ds,qs} \qquad (9.13)$$

$$T_e = \frac{3}{2} p_1 \left(\psi_d i_q - \psi_q i_d\right)$$

$$\frac{J}{p_1} \cdot \frac{d\omega_r}{dt} = T_e + T_{mech}; \quad T_e < 0 \text{ for generating}$$

V_d, V_q, V_{dr}, V_{qr}, ψ_d, ψ_q, ψ_{dr}, ψ_{qr}, L_s, L_r, L_m, R_s, R_r, J, T_e, T_{mech}, p_1 are stator voltages, rotor voltages, stator and rotor flux linkages, inductances, resistances, inertia, electromagnetic and mechanical torque, and pole pairs.

Aligning the system of coordinates to the stator flux ψ_s is beneficial because the latter varies less, except in the case for faults. Consequently,

$$\underline{\psi}_s = \psi_s = \psi_d, \quad \psi_q = 0, \quad \frac{d\psi_q}{dt} = 0 \tag{9.14}$$

$\left(\dfrac{d\psi_d}{dt} = 0 \right)$ and zero resistance ($R_s = 0$),

$$V_d \approx 0, \quad V_q \approx -\omega_1 \psi_d, \quad \psi_q = 0 = L_m i_{qr} + L_s i_q \tag{9.15}$$

Consequently, the stator active and reactive powers P_s, Q_s become

$$P_s = \frac{3}{2}\left(V_d i_d + V_q i_q\right) = \frac{3}{2}V_q i_q = \frac{3}{2}\omega_1 \psi_d \frac{L_m}{L_s} i_{qr}$$

$$Q_s = \frac{3}{2}\left(V_q i_d - V_d i_q\right) = \frac{3}{2}V_q i_d = -\frac{3}{2}\omega_1 \frac{\psi_d}{L_s}\left(\psi_d - L_m i_{dr}\right) \tag{9.16}$$

It becomes clear that active and reactive stator power control can be accomplished through the rotor currents i_{qr}, i_{dr}. The rotor voltages for steady state are from (9.13) and (9.16):

$$V_{dr}^* = -R_r i_{dr} + L_{sc} S \omega_1 i_{qr}$$

$$V_{qr}^* = -R_r i_{qr} - S\omega_1 \left(\frac{L_m}{L_s} \psi_d + L_{sc} i_{dr} \right) \tag{9.17}$$

The current controllers in rotor coordinates for i_{dr} and i_{qr} have to be accompanied by the compensation of back-emfs (the second terms in [9.17]). Figure 9.8 shows how field-oriented control (FOC) of the rotor-side PWM converter is accomplished.

As shown in Figure 9.9, a similar FOC aligned to the grid voltages vector (or to their virtual flux) can be used for the grid-side PWM converter.

A digital simulation code is used to study the transients of a 2 MW DFIG, when a three-phase short circuit occurs and the dq currents in the stator are limited to 50% $I_{s\,rated}(I_d)$ and 150% $I_{s\,rated}(I_q)$, respectively, with the results as shown in Figure 9.10 [4].

The machine data are $P_n = 2$ MW, $V_{snl} = 690$ V, $2p_1 = 4$, $f_{1n} = 50$ Hz, $l_m = 3.658$ pu, $l_{sl} = 0.0634$ pu, $l_{rl} = 0.08466$ pu, $r_s = 4.694 \times 10^{-3}$ pu, $r_r = 4.86 \times 10^{-3}$ pu, and h (inertia) = 3.611 s.

It is evident that the machine may stand the fault, and it also retains control. The speed increases during transients, because the active power cannot be transmitted to the grid during the short-circuit condition. With the current demanding grid standards for wind generators, DFIGs should remain online even during asymmetric dip voltage sags, while providing reactive current to be ready to participate in the voltage restoring right after the fault.

All these aspects require special treatment and additional measures of protection of the machine-side converter [2].

The DFIG has proved to be a cost-effective 3G transmission wind generator and is used up to 4.5 MW, despite the brush–copper ring power transfer to the rotor. There are ways to obtain 10 MW with 1G transmission system (down to 50 rpm) [12].

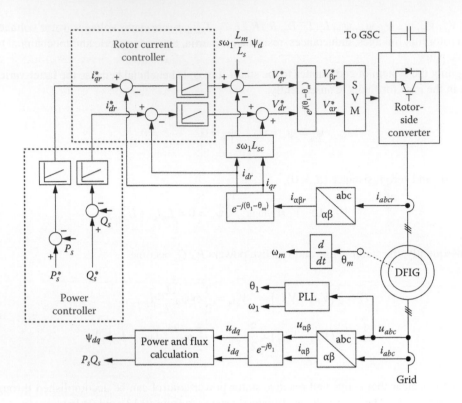

FIGURE 9.8 P_s and Q_s FOC of the rotor-side converter of DFIG.

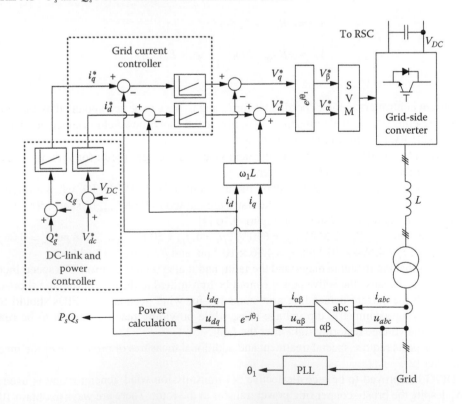

FIGURE 9.9 DC-link V_{DC} and Q_g FOC of the grid-side converter of DFIG.

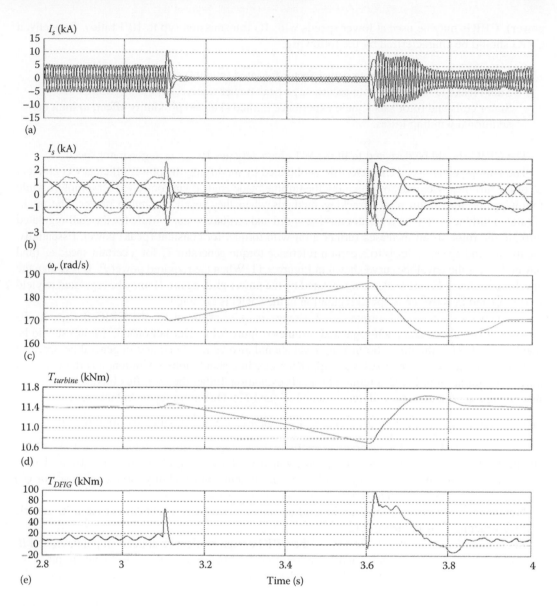

FIGURE 9.10 MW DFIG three-phase short circuit on the power grid with limitations on the rotor currents: (a) stator currents, (b) rotor currents, (c) speed, (d) turbine torque (Nm), and (e) DFIG electromagnetic torque.

9.3 CRIG

9.3.1 INTRODUCTION

Currently, CRIGs have been applied to 2.5 MW in 3G transmission drives with 6/8 pole machines, interfaced with full power bidirectional AC–DC–AC PWM converters. This can be done by using oversized DC-link capacitors to provide full active and reactive power AC–DC–AC converters [1, 14]. The robustness of the induction motor with aluminum or even copper bar cage rotor provides a solid ground for wind generator applications. So far, the CRIG was used only with 3G transmission and 2.5 MW. But, at the cost of higher kVA in the converter (due to lower factor power), CRIGs may be used at lower speeds with 1G transmission (up to 10/1 ratio). Typically, a CRIG should be characterized by the following:

- High efficiency and moderate power factor for rated (middle) speed.
- Small leakage inductances, to secure large peak power reserve.
- Stator coils and rotor bars should experience limited skin and proximity effect, to limit the additional winding losses.

The CRIG should be designed to fit the WT characteristics $P_t(\omega_r)$, ω_r—generator speed.

The flattop power/speed characteristic is obtained by modifying the power efficiency coefficient C_p of the turbine, essentially by changing the turbine blade angle β via controlled servomotors while following the maximum power tracking methods when they reach maximum power. The pitch-regulated turbine keeps the generator turbine speed and torque constant, while for the stall-regulated turbine, the rotor speed increases further a bit with torque, for constant (peak) power. Essentially, the input wind speed v_w leads to a certain reference torque generator T_e^* for a certain speed ω_r^* (and power P^*), for the envelope curve shown in Figure 9.11. When the required power P_e^* is less than the envelope power in Figure 9.11, being dictated by the power grid, again, P_e^* may be regulated to yield T_e^* (or ω_r^*) and thus torque or speed regulation in the CRIG is executed.

The generator speed and torque (ω_r^*, T_e^*) range is quite essential for the CRIG design. It differs in stall-regulated and pitch-regulated turbines (Figure 9.11). But not only T_e^* (or ω_r^*) should be regulated in CRIG, the magnetic flux ψ_m level should also be regulated and in general be operated with low total losses (core losses and copper losses). In a stand-alone operation, a certain level of magnetic saturation in CRIG is required for self-excitation. Either the rotor flux linkage ψ_r^* or stator flux linkage ψ_s^* can be regulated versus speed ω_r^* (and T_e^*). One more factor to be considered is the fact that the emf $\left(E_{1r} = \omega_1^* \dfrac{L_m}{L_r} \psi_r^* \text{ or } E_{1s} = \omega_1^* \psi_s^*; \ \omega_1^* = \omega_r^* + S\omega_1 \right)$ should vary slightly with power, to operate the machine-side converter optimally in the voltage-boosting mode. The DC-link voltage should slightly increase with power (speed), which is above the grid voltage in order to deliver power to the grid with low total system losses and to help controllability. So planning ψ_r^* (or ψ_s^*) versus speed ω_r^* and P_e^* is not a trivial task in the case of CRIGs for wind power.

Now if the CRIG is connected to the grid through an AC–DC–AC PWM converter, its generic control structure is as in Figure 9.12.

In essence, for active stall control WTs, the CRIG design bottleneck appears in points A_S and B_S and for pitch control in points A_P and B_P in Figure 9.11. The active stall control tends to place the blades like "a wall in the wind," while the pitch control leads to "a flag in the wind" situation, with

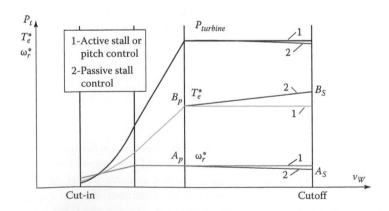

FIGURE 9.11 Generic WT characteristics: generator power, speed, and torque versus wind speed.

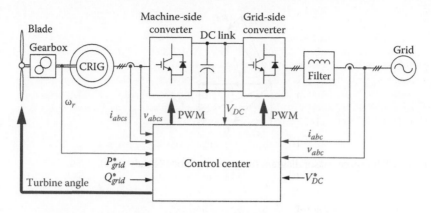

FIGURE 9.12   CRIG controlled to the grid through an AC–DC–AC converter.

a notably lower stress on the wind tower. The stall control may have an advantage in terms of lower complexity and drive cost but requires the CRIG to be designed for maximum power at lower speed and higher torque.

9.3.2   CRIG CIRCUIT MODEL AND PERFORMANCE

In order to model the CRIG, the space phasor of the model is considered for a three-phase CRIG in general coordinates (for motor mode consideration of signs) [3]:

$$
\frac{d\overline{\psi}_s}{dt} = R_s\overline{i}_s - \overline{V}_s + j\omega_b\overline{\psi}_s; \quad \overline{\psi}_s = L_s\overline{i}_s + L_m\overline{i}_r; \quad L_s = L_{sl} + L_m
$$
$$
\frac{d\overline{\psi}_r}{dt} = R_r\overline{i}_r - j(\omega_b - \omega_r)\overline{\psi}_r; \quad \overline{\psi}_r = L_r\overline{i}_r + L_m\overline{i}_s; \quad L_r = L_{rl} + L_m
$$
(9.18)

$$
T_e = \frac{3}{2}p_1 \operatorname{Im}\left(\overline{\psi}_s\overline{I}_s^*\right) < 0 \quad \text{for generating}
$$

In dq synchronous coordinates ($\omega_b = \omega_1$), for rotor flux orientation, (9.18) becomes

$$
\left(R_s + \left(-p + j\omega_1\right)L_{sc}\right)\overline{i}_s - \overline{V}_s = \left(-j\omega_1 + p\right)\frac{L_m}{L_r}\overline{\psi}_r; \quad \overline{\psi}_s = L_s i_d + jL_{sc}i_q; \quad \overline{i}_s = i_d + ji_q
$$
$$
R_r\frac{L_m}{L_r}\overline{i}_s = \left(-jS\omega_1\frac{L_m}{L_r} - p + \frac{R_r}{L_r}\right)\overline{\psi}_r; \quad T_e = \frac{3}{2}p_1\left(L_s - L_{sc}\right)i_d i_q; \quad \omega_1 = \omega_r + S\omega_1
$$
(9.19)

where i_d, i_q are stator current components in the rotor flux coordinates. The vector (space phasor) diagram for the steady state is illustrated in Figure 9.13.

Also, for steady state,

$$
\overline{V}_s = I_sR_s + j\omega_1\overline{\psi}_s; \quad \psi_r = L_m i_d; \quad i_q = -S\omega_1\frac{L_r}{R_r}i_d
$$
$$
0 = R_r\overline{i}_r + jS\omega_1\overline{\psi}_r
$$
(9.20)

In rotor flux coordinates, $i_s = i_d + ji_q$; $\overline{\psi}_r = \psi_{dr}$; $\psi_{qr} = 0$.

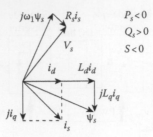

FIGURE 9.13   CRIG space phasor diagram for generator mode (steady state).

With $\psi_s = L_s i_d + jL_{sc}i_q$ and neglecting the stator copper losses,

$$P_s \approx \frac{3}{2}\mathrm{Re}\left(\overline{V}_s \overline{I}_s^*\right) = \frac{3}{2}\omega_1 \mathrm{Im}\left(\overline{\psi}_s \overline{I}_s^*\right) = \frac{3}{2}\omega_1\left(L_s - L_{sc}\right)i_d i_q$$

$$Q_s = \frac{3}{2}\mathrm{Im}\left(\overline{V}_s \overline{I}_s^*\right) = \frac{3}{2}\omega_1 \mathrm{Re}\left(\overline{\psi}_s \overline{I}_s^*\right) = 3\omega_1\left(L_s i_d^2 + L_{sc}i_q^2\right) \tag{9.21}$$

Both ideal powers may be controlled by the variation $i_d > 0$ and $i_q < 0$ (generating), but as demonstrated in FOC of the cage rotor for IMs, i_d and i_q may be varied separately (decoupled). So P_s is controlled by i_q control and Q_s by i_d control. Besides FOC, direct torque (power) and flux control may be applied to CRIG as done earlier for IM drives [16].

Example 9.2

Let us consider a large power CRIG with the following data $V_n = 3200$ V (Y), $S_n = 3$ MVA, $r_s = 0.015$ pu, $r_r = 0.0125$ pu, $l_{sl} = l_{rl} = 0.05$ pu, and $l_m = 3$ pu, with six poles generating at 50 Hz with a slip $S = -0.015$. Calculate the resistances and reactances in Ω, stator flux, i_d and i_q, and electromagnetic torque and powers.

Solution

Approximately $\psi_s \approx \dfrac{V_s}{\omega_1} = \dfrac{3200\sqrt{2}/\sqrt{3}}{2\pi 50} = 8.306$ Wb.

The nominal reactance $X_n = \dfrac{V_{1n}/\sqrt{3}}{I_n} = \dfrac{V_{1n}}{S_n} = \dfrac{3200^2}{3\times 10^6} = 3.41\,\Omega.$

$R_s = 0.015 \times X_n = 0.0512\,\Omega,\ R_r = 0.0427\,\Omega,\ X_{sl} = X_{rl} = .05\times 3.41 = 0.17\,\Omega,\ X_m = 10.24\,\Omega,$
$L_m = 0.0326$ H, $L_s = L_r = 0.033$ H, $L_{sc} = 1.087$ mH

but in rotor coordinates $\psi_s = \sqrt{\left(L_s i_d\right)^2 + \left(L_{sc}i_q\right)^2}$ and $i_q = -S\omega_1 \dfrac{L_r}{R_r}i_d.$

From the previous two equations, we may calculate i_d and i_q as

$$i_d = \frac{\psi_s}{\sqrt{L_s^2 + L_{sc}^2\left(s\omega_1\frac{L_r}{R_r}\right)^2}} = \frac{8.306}{\sqrt{0.033^2 + \left(1.087\times 10^{-3}\right)^2\left(-0.015\times 314\frac{0.032}{0.0427}\right)^2}} = \frac{8.306}{0.033} = 250\ \mathrm{A}$$

Now $i_q = -S\omega_1 \dfrac{L_r i_d}{R_r} = 0.015\times 314\times \dfrac{0.033}{0.0427}\times 250\ \mathrm{A} = 910.86\ \mathrm{A}.$

The electromagnetic torque T_e is

$$T_e = \frac{3}{2} p_1 \left(L_s - L_{sc} \right) i_d i_q = \frac{3}{2} \times 3 \left(0.03428 - 0.001087 \right) \times 250 \times \left(910.86 \right) = 34013 \text{ Nm}$$

The delivered active power P_1 is

$$P_1 = T_e \frac{\omega_1}{p_1} - 3R_s i_s^2 = 34,013 \times \frac{314}{3} - 3 \times 0.0512 \left(250^2 + 910.86^2 \right) = 3.42 \text{ MW}$$

The reactive power Q_s is

$$Q_s = -3\omega_1 \left(L_s i_d^2 + L_{sc} i_q^2 \right) = -3 \times 314 \left(0.03428 \times 250^2 + 1.087 \cdot 10^{-3} \times 910.86^2 \right) = -2.88 \text{ MVAR}$$

This leads to a power factor $\cos \varphi_1 = 0.767$, which corresponds to a 6-pole CRIG at 50 Hz, delivering 3.4 MW, while the rated MVA was stated at 3 MVA. The efficiency η_{ei} is, if the iron and mechanical losses are neglected,

$$\eta_{ei} = \frac{P_1}{P_1 + \frac{3}{2} R_s \left(i_d^2 + i_q^2 \right) + \frac{3}{2} R_r \left(\frac{L_m}{L_r} i_q \right)^2}$$

$$= \frac{3.42}{3.42 + \frac{3}{2} 0.052 \left(250^2 + 910.86^2 \right) + \frac{3}{2} 0.04266 \left(\frac{0.0326}{0.033} 910.86 \right)^2}$$

$$= 96.01\%$$

The efficiency is rather acceptable, but when adding the iron and mechanical losses, the efficiency can drop down to around 95%. A copper cage could bring down the slip for the same power to around (less than) 1%, and thus, the efficiency would increase. The leakage inductances have been considered rather large, and they may be reduced to some extent by design. As no real design is behind this numerical example, an optimal design may bring notable CRIG improvements in performance per cost.

9.3.3 CRIG CONTROL

Scalar (*V/f* with stabilizing loops), FOC, and direct power control may be implemented for controlling a CRIG [1–4].

A rather straightforward FOC of CRIG may be implemented in the synchronous coordinates, as in direct and indirect versions, based on the equation obtained from (9.19):

$$\overline{\psi}_r = \left[\int \left(\overline{V}_s - R_s \overline{i}_s \right) dt - L_{sc} \overline{i}_s \right] \cdot \frac{L_r}{L_m} \tag{9.21}$$

The integral may be "translated" into a filter to allow safe operation down to 4–5 Hz, or a combined voltage and current observer may be used [16]. For the indirect FOC, current decoupling equations need to be used. They use the following expressions from (9.19) and (9.20):

$$\overline{\psi}_r \left(1 + p \frac{L_r}{R_r} \right) = L_m i_d; \quad S\omega_1 \frac{L_r}{R_r} \psi_r = L_m i_q \tag{9.22}$$

It is also useful to perform emf compensation by adding

$$V_{DC}^* = -\omega_1^* L_{sc} i_q; \quad \omega_1^* = \omega_r + \left(s\omega_1\right)^*; \quad V_{qc}^* = \omega_1^* L_s i_d \tag{9.23}$$

and thus allow for lower gain in the dq current controllers.

Figure 9.14 illustrates both the indirect and direct FOC with dq current controllers and emf compensator for the CRIG-side converter. The reference T_e^* originates in the WT maximum power point tracking (MPPT), based on wind or generator speed (v_w or ω_r).

The grid-side converter controls the DC-link voltage and reactive power Q_g to the power grid. It has to be emphasized that the reactive power source is the DC-link capacitor, where the latter also magnetizes the CRIG and eventually sends reactive power to the grid (by pertinent overrating). V_{DC}^* may be slightly adapted (increased) in relation to grid active and reactive power delivery levels.

Simulation results on the direct and indirect FOC of the MW range CRIG and direct power control are given in [1], while a direct torque and flux control (DTFC) sensorless IM control system, applicable to CRIG, is presented in [4]. The reactive power delivery can be controlled to the desired value.

The synchronization process is rather straightforward; at first the machine-side converter is idle while the grid-side converter charges the DC-link capacitor to the desired V_{DC}^*. Then the machine-side converter starts working, after the WT reaches the cut-in speed. Starting in stand-alone applications implies an additional low-voltage battery and diode to initiate self-excitation of DC-link voltage bus. Operation under asymmetric voltage sags, harmonics, and switching from grid to stand-alone operation (when required) and back should also be considered when using CRIG.

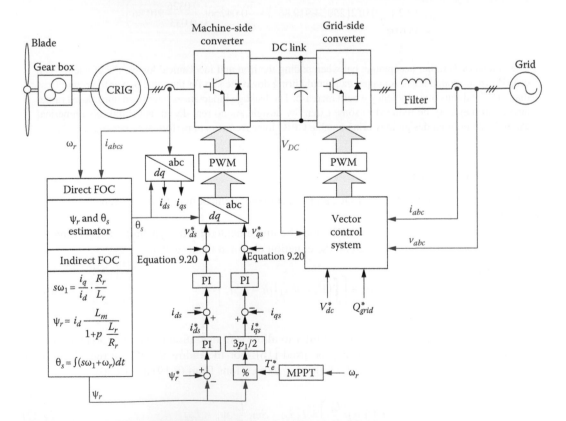

FIGURE 9.14 Indirect or direct FOC of CRIG with emf compensation.

9.4 PMSG

9.4.1 INTRODUCTION

Three-phase PMSGs with three slots/pole, single-layer winding, and surface PM rotor poles have been introduced for high power (torque) WTs (e.g., 3 MW, 15 rpm). Other configurations of PMSG such as Vernier dual stator type (Figure 9.15), transverse flux type with circular shape coils (Figure 9.16), and flux reversal types (with tooth-wound coils or multiple pole coils) (Figure 9.17) have also recently been investigated [8].

All these new configurations, which allow higher torque density, may certainly be suitable for low-speed/high-torque generators. But, like "flux modulation" machines, they show a smaller power factor than surface PM rotor distributed winding PMSGs. This simply means a machine converter kVA overrating: $\varphi = \tan^{-1}\left(\dfrac{X_q I_{rated}}{E_{PM}}\right)$; φ—power factor angle.

A brushless DC multiphase reluctance machine, which mimics the operation of an exciter brush DC machine that brushes off the neutral axis, is shown in Figure 9.18. The current is bipolar and has two values for the excitation and torque operation. The advantages of this machine include a

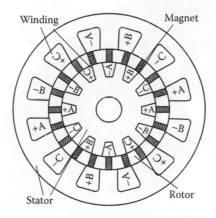

FIGURE 9.15 Vernier PMSG.

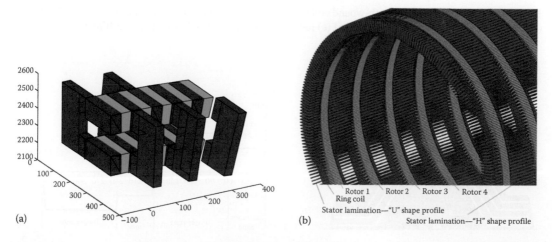

FIGURE 9.16 Transverse flux axial air-gap PMSM (one phase).

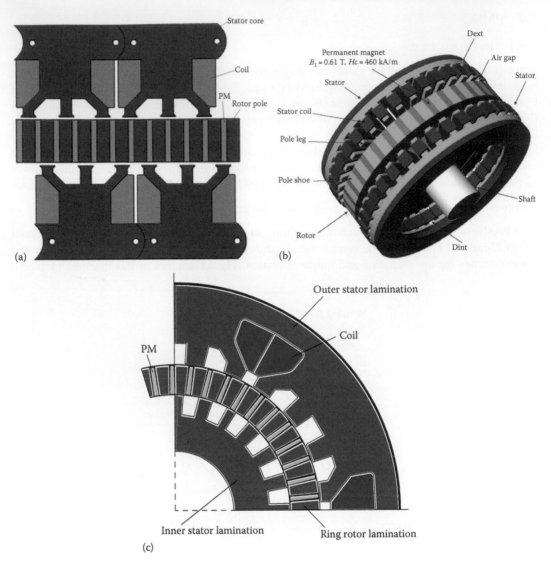

FIGURE 9.17 Flux reversal tooth-wound coil PMSG: (a) axial air-gap flux reversal machine, with IPM rotor, (b) tooth-wound axial air-gap IPMSM, and (c) radial-air-gap flux reversal machine, with IPM rotor.

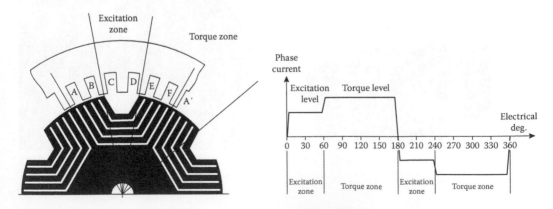

FIGURE 9.18 Brushless DC multiphase reluctance machine—basic structure and phase current in the windings.

relatively large air gap, the absence of PMs or brushes, and good utilization of the active material as all phases are active simultaneously [28].

The following will be discussed:

- Optimal design of a surface PM rotor PMSG (8 MW, 3.2 kV, 480 rpm)
- Advanced control aspects of PMSGs with an AC–DC–AC converter connected to the grid

9.4.2 OPTIMAL DESIGN OF PMSGS

Optimal design algorithms (ODAs) have to search for a set of parameters (variables) grouped in a vector $\overline{X}$, which minimizes (or maximizes) an objective (or fitting or cost) function $F_{ob}(X)$ and fulfills some constraints.

The ODA presupposes

- A set of specifications, say, for the PMSG
- A model for the object (PMSG), analytical or (and) numerical, which links the parameters (variables) to the objective (fitting) function
- A mathematical optimization (search) method that investigates part of the parameter vector space (range), to yield with high probability a global optimum solution, within a reasonable computational time

ODAs may be in general deterministic or metaheuristic (evolutionary) [19, 20]. Some of the metaheuristic methods are enumerated here:

- Black hole–based optimization (BHBO)
- Particle swarm optimization (PSO)
- Gravitational search algorithm (GSA)
- Particle swarm optimization gravitational search algorithm (PSOGSA)

A comparison of their results on a benchmark is shown in Table 9.3. It suggests that PSO-M (modified) is overall better. But no generalization is allowed.

Optimal design methods such as modified Hooke–Jeeves algorithm, genetic algorithm, bee colony algorithm [17], and differential evolution (DE) have also been used successfully to design electric machines.

TABLE 9.3

Optimization Algorithms on a Benchmark [28]

		DeJong	Rosenbrock	Ackley	Rastring
BHBO	Time (s)	19.4	41.5	25.5	25.2
	Best fitting val.	9.5×10^{-6}	26.2	1.65	28.87
PSO_O	Time (s)	21.8	43.1	25.9	25.1
	Best fitting val.	6.3×10^{-16}	12.4	9.3×10^{-7}	24.9
PSO_M	Time (s)	25.5	45.1	27.51	28.8
	Best fitting val.	2.5×10^{-27}	12.5	9.5×10^{-13}	19.8
GSA	Time (s)	897.2	953.4	902.9	899.2
	Best fitting val.	1.5×10^{-18}	25.7	1.1×10^{-9}	3.97
PSOGSA	Time (s)	926.8	933	948	927.7
	Best fitting val.	4.8×10^{-8}	15.4	0.037	41

Single-composite or multiple-objective function design optimization methods are now in use. In a single-composite objective function optimization algorithm, more than one term may be included (but normed to money/cost), such as active and passive material cost and capitalized cost of losses, both of the wind generator plus transmission (in this case) and also that of the interfacing PWM converter.

In a multiple-objective (e.g., DE) optimization algorithm, each objective function (say, minimum losses) is followed through optimization versus [18, 19], say, initial cost or weight to build the Pareto clouds and thus allow the designer more ways to choose the final solution while explaining also the sensitivity/robustness of the design to various variables. In what follows, in short, the modified Hooke–Jeeves algorithm is described, which is used with a single-composite objective function [20].

The objective function as a cost function (in U.S. $) C_t is expressed as

$$C_t = C_a + C_e + C_f + C_p \tag{9.24}$$

where
    C_a is the active material cost (lamination, copper, PMs)
    C_e is the cost of energy loss during the expected life of PMSG for certain average-daily duty cycle
    C_f is the frame, bearings, shaft, etc., costs
    C_p is the penalty function cost

The cost of energy loss C_e (here only for the generator) is

$$C_e \approx P_e\left(1 - \frac{1}{\eta_n}\right)n_{hy}n_y p_e \tag{9.25}$$

where
    n_{hy} is the hours/year
    n_y is the year
    p_e is the energy price USD/kWh
    P_e and η_n are equivalent average power and efficiency

The energy loss cost C_e proves to be dominant in the objective function. It is thus feasible to modify the energy loss cost C_e as

$$C_e = n_{hy}n_y p_e P_i; \quad P_i = \begin{cases} P_n\left(\dfrac{1}{\eta_0} - \dfrac{1}{\eta_n}\right) & \text{if } \eta_n < \eta_0 \\ 0 & \text{if } \eta_n \geq \eta_0 \end{cases} \tag{9.26}$$

when η_0 is the lower limit of the targeted efficiency.

In this way, the design is more flexible in computing cost versus efficiency while selecting apparently the best cost solution above a certain efficiency level.

The frame cost C_f is $C_f \approx p_f m_a$, where

- m_a—total mass of active materials,
- p_f—average price of frame and auxiliary components/kg.

The penalty component C_p deals mainly with overtemperature T_{max}:

$$C_{Temp} = \begin{cases} k_T (T - T_{max}) C_i; & \text{if } T > T_{max} \\ 0 & \text{if } T < T_{max} \end{cases} \quad (9.27)$$

where $C_i = C_a + C_f$ and k_T is the penalty coefficient.

Penalty components for PM demagnetization may be added for most critical situations. In addition to the initial cost, energy cost, cost of PWM converter (per KVA), and the cost of converter losses may be added if so desired. To simplify the case, these additional terms are not included in the study reported here.

The evolution of C_t during the optimization design process and power losses, efficiency, weight, and material cost components are all illustrated in Figures 9.19 and 9.20 for an 8 MW, 480 rpm PMSG.

Despite the rather large number of iterations, the total computational time was less than 100 s on a standard dual-core 2.4 GHz desktop computer. Even with 20 of such runs, from randomly different initial variable vectors, to better secure a global optimum, the computational time would be less than 2000 s (around 33 min). Finite element method (FEM) is embedded in the optimization algorithms after each entire optimization run, in a few points, in order to check the average torque T_e and inductances L_d and L_q and to correct the latter's analytical expressions, by under relaxation fudge factors until sufficient convergence is met. This would increase the computational time by up to 30 times, meaning 6 h on a standard contemporary desktop computer. The same optimal design methodology is applied for the same power, 8 MW, but at 1500 rpm, 3.6 kV. Selected comparative results are shown in Table 9.4.

The initial cost, the PM, and the total weight are smaller for the 1500 rpm generator, but the efficiency is a bit lower because the mechanical losses are notably higher than the stator electrical losses at 1500 rpm.

For this case, the efficiency, cost, and weight of the mechanical transmission system for the two speeds are not introduced in the optimal design.

9.4.3 CIRCUIT MODELING AND CONTROL OF PMSG

The *dq* model is suitable for PMSG investigation for transients and its control:

$$\bar{i}_s R_s - \bar{V}_s = -\frac{d\bar{\psi}_s}{dt} - j\omega_r \bar{\psi}_s; \quad \bar{\psi}_s = \psi_d + j\psi_q$$

$$\psi_d = L_d i_d + \psi_{PM}; \quad \psi_q = L_q i_q; \quad \bar{i}_s = i_d + j i_q \quad (9.28)$$

$$T_e = \frac{3}{2} p_1 (\psi_d i_q - \psi_q i_d) < 0 \quad \text{for generating}$$

where
  V_s, ψ_s, and i_s are voltage, flux linkage, and current space vectors, respectively, with their *dq* components
  T_e is the electromagnetic torque

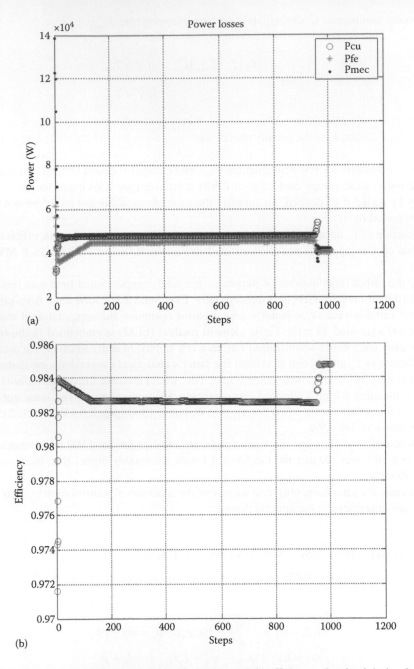

FIGURE 9.19 Evolution of (a) losses during optimization and (b) efficiency of optimal design for an 8 MW, 480 rpm PMSG (Pcu, Pfe, and Pmec are copper, iron, and mechanical losses, respectively).

For surface-mounted PMSGs, $L_d = L_q = L_s$ and, in general of maximum torque per losses, $i_d = 0$ and thus pure i_q control is adequate (no reluctance torque is available anyway). Pure i_q control leads to an increase in reactive power Q_e, which is

$$P_e = \frac{3}{2}V_q i_q < 0; \quad Q_e = \frac{3}{2}\omega_1 L_q I_q^2 > 0; \quad i_d = 0 \tag{9.29}$$

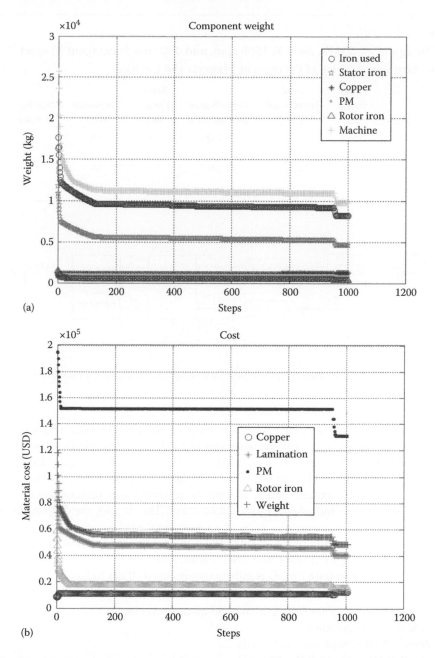

(a)

(b)

FIGURE 9.20 Evolution of (a) weight and (b) material cost for different components during optimization design for an 8 MW, 480 rpm PMSG.

This leads automatically to an "underexcited" operation mode (generated voltage is higher than the terminal voltage), which is favorable for the voltage-boosting operation of the machine-side PWM voltage source converter, with V_{DC} being the DC-link voltage (the DC-link voltage is decided by the grid voltage) in steady state:

$$V_d = -\omega_1 L_q i_q; \quad V_q = \omega_r \psi_{PM} + i_q R_s; \quad V_s = \sqrt{V_d^2 + V_q^2}; \quad V_s < \frac{\pi}{3\sqrt{3}} V_{DC} \qquad (9.30)$$

TABLE 9.4

Optimized Design of an 8 MW, 3600 V, 1500 rpm, and 480 rpm Permanent Magnet Synchronous Generator (Price of Permanent Magnet: 150 USD/kg)

Generator	Efficiency	Total Weight (kg)	Total Initial Cost (USD)	Outer Stator Diameter (m)	Stack Length (m)	Frequency (Hz)	Number of Poles	PM Weight (kg)
8 MW, 1500 rpm	0.98296	8560	197,027	1.252	3.5	75	6	629
8 MW, 480 rpm	0.98475	9799	249,376	2.177	3	56	14	875

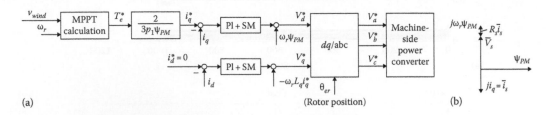

FIGURE 9.21 Generic vector control of PMSG: (a) control diagram and (b) vector orientation.

Neglecting R_s in (9.31), it follows that V_s increases proportionally to the speed and less than with load (I_q). At maximum speed ω_{rmax}, still, the inequality (9.31) has to be fulfilled to allow the control of the machine-side PWM voltage source converter. With ω_{rmax}, and P_n known, the design methodology is rather straightforward [4], and its details are skipped here, but they have been included in the optimal design code discussed earlier. To limit the reactive power requirement (the voltage-boosting ratio), the machine inductance L_s should be small. This is why the distributed winding was chosen.

With $i_d^* = 0$, the generic control of the PMSG resides in knowing the reference torque T_e^* versus wind or generator speed (v_{wind}, ω_r) as for the CRIG (Figure 9.21).

A practical control system for a PMSG is rather complicated as it has to handle

- Bidirectional power transfer from grid to stand-alone operation
- Motion-sensorless operation
- Asymmetric deep voltage sags and swells
- Filtering of output voltage at the grid side

When a wind park is controlled, participant PMSGs may be controlled in series or parallel on a common DC voltage bus through their machine-side converter. From there, a single (larger) inverter interfaces the local AC power grid. This latter aspect is beyond our scope here.

A typical PMSG sensorless vector control system for the machine-side converter with bidirectional stand-alone–grid modes is shown in Figure 9.22 [21]. Voltage harmonics caused due to non-linear loads such as a diode bridge rectifier can be filtered out. Single-phase voltage sags cause DC-link voltage pulsations that allow rather asymmetric PMSG stator currents. A smooth transition from stand-alone to grid mode is possible in this machine.

More details on PMSG design and control can be seen in [1, 4].

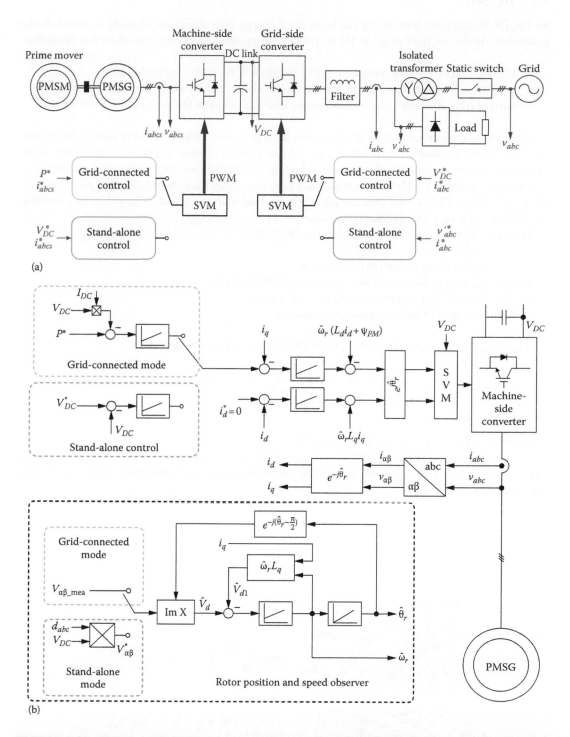

FIGURE 9.22 PMSG side converter with (a) grid and stand-alone operation; (b) rotor position and speed observer. (From Tutelea, L. and Boldea, I., *2010 IEEE International Symposium on Industrial Electronics*, 2010, pp. 1504–1509, Bari, Italy.)

9.5 DCE-SG

So far, DCE-SGs have been applied in large wind energy conversion units mainly as direct-driven generators. However, their usage in 1G or 3(4)G transmission drives may also show merits such as

- Elimination of PMs (or reduction of PM weight/kW by hybrid excitation [8])
- Opportunity to use only a diode rectifier in the machine side, where the DC overexcitation current controller alone provides the power flow to a common DC voltage bus of a wind park

Although the superconducting SC–SGs have been proposed for up to 10 MW (direct driven) [8], they may be more suitable for 1G transmission drives in order to allow a reasonable frequency in the generator and the power converter. Superconducting SC–SCGs have been built with iron-back core (operating frequency 2.2 Hz) and with air-back core (operating frequency 0.833 Hz), giving an efficiency of 96.4%. The SC–SG does not allow any no-load voltage control. It may thus be controlled as a PMSG with very low internal inductance.

In what follows, we will treat, in short, the DCE-SG:

- The phase circuit model for steady state
- Optimal design for unity power factor for 8 MW, 3.6 kV generator
- The *dq* model for transient simulation and analysis
- Vector control of the machine-side converter

The steady-state phase circuit model of a DCE-SG is rather standard:

$$i_s^* R_s + \underline{V}_s = -j\omega_r \underline{\psi}_s; \quad \underline{\psi}_s = L_{dm}\underline{i}_F + L_d\underline{i}_d + L_q\underline{i}_q$$
$$T_e = 3p_1\left(L_d i_F + \left(L_d - L_q\right)i_d\right)i_q,$$
(9.31)

with a phasor diagram as shown in Figure 9.23.

For simplicity, the additional PMs between rotor poles, as shown in Figure 9.23a, are not considered in (9.32). As the equations are written for the generator mode, the active power and torque are positive:

$$P_e \approx 3\frac{\omega_1}{p_1}T_e > 0; \quad Q_e = -3\omega_1 L_{dm}i_F i_d - \frac{3}{2}\omega_1\left(L_d i_d^2 + L_q i_q^2\right)$$
(9.32)

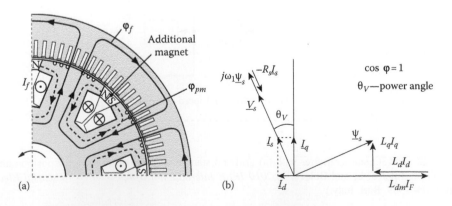

(a) (b)

FIGURE 9.23 DCE-SG: (a) topology layout and (b) phasor diagram.

When $Q_e > 0$, the machine may "deliver" reactive power ($i_d < 0$), while $Q_e < 0$ means "absorbing" reactive power. Now operating at about unity power factor (in the case of the diode rectifier machine interface, the fundamental power factor is unity) is rather simple, but the reluctance torque is negative as $L_d - L_q > 0$ and $i_d < 0$ ($Q_e = 0$), while $i_q > 0$ for the generating mode. The bottleneck in the design is to provide sufficient voltage under load at minimum speed to deliver the reference power at that speed. So the highest excitation current is required at lowest speed, and thus, the design methodology has to carefully follow this aspect. If a PWM voltage source converter (rectifier) is used on the machine stator side, this constraint is not so severe by using some voltage boosting, corroborated with excitation current rise at low speeds, so the final solution may lead to a higher efficiency system (excitation losses in the generator are reduced).

If the speed range is small (±10%), the generator excitation losses to cover the power control for a diode rectified output may be acceptable and result in a lower-cost solution. However, the motoring is not possible.

9.5.1 Optimal Design of DCE-SG

The general design methodology of the DCE-SG model is rather similar to that of standard SGs [4], and it contains

- *The specifications* (short-circuit ratio, transient inductance, reactive–active power capability and power factor, excitation system and its voltage ceiling, voltage and frequency [speed] control range, negative sequence voltages and currents, harmonics of generator at no-load voltages, temperature ratings, start–stop cycles, forces, armature voltage, runaway speed)
- *Design issues* (output coefficient and base stator geometry; number of stator slots; design of stator winding; design of stator core; salient pole rotor design; open-circuit saturation curve; field current at full load and lowest assigned power factor; stator resistance; leakage inductance; and synchronous inductance L_d, L_q; calculated losses and efficiency; time constants and subtransient (transient) inductance; cooling system and thermal design; design of brushes and slip rings [if any], design of bearings, frame, etc.; brakes and jacks design; exciter design)

The SG design heritage is incorporated into an optimal design of a 7.6 MW, 3.6 kV, $n = 11$ rpm wind generator [4], based on modified Hooke–Jeeves method [20]. A parameter variable is left to choose the number of pole pairs p, which defines the stator frequency:

$$f_s = p\,f_n \qquad\qquad (9.33)$$

A higher number of poles lead to a better efficiency, and lower weight up to a point, but, at the cost of a large outer diameter of the stator, can be expected, as shown in Figure 9.24.

From Figure 9.24, more illustrative data are selected, and they are shown in Table 9.5.

The result for the fifth optimal design case (180 poles, 12 m outer stator diameter) in Table 9.5 is followed in detail.

A few cost function components are considered:

- Active material cost
- Overtemperature penalty
- Energy loss
- Initial generator cost
- Total cost function

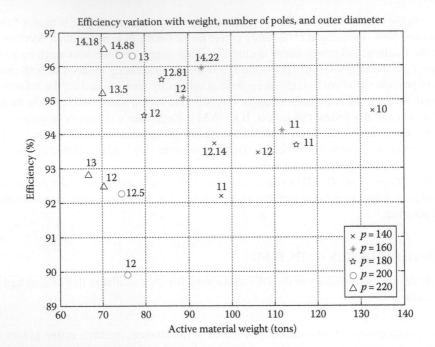

FIGURE 9.24 Rated efficiency of DCE-SG (7.6 MW, 11 rpm) versus active material weight (tons) at various pole pairs and thus various outer stator diameters.

TABLE 9.5

Design of a 7.6 MW, 11 rpm, DC-Excited Synchronous Generator

No.	Poles	Outer Diameter (m)	Efficiency (%)	Weight (tons)	Comments
1	220	14.18	96.53	70.67	Better efficiency
2	220	13	92.81	66.81	Lowest weight
3	140	10	94.68	132.9	Lowest diameter
4	200	13	96.32	77.02	Optimum 1
5	180	12	94.54	80.0	Optimum 2

The evolution of the cost function and its components (similar to the definitions for those of the PMSG) is shown in Figure 9.25.

As shown in Figure 9.25, it should be noticed that single-composite cost function decreases early by 1000 times, mainly due to the overtemperature penalty cost sudden reduction in the first computation steps. After this stage, the reduction of composite cost function is rather small and 25 iterations seem enough to produce an optimal design. The electric and magnetic loss evolution (Figure 9.26) shows a mild reduction as initial cost is an important component of the composite cost. The mechanical losses were assigned 2.92 kW (due to small peripheral speed and notably large air gap).

The total computational time was around 70 s on a "core 2 duo" 2.4 GHz desk-top computer. With 20 runs from randomly different initial variable vectors, to better secure a global optimum, the computational time would be 1400 s. Now with FEM embedded for torque and inductances computation, the computational time would be raised by about 30 times to 25 h on a standard contemporary desktop computer.

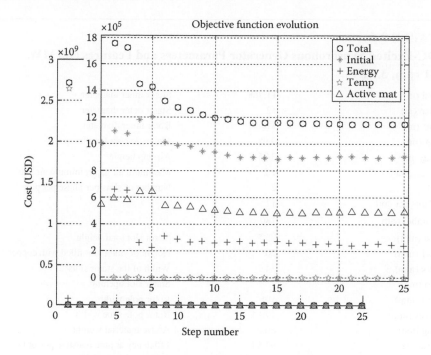

FIGURE 9.25 Cost function and its component evolution for the optimal design of DCE-SG 7.6 MW of 11 rpm.

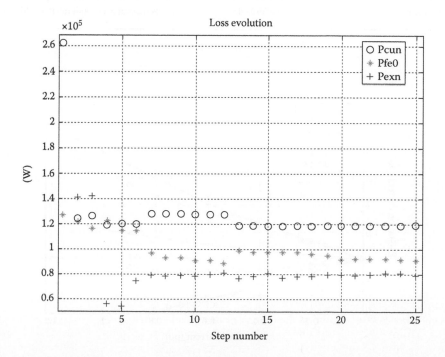

FIGURE 9.26 Loss evolution for a 7.6 MW, 11 rpm, cos φ = 1, DCE-SG optimal design (Pcun, Pfe0, and Pexn are copper, iron, and excitation losses, respectively).

TABLE 9.6

DC-Excited Synchronous Generator Parameters and Features (7.6 MW, 11 rpm, 3.6 kV)

Dsi (m)	11.768	Stator inner diameter
Dso (m)	12	Stator outer diameter
Dri (m)	11.491	Rotor inner diameter
Lg (m)	0.98	Magnetic core length
hag (mm)	3.25	Air-gap height
hag_{max}/hag	1.2	Maximum air gap/minimum air gap
q1	2	Number of slots per phase per pole
Turns per coil	12	
Parallel path	180	
hst (mm)	25.5	Stator slot total height
hs1 (mm)	20	Height of stator slot filled with copper
Ws (mm)	14.7	Width of the stator slots
So (mm)	7.	Stator slot opening
wrt (mm)	105	Rotor pole width
wrps (mm)	170.4	Rotor pole shoe width
mg (ton)	80.0	Active material weight
Etan (%)	94.54	Efficiency at pure resistive power factor
Pcun (kW)	237.61	Stator copper losses
Pex0n (kW)	34.68	Excitation, load losses
Pexn (kW)	84.13	Excitation losses at rated resistive load
Pfen (kW)	96.69	Iron-rated losses
Stator linkage flux (Wb)	2.96/2.79	Analytical/FEM
Xd (pu)	0.878/0.824	d-axis reactance per unit (analytical/fem)
Xq (pu)	0.586/0.443	q-axis reactance per unit (analytical/fem)

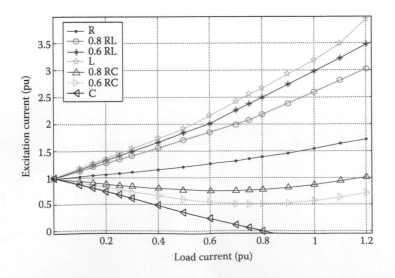

FIGURE 9.27 Excitation current versus load current of DCE-SG (R, L, and C are resistive, inductive, and capacitive loads, and RL and RC are resistive inductive and resistive capacitive loads).

Note: The multiobjective optimal design of PMSG or DCE-SG could be attempted by multiobjective evolutionary, differential evolutionary, and optimization methodology where FEM machine modeling is exclusively used but only in 10–12 points per electrical period to reduce the computational time. The computational time is higher (at least 50 h on a strong multi-desk computer crew), but much more information on the design and its robustness may be collected [18, 19].

More data on the optimized 7.6 MW, 11 rpm DCE-SG design are summarized in Table 9.6 and in Figures 9.27 through 9.30.

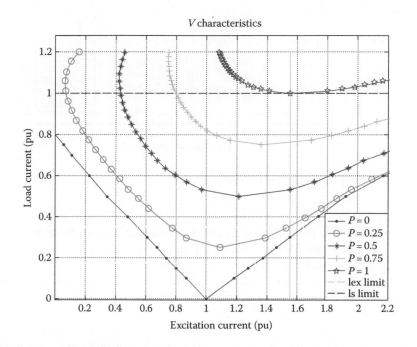

FIGURE 9.28 V curves DCE-SG.

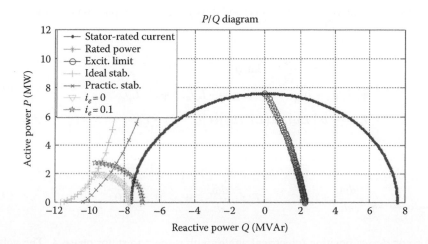

FIGURE 9.29 Active–reactive power capabilities of DCE-SG.

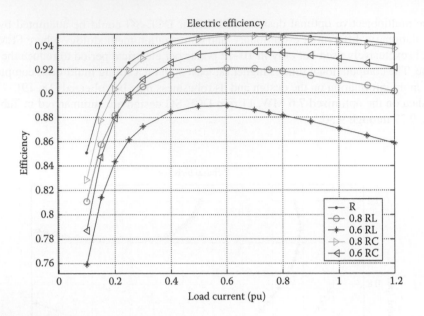

FIGURE 9.30 Efficiency versus load of DCE-SG.

The compromise between the initial cost and the loss cost leads to a moderate efficiency (94.5%) at unity power factor load. Besides, the design for unity power factor leads to moderate active–reactive power capabilities; see the curve from Figure 9.29.

Some key 2D FEM validation simulation checks have been run for the optimal analytical design to check the air-gap flux linkage and flux distribution at full load (i_F, i_d, i_q given) to check the torque realization in a saturated core machine (Figure 9.31).

With all circuit parameters, R_s, R_F, L_F, L_{dm}, L_{qm}, and L_{sl}, calculated analytically and validated or directly calculated from FEM, the *dq* model for exploring the transients and control may be used directly.

9.5.2 Active Flux–Based Sensorless Vector Control of DCE-SG

For simplicity, let us ignore the rotor cage (if any) and thus write the *dq* (space phasor) model of the DCE-SG in rotor coordinates as [4]

$$
\bar{i}_s R_s - \bar{V}_s = -\frac{\partial \overline{\psi}_s}{\partial t} - j\omega_r \overline{\psi}_s
$$
$$
\overline{\psi}_s = \psi_d + j\psi_q; \quad \psi_d = L_{dm}i_F + L_d i_d; \quad \psi_q = L_q i_q \tag{9.34}
$$
$$
T_{eg} = \frac{3}{2} p_1 \left[L_{dm}i_F + \left(L_d - L_q \right)i_d \right] i_q; \quad L_d > L_q
$$

where T_{eg} is negative for generating mode ($i_q < 0$), so the motion equations are

$$
\frac{J}{p_1} \cdot \frac{d\omega_r}{dt} = T_{prim\,mover} - T_{eg}; \quad \frac{d\omega_r}{dt} = \theta_{er}; \quad \theta_{er} = p_1 \theta_r \tag{9.35}
$$

where θ_{er} is the rotor *d*-axes position (electrical angle) with respect to stator phase *A*. The voltage and current phasor are

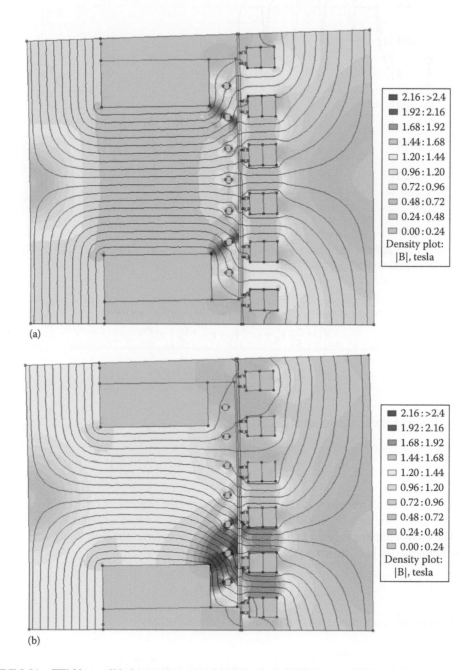

(a)

(b)

FIGURE 9.31 FEM key validation on the optimal design of a 7.6 MW, 11 rpm DCE-SG: (a) field distribution at no load, (b) field distribution at full load cos φ = 1. (*Continued*)

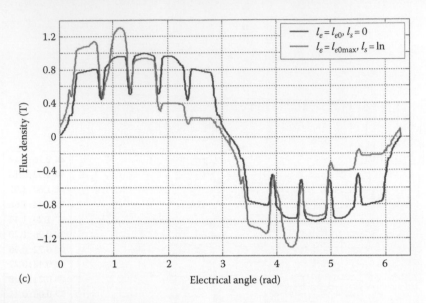

(c)

FIGURE 9.31 (*Continued*)   FEM key validation on optimal design of 7.6 MW, 11 rpm DCE-SG: (c) radial air-gap flux density (at no load and full load).

$$\overline{V}_s = \frac{2}{3}\left(V_a + e^{j\frac{2\pi}{3}} V_b + e^{-j\frac{2\pi}{3}} V_c \right) \cdot e^{-j\theta_{er}}$$

$$\overline{I}_s = \frac{2}{3}\left(I_a + e^{j\frac{2\pi}{3}} I_b + e^{-j\frac{2\pi}{3}} I_c \right) \cdot e^{-j\theta_{er}}$$

(9.36)

Finally, the field circuit equations are

$$i_F^s R_F - V_F^s = -\frac{d\psi_F^s}{dt}; \quad \psi_F^s = L_{lF}^s i_F + L_{dm}\left(i_d + i_F^s \right)$$

(9.37)

$V_F^s, R_F^s, L_{lF}^s, i_F^s, \psi_F^s$ are all reduced to the stator by a transformation (equivalent turns ratio) coefficient k_F.

Magnetic saturation may be modeled mainly by L_{dm} and L_{qm}, which depend on $i_{dm} = i_d + i_F^s$ and $i_{qm} = i_q$. The cross-coupling saturation has to be carefully considered for both steady-state and transient operation modes [4]. The active flux $\overline{\psi}_d^a$ is [23]

$$\overline{\psi}_d^a = \overline{\psi}_s - L_q \overline{i}_s$$

(9.38)

It turns to be the multiplier to of i_q in the torque expression [23]

$$\overline{\psi}_d^a = \left[L_{dm} i_F - \left(L_d - L_q \right) \right] i_d; \quad T_e = \frac{3}{2} p_1 \left| \overline{\psi}_d^a \right| i_q$$

(9.39)

But this flux linkage is aligned along the d-axis of the rotor, and thus, its position is θ_{er} and its speed $\hat{\omega}_{\psi_d} = \hat{\omega}_r$ at all loads. Cross-coupling saturation introduces an error to θ_{er}, when estimated, which may be handled by adding a correction angle $\Delta\theta_{er}$. For unity power factor, the steady-state equations and the vector diagram (Figure 9.32) are simplified to

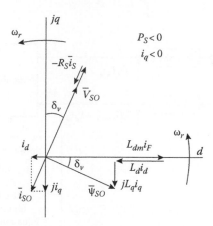

FIGURE 9.32 Vector diagram of DCE-SG at cos φ = 1 (generator mode).

$$i_{so}R_s + V_{so} = \omega_r\psi_{so}; \quad T_e = \frac{3}{2}p_1\left|\overline{\psi}_d^a\right|i_q = -\frac{3}{2}\psi_s i_s; \quad i_q < 0 \tag{9.40}$$

Figure 9.32 illustrates (9.41).

From (9.41) and Figure 9.32 neglecting R_s ($R_s = 0$), the following can be obtained:

$$\sin(\delta_v) = \frac{L_q i_q}{\psi_s}; \quad \psi_s \approx \frac{V_{so}}{\omega_r} = \sqrt{\left(L_{dm}i_{Fo} + L_d i_d\right)^2 + L_q^2 i_q^2} \tag{9.41}$$

$$\tan(\delta_v) = \frac{i_{do}^s}{i_{qo}}; \quad (i_d < 0, i_q < 0) \quad T_e = \frac{3}{2}p_1\psi_{so}i_{so}; \quad i_{so} = \sqrt{i_{do}^2 + i_{qo}^2} \tag{9.42}$$

From (9.42) and (9.43), for given $\omega_1^*, V_{so}^*, T_e^*$, first, ψ_{so}^*, then i_{so}^*, then i_d^*, i_q^*, and then finally, i_F^*, for unity power factor (cos $\varphi_1 = 1$), can be calculated. The reference value i_F^* for cos $\varphi_1 = 1$ may be corrected to parameter detuning by a correction loop based on the power factor angle error ($\varphi_1^* - \varphi_1$), in steady state ($\varphi_1^* = 0$). Now the thing missing for vector control is only the rotor position θ_{er} and speed ω_r^*. A stator flux $\widehat{\psi}_s$ estimator based on a voltage model may be used as an operation when very low speed is not needed (if it is, $V_{comp} \neq 0$, then, a combined voltage and current model estimator will do like (9.43) through (9.45)).

$$\widehat{\psi}_s = \int\left(\overline{V}_s - R_s\overline{i}_s + \overline{V}_{comp}\right)dt; \quad \widehat{\psi}_d^a = \widehat{\psi}_s - L_q\overline{i}_s \tag{9.43}$$

In the stator coordinates,

$$\widehat{\psi}_d^a = \psi_{d\alpha}^a + j\psi_{d\beta}^a \tag{9.44}$$

$$\widehat{\theta}_{er} = \tan^{-1}\left(\frac{\psi_{d\beta}}{\psi_{d\alpha}}\right) \tag{9.45}$$

A PLL observer is used to get the refinement values of $\widehat{\theta}_{er}$ and $\widehat{\omega}_r$. Finally, a vector control system as shown in Figure 9.33 may be obtained.

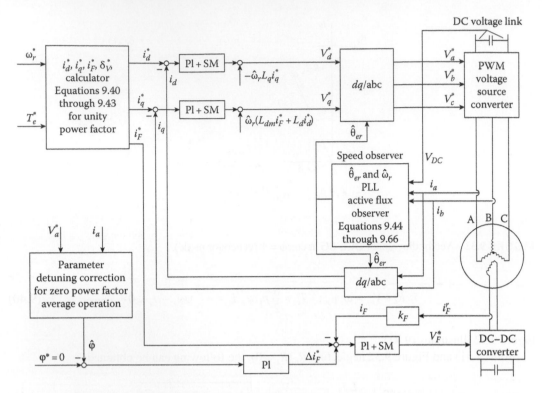

FIGURE 9.33 Generic unity power factor vector sensorless control of DCE-SG with active flux–based rotor position and speed observer (PI+SM means proportional integral and sliding mode controller).

The typical operation of such a drive in the motoring mode sensorless acceleration under full load is shown in Figure 9.33 [24]. The drive can operate in four quadrants. Though a few companies have a few DCE-SGs with WTs up to around 8 MW at 11 rpm in operation, no deep investigation into their control is available up to now.

As it can be inferred from the optimal design and from the control paragraph, only unity power factor control was attempted. For a given DC voltage power bus, the inverter always needs a bit of voltage boosting, so operation close to unity power factor (a bit lagging) is required. Alternatively, if a diode rectifier is used, unity power factor (for the fundamental components) is almost implicit, and then only the field-winding converter should be controlled, based on required speed ω_r^*, torque T_e^*, and (power P_e^*) desired for MPPT. So, the burden to "produce" reactive power resides in the converter with the DC-link capacitor. Alternatively, a dedicated active parallel power filter capable of "producing" reactive power and filter the harmonics in the AC power grid may be added. More work is envisaged in this field, especially with the maturing of HVDC power transmission lines for transmitting large power from wind farms.

9.6 MODELING OF ELECTRIC GENERATORS BY FINITE ELEMENT ANALYSIS (FEA)

As previously mentioned and illustrated in this chapter, finite element analysis (FEA) is employed for the study of the electromagnetic field in electrical machines, such as the generators used in wind turbines. This approach is required due to the detailed geometrical features of the devices and the nonlinear characteristics related to the ferromagnetic materials. Figure 9.34 illustrates the main steps of typical FEA for a PMSG example with 12 poles mounted on the rotor surface and a stator employing a winding distributed in a two slots per pole and phase configuration.

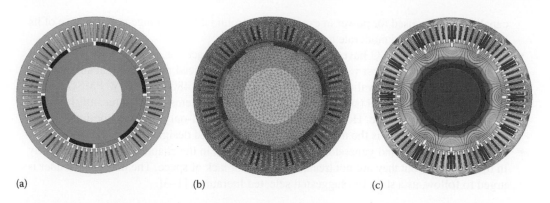

(a)   (b)   (c)

FIGURE 9.34 Example of finite element analysis (FEA) for a PMSG using the ANSYS Maxwell® software: (a) geometric model showing the alternating north and south rotor magnetic poles and the stator winding phase distribution, red–yellow–blue, (b) FE mesh with triangular elements, and (c) magnetic field plot including flux lines and the colored map of flux density.

In a first step of the analysis, a geometrical model is created. A parametric approach, in which the main independent design variables, such as magnet length in the direction of magnetization and pole arc, the stator tooth width, back iron width, etc., are parametrized, is highly recommended as the model can then be used for a variety of design studies, as well as for automated optimization. An FE mesh, typically a triangulation of first or multiple order elements, is generated to approximate, as closely as possible, the geometry of the computational domain, which typically for the radial flux machines, is represented by the 2D cross section, while special topologies, such as axial flux configurations, may require 3D analysis.

The electromagnetic FEA yields the field distribution, as illustrated in Figure 9.34 for the magnetic flux lines and the colored map of the magnetic flux density. For clarity, the figure depicts the entire cross section of the generator, but in order to save resources, the actual computations can be performed on the smallest domain that has periodical repeatability, which in the case of the considered example is one pole region of the electric machine. More details on FEA and its practical applications are available, for example, in the ANSYS Maxwell® software documentation [29].

For PMSG, based on the electromagnetic field results from FE simulations in the *abc* reference frame, which consider through time-stepping the variation of the stator phase currents and the rotor movement, the torque in the air gap can be calculated, as well as the losses in the ferromagnetic core and in the PM. Equivalent circuit parameters for the *dq* model, namely, back emf and inductances, can be derived from such *abc* analysis, or through other FE-based approximations, for use in simulations of drive system including the controllers. Coupled simulations of power electronics circuits and electric machine fields are also possible, but the use of extremely small time steps in order to account for high PWM switching frequencies requires substantial computational resources and therefore is typically employed in the last stages of design verification.

9.7 SUMMARY

- The present chapter has introduced in notable detail the main up-to-date commercial variable speed high power wind generators (up to 8 MW) in terms of the modeling, performance, optimal design, and control of the machine-side converter.
- The DFIG was treated first, with a numerical example, since it covers around 50% of all the more than 350 GW in operation in the year 2015.
- The CRIG was given a rather detailed treatment in terms of modeling, design performance, and control, and it requires a full power AC–DC–AC PWM converter interface, which is

capable of being used for power in excess of the around 2.5 MW range unit, because of its higher robustness and moderate total system costs.

- The PMSG and DCE-SG have been given even more space for an optimal design, based on a single-composite cost (objective) function and advanced analytical machine models with only trial key FEM validations. The case studies on a 480 rpm 8 MW PMSG and an 11 rpm, 7.6 MW DCE-SG (similar to the largest power systems installed recently) showed good overall performance. However, a multiobjective FEM-only modeling optimal design based on DE algorithm is the next logic step for a finalized design.
- New topologies of wind generators are also mentioned in the chapter, introduced recently in the literature, but they are not treated in detail for lack of space. The interested reader is urged to follow, as a start, the suggested selected literatures [1–3].

REFERENCES

1. B. Wu, Y. Lang, N. Zargari, and S. Kouro, *Power Conversion and Control of Wind Energy Systems*, Wiley-IEEE Press eBook Chapters, Hoboken, NJ, 2011.
2. G. Abad, J. Lopez, M. Rodriguez, L. Marroyo, and G. Iwanski, *Doubly Fed Induction Machine: Modeling and Control for Wind Energy Generation*, Wiley, Oxford, U.K., 2011.
3. M.G. Simoes and F.A. Farret, *Renewable Energy Systems: Design and Analysis with Induction Generators*, CRC Press/Taylor & Francis Group, New York, 2004.
4. I. Boldea, *Electric Generators Handbook*, vols. 1 and 2, CRC Press/Taylor & Francis Group, New York, 2016.
5. D. Allaei, Y. Andreopoulos, and A. M. Sadegh, New wind power technology: INVELOX, McGraw-Hill, *Yearbook of Science and Technology*, 2015, pp. 203–207.
6. Vestas Wind System A/S, V164 8MW, 10/2012-EN, Available: http://nozebra.ipapercms.dk/Vestas/Communication/Productbrochure/V16480MW/V16480MW/, accessed November 2016.
7. Enercon Gmbh, E-126, 2016, Available: http://www.enercon.de/en/products/ep-8/e-126/, accessed November 2016.
8. K. Ma, L. Tutelea, I. Boldea, D. Ionel, and F. Blabjerg, Power electronic drives, controls, and electric generators for large wind turbines-an overview, *EPCS*, 43, 1406–1421, 2015.
9. H. Polinder, F.F.A.v.d. Pijl, G.J.d. Vilder, and P. Tavner, Comparison of direct-drive and geared generator concepts for wind turbines, in *IEEE International Conference on Electric Machines and Drives, 2005*, pp. 543–550, 2005.
10. H. Polinder, D. Bang, R.P.J.O.M.v. Rooij, A.S. McDonald, and M.A. Mueller, 10 MW wind turbine direct-drive generator design with pitch or active speed stall control, in *2007 IEEE International Electric Machines and Drives Conference*, pp. 1390–1395, 2007.
11. L. Hui and C. Zhe, Design optimization and evaluation of different wind generator systems, in *International Conference on Electrical Machines and Systems, 2008 (ICEMS 2008)*, pp. 2396–2401, 2008.
12. Vestas Wind Systems A/S, 12/05 UK, V120-4.5 MW, Available: www.nrg-systems.hu/dok/en/V120_UK.pdf, accessed November 2016.
13. V.D. Colli, F. Marignetti, and C. Attaianese, Analytical and multiphysics approach to the optimal design of a 10-MW DFIG for direct-drive wind turbines, *IEEE Transactions on Industrial Electronics*, 59, 2791–2799, 2012.
14. R.A. McMahon, P.C. Roberts, X. Wang, and P.J. Tavner, Performance of BDFM as generator and motor, *IEE Proceedings: Electric Power Applications*, 153, 289–299, 2006.
15. L. Mihet-Popa and I. Boldea, Variable speed wind turbines using induction generators connected to the grid: Digital simulation versus test results, in *Rec. of IEEE OPTIM 2006 (IEEExplore)*, pp. 287–294, Brasov, Romania, 2006.
16. Siemens, The right Siemens wind turbine for all conditions, Available: http://www.siemens.com/global/en/home/markets/wind/turbines.html, accessed November 2016.
17. I. Boldea and S.A. Nasar, *Electric Drives*, CRC Press/Taylor & Francis Group, New York, 2016.
18. I. Boldea and L. Tutelea, *Electric Machines*, CRC Press/Taylor & Francis Group, New York, 2010.
19. B. Novakovic, Y. Duan, M. Solveson, A. Nasiri, and D.M. Ionel, Comprehensive modeling of turbine systems from wind to electric grid, in *2013 IEEE Energy Conversion Congress and Exposition*, pp. 2627–2634, Denver, CO, 2013.

20. A. Fatemi, D.M. Ionel, N.A.O. Demerdash, and T.W. Nehl, Fast multi-objective CMODE-type optimization of electric machines for multicore desktop computers, in *2015 IEEE Energy Conversion Congress and Exposition (ECCE)*, pp. 5593–5600, 2015.

21. L. Tutelea and I. Boldea, Surface permanent magnet synchronous motor optimization design: Hooke Jeeves method versus genetic algorithms, in *2010 IEEE International Symposium on Industrial Electronics*, pp. 1504–1509, Bari, Italy, 2010.

22. M. Fatu, F. Blaabjerg, and I. Boldea, Grid to standalone transition motion-sensorless dual-inverter control of PMSG with asymmetrical grid voltage sags and harmonics filtering, *IEEE Transactions on Power Electronics*, 29, 3463–3472, 2014.

23. I. Boldea and S.C. Agarlita, The active flux concept for motion-sensorless unified AC drives: A review, in *International Aegean Conference on Electrical Machines and Power Electronics and Electromotion, Joint Conference*, pp. 1–16, Istanbul, Turkey, 2011.

24. I. Boldea, G.D. Andreescu, C. Rossi, A. Pilati, and D. Casadei, Active flux based motion-sensorless vector control of DC-excited synchronous machines, in *2009 IEEE Energy Conversion Congress and Exposition*, pp. 2496–2503, San Jose, CA, 2009.

25. S. Krohn (ed.), P.E. Morthorst, and S. Awerbuch, *The Economics of Wind Energy*, The European Wind Energy Association, 2009.

26. Havøygavlen Windmill Park, Norway, Power-Technology.com, Available: http://www.power-technology.com/projects/havoygavlen/havoygavlen5.html, accessed November 2016.

27. H. Zhong, L. Zhao, and X. Li, Design and analysis of a three-phase rotary transformer for doubly fed induction generators, *IEEE Transactions on Industry Applications*, 51, 2791–2796, 2015.

28. D. Ursu, Brushless DC multiphase reluctance machines and drives, PhD, University Politehnica Timisoara, Timişoara, Romania, 2014.

29. ANSYS, *Maxwell® 2D/3D Field Simulator, User's Manual*, Vol. 17, Pittsburgh, PA, 2016.

10 Design Considerations for Wind Turbine Systems

Marcelo Godoy Simões, S.M. Muyeen, and Ahmed Al-Durra

CONTENTS

Abstract...251
10.1 Introduction..251
10.2 Wind Turbine Components..252
10.3 Wind Turbine Characteristics..254
10.4 Wind Energy Generator Systems...256
    10.4.1 Fixed Speed Wind Turbines..257
    10.4.2 Variable Speed Wind Turbines..257
10.5 Wind Speed And Generator Performance...259
10.6 Wind Turbine Controls 261
    10.6.1 Mechanical Control...262
    10.6.2 Electrical Control..263
10.7 Summary...264
References...264
Further Reading 265

ABSTRACT

In this chapter, the fundamentals of wind energy conversion system including wind turbine components, drive train, wind turbine characteristics, and wind generators are explained at the beginning. Then the design considerations of a small wind turbine generator system such as wind speed probability distribution, component losses, system efficiency, etc., are explained in detail. Different types of wind turbine generator system topologies including fixed and variable speed systems are presented. Finally, mechanical, electrical, and overall control schemes of wind energy conversion system are demonstrated.

10.1 INTRODUCTION

The wind energy conversion has some simple foundations: the kinetic energy of the wind is converted into mechanical shaft movement with a turbine, then it is converted into electrical energy with a coupled electrical generator. Therefore, the wind propels the blades of a turbine, rotating a shaft connected to the generator rotor and in turn producing electricity.

The global installed wind power had a fairly linear increasing capacity from 6 GW in 1996 to nearly 121 GW in 2008. After that, the growth resembled more an exponential curve, because in 2014 it was estimated at a total worldwide power capacity of 367 GW. There are many potential applications for small-scale wind users—typically residential, rural, or commercial applications— either totally isolated or connected to the power grid. When a wind energy system is connected to the public grid, advanced smart inverters may provide low-voltage support and reactive power compensation if necessary. Small power user (prosumer) motivation is always based on income, in order to offset their initial investments, by either saving money with their own microgrid or selling power to the utility via net metering or a contract. Utilities on the other hand might be motivated in dealing with those small prosumers willing to postpone their grid reinforcement infrastructure investments or displacing reactive power required hardware or improving power factor in certain zones of the distribution grid. Therefore, it is very important to understand the technology and economic basis for small wind-based power systems.

In order to understand how to design a small wind power system, it is initially necessary to assess (on an hourly basis) the typical load flow in the installation and all sources of energy that must meet an average energy balance. Then, storage compensation may have to provide the instantaneous power balance, typically a battery is used for such instantaneous compensation or a connection to the utility grid. In parallel, other devices may have longer time needs, such as a photovoltaic power system, pumped hydro, compressed air, diesel generator, or fuel cell. There are several other parameters to consider: how the power quality impacts the overall performance, how the system protection requirement for islanding and reconnection is, how integrated must the battery and inverter controllers be, and how the electromechanical protections against strong and unsafe wind speeds are. It is necessary to understand how the performance is affected by local conditions, that is, the random nature of the wind, nearby obstructions, power-demand profiles, turbine-related factors, and deterioration due to aging.

The objective of this chapter is to provide some understanding of a few important topics for designing a small wind power system with a tutorial-like educational approach in the control and design characteristics for small wind energy projects. The topics discussed in this chapter are concerned with wind turbine components and their characteristics, typical wind turbine generators, and the overall wind energy design. There is an example of an energy study with understanding of the wind turbine control, explaining both mechanical and electrical aspects.

10.2   WIND TURBINE COMPONENTS

Wind turbines are categorized based on the orientation of their spin axis into horizontal-axis wind turbines (HAWT) and vertical-axis wind turbines (VAWT) [1]. VAWT have their rotor shaft transverse to the wind (but not necessarily vertical), and the main components are located at the base of the turbine; that is, the generator and gearbox are located close to the ground, facilitating service and repair. VAWT capture wind from any direction without being pointed towards the main wind direction. Typical designs such as *Savonius*, *Darrieus*, and *Giromill* have significant torque variation during each revolution causing bending moments on the blades with consequent mechanical stress and material fatigue. VAWT have been studied and deployed in the past few years, but the recent commercial success of wind turbines is really due to HAWT. Figure 10.1 shows the typical wind turbine components in a HAWT. There are three categories of components: mechanical, electrical, and control. The following is a brief description of the main components:

- *Tower* is the physical structure that holds the wind turbine. It supports the rotor, nacelle, blades, and other wind turbine equipment. Typical commercial wind towers are usually 50–120 m long and they are constructed from concrete or reinforced steel.
- *Blades* are physical structures, which are aerodynamically optimized to help capture the maximum power from the wind in normal operation with a wind speed in the range of about 3–15 m/s. Each blade is usually 20 m or more in length, depending on the power level.
- *Nacelle* is the enclosure of the wind turbine generator, gearbox, and internal equipment. It protects the turbine's internal components from the surrounding environment.
- *Rotor* is the rotating part of the wind turbine. It transfers the energy in the wind to the shaft. The rotor hub holds the wind turbine blades while connected to the gearbox via the low-speed shaft.
- *Pitch* is the mechanism of adjusting the angle of attack of the rotor blades. Blades are turned in their longitudinal axis to change the angle of attack according to the wind directions.

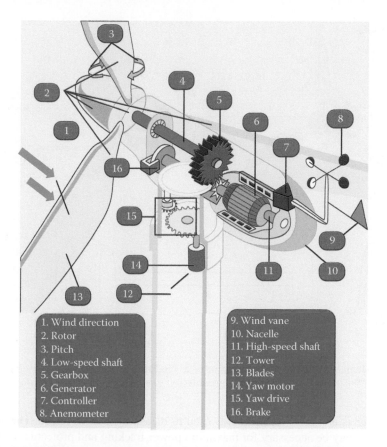

1. Wind direction
2. Rotor
3. Pitch
4. Low-speed shaft
5. Gearbox
6. Generator
7. Controller
8. Anemometer
9. Wind vane
10. Nacelle
11. High-speed shaft
12. Tower
13. Blades
14. Yaw motor
15. Yaw drive
16. Brake

FIGURE 10.1 Wind turbine components. (Adapted from National Renewable Energy Laboratory [NREL] and Bureau of Ocean Energy Management [BOEM], Wind turbine nacelle schematic, Offshore wind energy, Available: https://www.boem.gov/Offshore-Wind-Energy/, Retrieved on January 18, 2017.)

- *Shaft* is divided into two types: low and high speed. The low-speed shaft transfers mechanical energy from rotor to gearbox, while the high-speed shaft transfers mechanical energy from gearbox to generator.
- *Yaw* is the horizontal moving part of the turbine. It turns clockwise or anticlockwise to face the wind. The yaw has two main parts: the yaw motor and the yaw drive. The yaw drive keeps the rotor facing the wind when the wind direction varies. The yaw motor is used to move the yaw.
- *Brake* is a mechanical part connected to the high-speed shaft in order to reduce the rotational speed or stop the wind turbine overspeeding or during emergency conditions.
- *Gearbox* is a mechanical component that is used to increase or decrease the rotational speed. In wind turbines, the gearbox is used to control the rotational speed of the generator.
- *Generator* is the component that converts the mechanical energy from the rotor to electrical energy. The most common electrical generators used in wind turbines are induction generators (IGs), doubly fed induction generators (DFIGs), and permanent magnet synchronous generators (PMSGs).
- *Controller* is the brain of the wind turbine. It monitors constantly the condition of the wind turbine and controls the pitch and yaw systems to extract optimum power from the wind.

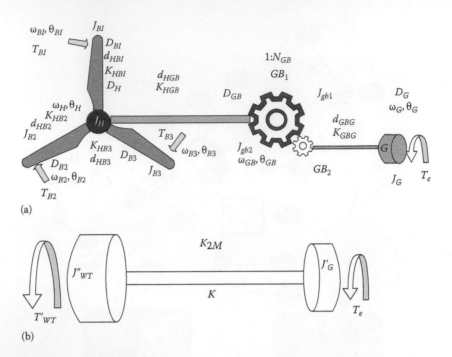

(a)

(b)

FIGURE 10.2 Mechanical model of a drivetrain of a wind turbine: (a) six-mass drivetrain, (b) two-mass equivalent model.

- *Anemometer* is a type of sensor that is used to measure the wind speed. The wind speed information may be necessary for maximum power tracking and protection in emergency cases.
- *Wind vane* is a type of sensor that is used to measure the wind direction. The wind direction information is important for the yaw control system to operate.

The drivetrain of a wind turbine is depicted in Figure 10.2a. It shows a six-mass drivetrain model with six inertias: three blade inertias ($JB1$, $JB2$, and $JB3$), hub inertia (JH), gearbox inertia (JGB), and generator inertia (JG). The angular positions for blades, hub, gearbox, and generator are represented by θ_{B1}, θ_{B2}, θ_{B3}, θ_{H}, θ_{GB}, and θ_{G}, while their corresponding angular speeds are ω_{B1}, ω_{B2}, ω_{B3}, ω_{H}, ω_{GB}, and ω_{G}. The elasticity between adjacent masses is expressed by their spring constants K_{HB1}, K_{HB2}, K_{HB3}, K_{HGB}, and K_{GBG} and the mutual damping between adjacent masses is expressed by d_{HB1}, d_{HB2}, d_{HB3}, d_{HGB}, and d_{GBG}. There are torque losses because of external damping elements of individual masses, represented by D_{B1}, D_{B2}, D_{B3}, D_H, D_{GB}, and D_G. This model requires a generator torque (T_E) and three individual aerodynamic torques acting on each blade (T_{B1}, T_{B2}, T_{B3}). The sum of the blade torques develops the turbine torque, T_{WT}. The aerodynamic torques acting on the hub and gearbox are assumed to be zero. Because the six-mass model is very complex, usually an equivalent two-mass model is used with equivalent shaft stiffness K_{2M}, that is, the masses of turbine and gearbox are lumped together into one mass, such simplified model is represented in Figure 10.2b.

10.3 WIND TURBINE CHARACTERISTICS

In order to understand the characteristics of wind energy conversion and how a turbine works, it is possible to make a simplified assumption that the wind approaches the turbine with constant velocity, with homogeneous properties (such as temperature, density), and without turbulence. Suppose

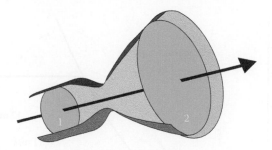

FIGURE 10.3   Control volume of wind as it travels through the turbine.

a perfect cylinder of wind arrives at the turbine (cross section 1), as indicated in Figure 10.3, and after passing through the turbine, there is an expansion of the volume (cross section 2). The turbine rotates, and a part of the energy is captured on the shaft so the wind will have a lower velocity just after crossing the turbine.

The fraction of power extracted from the wind by a real wind turbine can be defined by the symbol C_p, that is, the coefficient of performance or power coefficient. The actual mechanical power output P_m from a wind turbine is expressed in the following equation:

$$P_m = C_p \left(\frac{1}{2} \rho A v_w^3 \right) = \frac{1}{2} \rho \pi R^2 v_w^3 C_p (\lambda, \beta) \qquad (10.1)$$

where
 R is the blade radius of wind turbine (m)
 v_W is the wind speed (m/s)
 ρ is the air density (kg/m³)

The coefficient of performance varies with the wind speed, the rotational speed of the turbine, and turbine blade parameters, that is, blade pitch angle and angle of attack. Therefore, the power coefficient, C_p, is mainly a function of tip-speed ratio, λ, and blade pitch angle, β [deg]. The tip-speed ratio is defined in Equation 10.2 as:

$$\lambda = \frac{\omega_R R}{v_w} \qquad (10.2)$$

where
 ω_R is the mechanical angular velocity of the turbine rotor (rad/s)
 v_w is the wind speed (m/s)

The angular velocity, ω_R, is determined from the rotational speed, n (r/min), by Equation 10.3.

$$\omega_R = \frac{2\pi n}{60} \qquad (10.3)$$

The wind turbine characteristics can be found in the manufacturer datasheet. In order to calculate C_p for the given values of β and λ, the following numerical approximations can be used, as indicated in the following equations:

$$\lambda_i = \frac{1}{\dfrac{1}{\lambda + 0.02\beta} - \dfrac{0.003}{\beta^3 + 1}} \qquad (10.4)$$

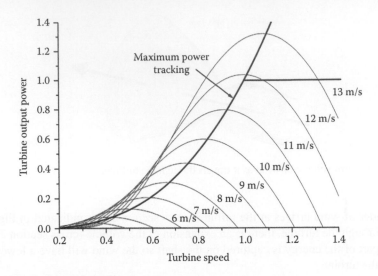

FIGURE 10.4   Wind turbine characteristics at different wind speeds.

$$C_p\left(\lambda,\beta\right) = 0.73\left[\frac{151}{\lambda_i} - 0.58\beta - 0.002\beta^{2.14} - 13.2\right]e^{\frac{-18.4}{\lambda_i}} \tag{10.5}$$

Using the wind turbine characteristics of Equations 10.1 through 10.5 and that *power = torque × angular speed*, a turbine output can be obtained at different wind speeds as shown in Figure 10.4. This figure shows the turbine rotation family of curves in per-unit, assuming that the turbine can have a margin for overpowering during high wind velocity, but typically 40% of the rated turbine speed is the maximum safety range, which is why the curves range up to 1.4 pu. A similar family of curves is plotted in Figure 7.12 for the discussion of maximum power tracking.

10.4 WIND ENERGY GENERATOR SYSTEMS

In order to select an electrical generator for small wind energy power application, it is necessary to study and define a wide set of variables such as voltage, frequency, speed, output power, slip factor (for induction machines), required source of reactive power, and field excitation along with other parameters.

Usually, a DC link is used between the generator and the load or the network to which the power has to be delivered, for example, an inverter or a rectifier on the generator side with an inverter on the utility side. There are other factors to consider, such as the capacity of the system, type of loads, availability of spare parts, voltage regulation, and cost. If loads are likely to be inductive or maybe with a low-power factor (such as phase-controlled converters, motors, and fluorescent lights), a synchronous generator could be a better choice than an IG (for larger power applications). On the other hand, for small power applications, an induction machine is easier to purchase or maybe a permanent magnet brushless AC generator would be available when choosing for inexpensive turbines. Selecting and sizing the generator is a very technical decision and viability studies must be conducted, particularly using power systems or power electronics software simulators. A gearbox will typically be used to convert the low-speed (typically 10–30 rpm) high-torque turbine shaft power to the shaft of the electrical generator. A gearbox is often used in small wind turbines, whereas multipole generator systems can be custom-made for multimegawatt systems. The most typical machines to convert the turbine shaft mechanical power into electrical power are

(1) PMSG, (2) self-excited induction generator (SEIG), (3) squirrel cage induction generator (SCIG), and (4) DFIG. Figure 10.5 shows the most typical commercially available electrical generator systems for wind energy applications.

The wind energy systems can be classified in several ways, but they broadly fit into two categories: (1) the fixed and (2) the variable speed wind generator systems.

10.4.1　Fixed Speed Wind Turbines

The SCIG can be used as a fixed speed wind generator, directly connected to the power grid using a step-up transformer. The rotor speed of an SCIG, which can be expressed in terms of the slip(s), varies with the amount of power generated. However, the rotor speed variation is very small, in the order of 2%–5% of the rated speed. This type of wind generator has many advantages, such as very little maintenance; rugged construction; low-cost, short-circuit resilience; and operational simplicity. On the other hand, as the rotor speed cannot be varied, fluctuations in the wind speed are translated directly into drivetrain torque fluctuations, causing higher structural loads than with a variable speed generator.

A very important factor for SCIGs directly connected to the grid is the need for reactive power to establish the rotating magnetic field of the stator. In case of large turbines and weak grids, a capacitor bank is connected at the terminal of the fixed speed wind generator. The capacitor bank is designed for rated conditions to maintain unity power factor operation, so the voltage fluctuation is natural in other operating conditions for fixed speed wind generator. Some turbines have two IGs connected to the grid, a small one and a large one, with different number of poles, and depending on the wind speed conditions, one of them will be selected to be connected to the grid.

10.4.2　Variable Speed Wind Turbines

The second type is the variable speed wind generator, which has again many subclassifications. The commercially available variable speed wind generators can be IGs with front-end converters, DFIGs, wound field synchronous generator (WFSG), or PMSG. The superior advantage of variable speed operation over that of the fixed speed is that more wind energy can be extracted for a specific wind speed regime. These variable speed wind generators, in general, are equipped with power electronic converters, rated partially or fully. Therefore, there are some losses in the power electronic converters, but the aerodynamic efficiency increases due to variable speed operation increasing the overall efficiency. The aerodynamic efficiency gain can exceed the electrical efficiency loss, resulting in higher overall efficiency. The mechanical stress is less with variable speed operation; the cost of the variable speed wind generator is usually higher but can be paid by the extra energy production. Further, it can provide an important ancillary service, that is, reactive power compensation for weak grids along with fault ride-through capability. Nowadays, most of the wind turbines are using the variable speed technology.

The modern variable speed wind turbine generator system topologies are shown in Figure 10.5. For the DFIG, a frequency converter, which consists of two partial-rated (30%–35% of nominal power of the generator) back-to-back voltage source converters, feeding the three-phase rotor winding through brushes and slip rings. Therefore, the mechanical and electrical rotor frequencies are decoupled. This wind generator requires a gearbox as the DFIG is a high-speed machine and limited to a few number of poles. On the other hand, with the direct-drive synchronous generator system (PMSG or WFSG), the generator is completely decoupled from the grid by a fully rated frequency converter, composed of back-to-back voltage source converters. The generator-side converter can be either a voltage source converter or a diode rectifier for small power applications. The generator is excited using either an excitation winding (in the case of WFSG) or permanent magnets (in the case of PMSG).

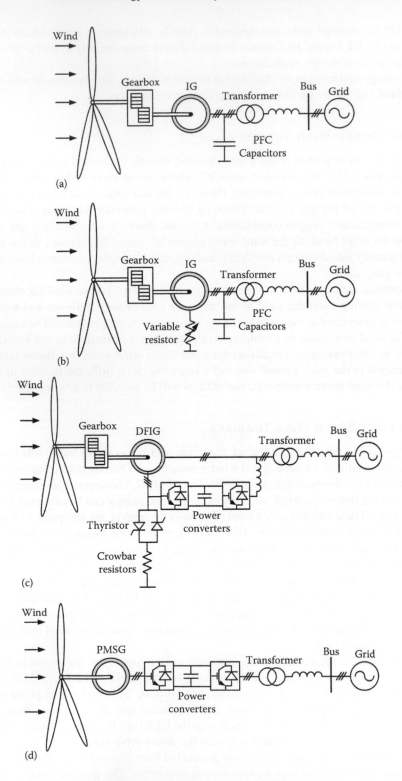

FIGURE 10.5 Commercially available wind turbine systems. (a) Fixed speed induction generator based system. (b) Variable speed induction generator with resistance control based system. (c) Variable speed doubly fed induction generator based system. (d) Variable speed back-to-back double pwm front end based system for permanet magnet synchronous generators.

Other possible variations are either an SCIG or a conventional synchronous generator connected to the wind turbine through a gearbox and to the grid by a power electronics converter of the full rating of the generator. Recently, variable speed wind generator using switched reluctance generators have also been developed as they can provide multiple poles [2].

10.5   WIND SPEED AND GENERATOR PERFORMANCE

The energy captured through the shaft of a wind turbine and converted into electrical energy can be evaluated by data analysis of the historical wind power intensity (in W/m²) in order to access the economic viability of a potential site; please refer to Chapter 7. It is appropriate to define local wind power as proportional to the distribution of wind speed occurrence. A statistics-based design should consider the different sites where the turbines will be installed. Thus, with the same annual average speed, very distinct wind power characteristics may affect the optimal design of the generator. Figure 7.4 displays a typical curve of wind speed distribution for a given site. If the wind speed is lower than 3 m/s (denominated by "calm periods"), the power becomes very low for the extraction of energy and the system is usually stopped. Calm periods will determine the necessary time for energy storage. Power distribution varies according to the intensity of the wind and with the power coefficient of the turbine. Then, a typical distribution curve of power may have the form as shown in Figure 7.5. Sites with high average wind speeds do not have calm periods, and there is no serious need for energy storage. However, high wind speed may cause structural problems in the system or in the turbine. The vertical axis of Figure 7.4 is given in percentage of hours/year per meter/second. For an optimal design on an electrical generator, the random nature of wind distribution in a particular site is considered in order to design the best operating range, defining electrical characteristics such as machine frequency and voltage ratings. The majority of losses occur because a gearbox is used to match generator speed with turbine speed.

An electrical generator used for a wind turbine system has an efficiency that is imposed by three main parts, (1) stator losses, (2) converter losses, and (3) gearbox losses, as depicted in Figure 7.2. Stator losses are considered by a proper design of the machine for the right operating range; converter losses are given by proper design of the power electronic circuits (on-state conduction losses of transistors and diodes, plus their frequency proportional switching losses, which may be neglected). One of the main factors responsible for a noticeable power loss is the use of a gearbox, as discussed in Chapter 7. The mechanical viscous losses due to a gearbox are proportional to the operating speed, as indicated by the following equation:

$$P_{gear} = P_{gear,rated}\frac{\eta}{\eta_{rated}} \tag{10.6}$$

where
$P_{gear,rated}$ is the loss in the gearbox at rated speed (in the order of 3% of rated power)
η is the rotor speed (r/min)
η_{rated} is the rated rotor speed (r/min)

Losses due to the use of a gearbox dominate the efficiency in most wind turbine systems, and simple calculations show that there occurs significant power dissipation in the generator system due to the gearbox. From the full energy available in the wind, just part of it can be extracted for energy generation, quantified by the power coefficient, C_p. The power coefficient is the relationship of the possible power extraction and the total amount of power contained in the wind. The turbine mechanical power P_t can be calculated by Equation 10.7.

$$P_t = \frac{\left(C_p \rho A V^3\right)}{2} \ \left(\text{in kg}\cdot\text{m/s}\right) \tag{10.7}$$

The air density ρ can be corrected by the gas law ($\rho = P/RT$) for every pressure (P) and temperature (T), where R is the ideal, or universal, gas constant, equal to the product of the Boltzmann constant and the Avogadro constant, with Equation 10.8.

$$\rho = 1.2929 \frac{273}{T} \frac{P}{760} \qquad (10.8)$$

where

P is the atmospheric pressure (in mm of mercury)

T is the Kelvin absolute temperature

Under normal conditions ($T = 296$ K and $P = 760$ mmHg), the value of ρ is 1.192 kg/m³ and at $T = 288\,\text{K}(15\,°\text{C})$ and $P = 760$ mmHg it is $1.225\,\text{kg/m}^3$. One can consider a temperature decrease of approximately one degree centigrade for every 150 m. The influence of humidity can be neglected. If just the altitude h (in meters) is known (10,000 ft = 3,048 m), the air density can be estimated by the two first terms of a series expansion as given in Equation 10.9.

$$\rho = \rho_0 e^{-\left(\frac{0.297}{3048}h\right)} \approx 1.225 - 1.194 \times 10^{-4} h \qquad (10.9)$$

So, the turbine torque is given by Equation 10.10.

$$T_t = P_t/\omega = \rho A R v_w{}^2 C_T/2 \qquad (10.10)$$

where the torque coefficient is defined as $C_T = C_p/\lambda$.

Suppose a very small wind turbine needs evaluation for a certain site. Assume $S = 1\,\text{m}^2$ and $\rho = 1.2929\,\text{kg/m}^3$, the maximum potential of wind can be obtained from the earlier equation (without taking into account the aerodynamic losses in the rotor, the wind speed variations at several points of the blade sweeping area, the rotor type, and so on) as calculated by Equation 10.11.

$$\frac{P}{A} = 0.5926 \times 0.6464 \times v_w^3 = 0.3831 \times v_w^3 \qquad (10.11)$$

where

P/A is the wind power per swept area (W/m²)

v_w is the wind speed (m/s)

Suppose a Rayleigh probability density function such as provided with the wind data in Figure 10.6. For a constant and steady wind speed of 6 m/s the available shaft turbine power is 260 W for such a small turbine. However, the wind is neither steady nor constant, and such instantaneous power calculation is not useful for sizing and for economic studies. In fact, the random nature of wind must also be considered, and a statistical analysis must be performed. Let us assume, for instance, that a wind turbine site with strong winds and with periods of low wind and some really good average speed can be defined as a Rayleigh probability density function. Equation 10.12 shows the calculation of the average cubic wind speed under Rayleigh conditions, and the shape factor (k) is related to the cubic average wind speed as given in Equation 10.12.

$$\left(v_w^3\right)_{avg} = \int_0^\infty v_w{}^3 h(v)\,dv = \int_0^\infty v_w{}^3 \cdot \frac{2v}{k^2}\exp\left[-\left(\frac{v_w}{k}\right)^2\right]dv = \frac{3}{4}k^3\sqrt{\pi} \qquad (10.12)$$

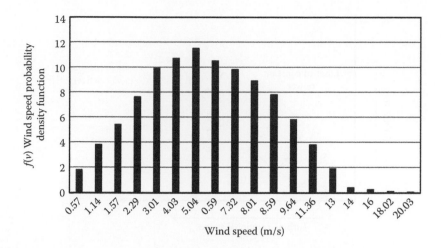

FIGURE 10.6 Rayleigh wind probability density function derived from experimental data for low wind speed, typical in inland and rural areas, estimated wind speed average = 6 m/s.

$$k = \frac{2}{\sqrt{\pi}} v_{wavg} \tag{10.13}$$

and the relationship of the average cubic wind speed with the average speed becomes (10.14)

$$\left(v_w^3 \right)_{avg} = \frac{3}{4} \sqrt{\pi} \left(\frac{2 v_{wavg}}{\sqrt{\pi}} \right)^3 = \frac{6}{\pi} \left(v_{wavg} \right)^3 \cong 1.91 \left(v_{wavg} \right)^3 \tag{10.14}$$

That is, given the Rayleigh wind speed probability distribution function, the average power extracted by the turbine shaft is 1.91 multiplied by the instantaneous power calculated in Equation 10.13. As an example, a Rayleigh distribution function with cubic average wind speed of 6 m/s as indicated gives 496 W for a 1 m blade radius, and of course, a larger-diameter turbine is required for typical machines used for small wind power systems.

The parameter c in the Rayleigh distribution can be evaluated from a set of N data points for the wind velocity. When experimental data are used to determine the parameters in the probability distributions, the computed result is called an estimate of the true parameter; Equation 10.15 gives the estimate of the true parameter $\hat{c}$.

$$\hat{c} = \sqrt{\frac{1}{2N} \sum_{k=1}^{N} v_w^2 (k)} \tag{10.15}$$

Since the turbine power rises with the cubic growth of wind speed and the square of the turbine radius, it is expected that for low wind Rayleigh site conditions (in the order of 8 m/s average cubic wind speed) it will need a turbine with a radius in the order of 3 m for a 1–2 kW generator.

10.6 WIND TURBINE CONTROLS

A typical wind turbine system has its operating range as illustrated in Figure 7.3, where it describes the control range for a typical wind turbine. For low-power wind turbines, permanent magnet generators and SEIGs have been used, for medium power, SCIGs are usually used, while high-power wind turbines have been used for doubly fed induction generators or synchronous machines. Small

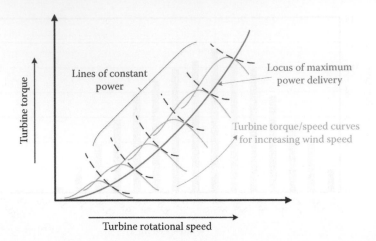

FIGURE 10.7   Family of a typical wind turbine torque versus their rotational speed for variable wind velocities.

wind turbine generators are usually based on inexpensive permanent magnet generators. Chapter 7 discusses a PMSG wind turbine controller with maximum power optimization and a back-to-back PWM grid-connected inverter in Figure 7.13.

Figure 10.7 shows a family of a typical wind turbine torque versus their rotational speed for variable wind velocity, where a change in the maximum power point can be observed when the wind velocity changes.

Large-scale wind turbines have built-in optimized speed and/or pitch control. However, the market for small-scale wind turbines does not accommodate expensive solutions. Gearboxes are often used in high-power wind turbines, but they significantly impact low-power wind turbines for two reasons: (1) they have low efficiency, since the viscous loss (1% of rated power per gearbox stage, which is proportional to the mechanical speed), and (2) they are the major factor (at least 19%) in the downtime in wind turbine generation due to their maintenance needs. Therefore, an optimized electric generator should be designed in order to have the best efficiency for a low wind velocity operating range, without a gearbox and a direct-drive system would be preferable. Small wind turbines have considerable cost constraints in order to be competitive in rural systems and applications for farms and villages. Typically, small turbine generators are made of permanent magnet machines, which are not optimized for capturing wind energy at the low wind speed range (up to about 7 m/s).

10.6.1   MECHANICAL CONTROL

A pitch controller is used in the wind turbine to control the mechanical power extraction from the turbine. As wind is intermittent and stochastic, there will be cases when the wind speed is higher than the rated wind speed for nominal power, and the wind generator will generate higher than power the rated conditions. This may lead to damages to the generator winding and/or may cause system instability. Therefore, the mechanical power extraction is limited from the turbine side using a pitch controller. A conventional pitch controller is shown in Figure 10.8, first the error signal

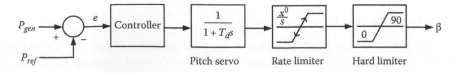

FIGURE 10.8   Block diagram of a pitch controller.

(between generated power, P_{gen}, and reference power, P_{ref}) is processed through a controller. Most of the time a simple PI controller is used and P_{ref} is considered as the rated power from the generator. The pitch servo is modeled with a first-order delay system with a time constant, T_d. As the pitch actuation system, in general, cannot respond instantly, a rate limiter is added to obtain a realistic response. Finally, a hard limiter is also added so that the pitch angle, β, is not capable of exceeding the design limit of the pitch system.

10.6.2 ELECTRICAL CONTROL

From a control point of view, the permanent magnet machine as discussed in Chapter 7 supports the design of control methodologies for several other wind energy systems. Figure 10.9 shows a diagram for a power electronics topology of a grid-connected wind turbine, as well as a stand-alone DC output without connection to the grid.

A wind electrical generator can be connected to a small-scale wind turbine and connected to either the distribution grid or to a DC load through power electronics interfaces. The control must be based on the load flow, which acts on the turbine rotation. As the rotor speed changes according to the wind intensity, the speed control of the turbine has to command low speed at low winds and high speed at high winds in order to follow the maximum power operating point. The maximum power tracking may require a hill-climbing type controller or maybe a fuzzy logic–based controller for commanding the speed of the generator [3, 4]. A general control scheme for variable speed wind turbine generator system using permanent magnet or wound rotor synchronous generator followed by full-rated back-to-back power converters is depicted in Figure 10.10.

The controller, in general, is developed based on a synchronously rotating reference frame concept where electrical quantities in the abc-reference frame are converted into a d–q reference frame [5–9]. Generator-side converter ensures maximum power transfer to the DC side as well as injecting no reactive power from the machine. Therefore, MPPT control is adopted at the generator side. On the other hand, grid-side converter is used mainly to maintain constant DC-link voltage along with unity power factor operation at the grid side. The grid-side inverter is often designed to compensate for reactive power at the point of common coupling in order to control the voltage. In that case, unity power factor is not maintained. A suitable filter must be designed to keep the harmonics within an acceptable range. The controllers shown in Figure 10.10 are typically PI controllers, but the use of different nonlinear and intelligent controllers is also possible.

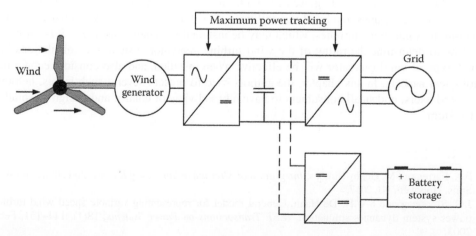

FIGURE 10.9 Power electronics topology for a small wind turbine with battery storage connected to the grid.

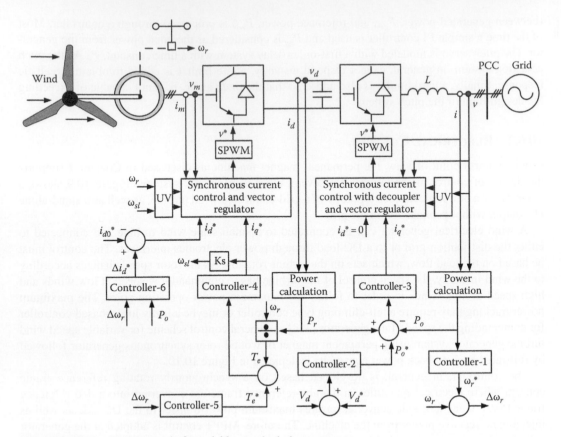

FIGURE 10.10   Basic control of a variable speed wind generator.

10.7 SUMMARY

Wind power is nowadays considered as an alternative to fossil fuel to generate electrical power. Several large-scale wind farms are already installed in many places throughout the world where good wind conditions are present. Apart from that, small-scale wind farms and stand-alone wind generators are also being installed in many parts of the world. Most of the good windy locations have already been occupied by large-scale wind farms populated with large wind turbines, offshore installations, but there are still a lot of possibilities for medium- and small-scale wind farms, as well as stand-alone small wind turbines, which may be installed by small consumers. It is important to know how to determine the rating of the wind turbine generator system, the calculations behind it, which type of wind generator will be effective when installed, and also consider economic and technological aspects. In this chapter, the authors tried to present several such issues, helping the readers, end users, and small consumers to have a broad view of control and design of small wind power systems.

REFERENCES

1. F.A. Farret and M. Godoy Simões, *Integration of Alternative Sources of Energy*, 2nd edn., John Wiley & Sons, Hoboken, NJ, 2017.
2. J.G. Slootweg and S.W.H. De Haan, General model for representing variable speed wind turbines in power system dynamics simulation, *IEEE Transactions on Power Systems*, 18(1), 144–151, February 2003.

3. M.G. Simoes, B.K. Bose, and R.J. Spiegel, Fuzzy logic based intelligent control of a variable speed cage machine wind generation system, *IEEE Transactions on Power Electronics*, 12(1), 87–95, January 1997.

4. M.G. Simoes, B.K. Bose, and R.J. Spiegel, Design and performance evaluation of a fuzzy-logic-based variable-speed wind generation system, *IEEE Transactions on Industry Applications*, 33(4), 956–965, July/August 1997.

5. E. Muljadi and C. Butterfield, Pitch-controlled variable-speed wind turbine generation, *IEEE Transactions on Industry Applications*, 37(1), 240–246, 2001.

6. B. Palle, M.G. Simoes, and F.A. Farret, Dynamic simulation and analysis of parallel self-excited induction generators for islanded wind farm systems, *IEEE Transactions on Industry Applications*, 41(4), 1099–1106, July–August 2005.

7. W. Qiao, X. Yang, and X. Gong, Wind speed and rotor position sensorless control for direct-drive PMG wind turbines, *IEEE Transactions on Industry Applications*, 48(1), 3–11, January–February 2012.

8. M. Rolak, R. Kot, M. Malinowski, Z. Goryca, and J.T. Szuster, Design of small wind turbine with maximum power point tracking algorithm, *2011 IEEE International Symposium on Industrial Electronics*, pp. 1023–1028, Gdansk, Poland, June 27–30, 2011.

9. G. Joos and J. Belanger, Real-time simulation of a wind turbine generator coupled with a battery supercapacitor energy storage system, *IEEE Transactions on Industrial Electronics*, 57(4), 1137–1145, April 2010.

FURTHER READING

P.M. Anderson and A. Bose, Stability simulation of wind turbine systems, *IEEE Transactions on Power Apparatus and Systems*, PAS-102(12), 3791–3795, December 1983.

J.G. Slootweg, Wind power: Modelling and impact on power system dynamics, PhD thesis, Delft University of Technology, Delft, the Netherlands, 2003.

S. Heier, *Grid Integration of Wind Energy Conversion System*, John Wiley & Sons Ltd., Chicester, U.K., 1998.

D.S. Zinger and E. Muljadi, Annualized wind energy improvement using variable speeds, *IEEE Transactions on Industry Applications*, 33(6), 1444–1447, November/December 1997.

R. Hoffmann and P. Mutschler, The influence of control strategies on the energy capture of wind turbines, *Conference Record of the 2000 IEEE Industry Applications Conference*, Vol. 2, pp. 886–893, Rome, Italy, October 8–12, 2000.

S.M. Muyeen, J. Tamura, and T. Murata, *Stability Augmentation of a Grid-connected Wind Farm*, Springer-Verlag, London, U.K., October 2008, ISBN 978-1-84800-315-6.

S.A. Papathanassiou et al., Mechanical stresses in fixed speed wind turbines due to network disturbances, *IEEE Transactions on Energy Conversion*, 16(4), 361–367, 2001.

B. Wu, Y. Lang, N. Zargari, and S. Kouro, *Power Conversion and Control of Wind Energy Systems*, John Wiley & Sons, Hoboken, NJ, 2011.

E. Hau, *Wind Turbines Fundamentals, Technologies, Application, Economics*, 3rd edn., Springer, Krailling, Germany, 2013.

G. Abad, J. Lopez, M. Rodreguez, L. Marroyo, and G. Iwanski, *Doubly Fed Induction Machine Modeling and Control for Wind Energy Generation*, John Wiley & Sons, Hoboken, NJ, 2011.

H. Polinder, F.F.A. van der Pijl, G.-J. de Vilder, and P.J. Tavner, Comparison of direct-drive and geared generator concepts for wind turbines, *IEEE Transactions on Energy Conversion*, 21(3), 725–733, September 2006.

E. Morgan, M. Lackner, R. Vogel, and L. Baise, Probability distributions for offshore wind speeds, *Elsevier Energy Conversion and Management*, 52, 15–26, 2011.

11 Marine and Hydrokinetic Power Generation and Power Plants

Eduard Muljadi and Yi-Hsiang Yu

CONTENTS

Abstract 267
11.1 Introduction 268
11.2 MHK Technologies 270
    11.2.1 WEC Technologies 270
    11.2.2 CEC Technologies 273
11.3 Electrical Generation 273
    11.3.1 MHK Generators 274
        11.3.1.1 Fixed-Speed Induction Generator: Type 1 MHK Generator 274
        11.3.1.2 Variable-Slip Wound-Rotor Induction Generator with Adjustable External Rotor Resistance: Type 2 MHK Generator 275
        11.3.1.3 Variable-Speed Doubly Fed Induction Generator–Partial-Size Power Converter: Type 3 MHK Generator 275
        11.3.1.4 Variable-Speed Full-Power Converter: Type 4 MHK Generator 276
    11.3.2 Prime Mover 276
    11.3.3 Direct Drive/Gearbox 277
    11.3.4 Control and Power Conversion 277
    11.3.5 MHK Power Plant 278
        11.3.5.1 Plant-Level Control 278
        11.3.5.2 Plant-Level Compensation 279
11.4 Energy Storage 280
    11.4.1 WEC Generator 280
    11.4.2 CEC Generator 281
11.5 Problems and Exercises 281
    11.5.1 WEC Generator 281
    11.5.2 CEC Generator 284
11.6 Homework Problems 287
11.7 Summary 288
Acknowledgment 288
References 289

ABSTRACT

Marine and hydrokinetic (MHK) power generation is a relatively new type of renewable generation. Predecessors such as wind power generation, hydropower plant generation, geothermal generation, photovoltaic generation, and solar thermal generation have gained a lot of attention because of their successful implementation. The successful integration of renewable generation into the electric power grid has energized global power system communities to take the lessons learned, innovations, and market structure to focus on the large potential of MHK to also contribute to the pool of renewable energy generation.

This chapter covers the broad spectrum of MHK generation. The state-of-the-art power take-off methods is discussed. The types of electrical generators and the options for implementation are presented.

11.1 INTRODUCTION

Marine and hydrokinetic (MHK) renewable energy has gained great interest in recent years because of its potential to provide a significant contribution to the electricity supply around the world. For example, studies have shown that the United States has a large theoretical MHK energy resource from the movements of tides, ocean and river currents, and waves; the total magnitude of the theoretical MHK resource is on the order of U.S. electricity demand (Figure 11.1). In particular, the U.S. West Coast, Hawaii, and Alaska have wave resources that are among the strongest. The magnitudes and locations of the MHK energy resources given based on a series of resource assessment studies [3–7] are plotted in Figure 11.1, and the total available resource from different types of MHK technologies is listed in Table 11.1. Note that the wave resource data were collected for a water depth of 200 m, and the 2012 U.S. electricity generation was estimated at 4054 TW-h/year [1, 2].

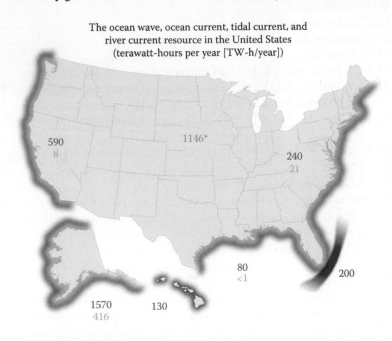

FIGURE 11.1   The theoretical MHK resources in the United States. (Courtesy of NREL, Denver, CO.)

TABLE 11.1
Total Available Resources and Equivalent Percentages of U.S. Electricity Generation in 2012

	Total Resource (TW-h/Year)	Equivalent % of 2012 Generation
Ocean wave (blue)	2640	65
Ocean current (red)	200	5
Tidal current (green)	445	11
River current	1381	34

A wide variety of MHK technologies have been proposed and tested by industry developers, particularly in oceanic countries such as Ireland, Denmark, Portugal, Sweden, the United Kingdom, and the United States. There are many examples of recent installations of wave energy conversion (WEC) devices. A Swedish company, Seabased, is currently installing a 10 MW demonstration plant off Sotenäs on the west coast of Sweden [8]. A U.S. company, Ocean Power Technologies, deployed a 150 kW system in Scotland in 2011 [9]. In the United Kingdom, Aquamarine tested their 315 kW Oyster wave energy device at the European Marine Energy Center's test site in Orkney in 2009 [10]. Northwest Energy Innovations tested their prototype Azura device off the coast of Oregon in 2012 and started a grid-connected demonstration project at the U.S. Navy's Wave Energy Test Site in Hawaii in 2015 [11]. Regarding current energy conversion (CEC) devices, in the United States, Ocean Renewable Power Company deployed a TidGen tidal power system in northern Maine [12], and Verdant Power completed a grid-connected demonstration of their prototype design in the East River in New York Harbor in 2009 [13]. In the United Kingdom, Marine Current Turbines successfully deployed a 1.2 MW SeaGen tidal energy system in Strangford Narrows in Northern Ireland in 2008 [14].

However, there are social, economic, regulatory, and environmental issues (e.g., marine mammal protection areas, fishing permit areas, and shipping routes) when considering MHK generation, and there are technical challenges, including the device power capture efficiency, the array/farm system performance, and the conversion efficiency of the power take-off system (PTOS). As a result, MHK technology is still emerging, and the cost of energy from MHK energy devices is not yet competitive with other forms of renewable energy. Studies from the Carbon Trust [15] and a reference model supported by the U.S. Department of Energy [16] showed that the levelized cost of energy (LCOE) is approximately 60–80 cents/kWh for a WEC array and 40–50 cents/kWh for a CEC farm in the commercial-demonstration-scale size of 10–30 MW. The Carbon Trust report also estimated the relative contribution of different subsystems to LCOE for typical WEC and CEC devices as summarized in Table 11.2.

A successful MHK energy converter design requires a balance between the design performance and cost. The cost of energy depends on the device's power output; the cost of manufacturing, deployment, and operations and maintenance (O&M); and environmental compliance (e.g., marine mammal protection areas and shipping lanes). The research community and industry developers will need to overcome technical and practical barriers to improve power generation performance and reduce LCOE to accelerate technology development to commercial readiness.

Cost reductions for MHK energy devices will come primarily from improving device power performance, optimizing array/farm layouts, reducing capital costs, and improving O&M strategies. In particular, the capability to apply an advanced control algorithm to improve power output, reduce loads, and improve the conversion efficiency of the PTOS is essential to improve the MHK device's power performance, especially for most WECs. The objective of this chapter is to review the status of current MHK renewable energy, particularly focusing on the generator and power conversion

---

TABLE 11.2

Estimated Percent Contributions to Levelized Cost of Energy for Early Commercial-Scale Wave Energy Conversion and Current Energy Conversion Systems

Technology	Installation (%)	Structural Costs (%)	Station Keeping (%)	Power Take-off (%)	Grid Connection (%)	O&M (%)
WEC	10.0	29.0	6.0	20.0	8	27.0
CEC	33.2	13.8	15.5	10.8	8.7	18.0

Source: Adapted from Carbon Trust, *Accelerating Marine Energy*, Carbon Trust, London, U.K., Tech. Rep., 2011.

system. After presenting the working principles of WEC and CEC systems, this chapter will reveal different types of generator designs, including linear and rotary systems, and types of generator systems for different MHK designs and energy storage.

11.2   MHK TECHNOLOGIES

MHK technology is still emerging, and different types of WEC and CEC designs have been proposed by developers. This section reviews WEC and CEC designs and their working principles.

A table to compare different MHK technologies in terms of their complexity, performance, reliability, and cost would be useful; unfortunately, in such a detail, the MHK devices and technologies are very broad and are not ready to be disseminated to the general public. Eventually, the winning concepts will find market and commercialization paths, and the MHK industry will reach the level of maturity as that of the present wind industry.

11.2.1   WEC TECHNOLOGIES

WEC devices extract energy contained within ocean surface waves and convert it into useful electrical power. To date, there are more than 100 prototypes of various WEC systems [17]. These devices are typically divided into point absorbers, terminators, attenuators, an oscillating water column (OWC), and overtopping designs [18–20]. The first three are often categorized as oscillating types of WEC designs and consist of one body or multiple bodies that are designed to directly convert wave energy into electrical power from the wave-induced relative translation motion and/or rotational motion between the body and a reference frame (e.g., seabed or another body) through the use of a linear or rotary PTOS. The overtopping devices and OWCs generate energy through the use of hydroturbines and air turbines, respectively. Figure 11.2 shows illustrations of the various types of WEC devices and illustrates the mechanisms through which the devices extract energy [20], and Figures 11.3 through 11.5 show the prototype WECs developed by the WEC industry.

Point absorbers are devices that are small with respect to the wavelength of incident waves. Point absorbers typically extract energy through a heaving or pitching motion, or a combination of both, as illustrated in Figure 11.2a. Generally, a point absorber is designed to have a system natural frequency close to the dominant frequency of waves to maximize its power output and is an attractive WEC concept because it is theoretically capable of absorbing energy from a wave front that is greater than the device's diameter or width [21].

Terminators and attenuators have dimensions that are on the same order of magnitude as the wavelength and have one dominant dimension. As shown in Figure 11.2b and c, terminators and attenuators are oriented with their dominant dimensions parallel and perpendicular to the incoming wave front, respectively. Theoretically, for a terminator that has a symmetrical body and is only allowed to oscillate in one mode of motion, it can absorb 50% of the energy from a wave front of the device's width. It is possible to absorb 100% of incoming wave energy if a nonsymmetrical body is implemented or the symmetrical body is allowed to oscillate in more than one degree of freedom. An attenuator, on the other hand, captures wave energy along its length from a large wave front length [21].

OWC and overtopping devices use turbines to generate electricity, as shown in Figure 11.2d and e. Overtopping devices gather water in a reservoir at a height higher than the mean free surface as waves pass over the top of the device. The resulting hydrostatic pressure difference that is created between the reservoir and the open ocean is used to drive a turbine that is similar to those used in conventional hydropower applications. OWC devices consist of a confined air chamber in which the pressure varies with water height within the chamber. As the water level rises and falls, air is driven into and out of an air turbine, which generates power.

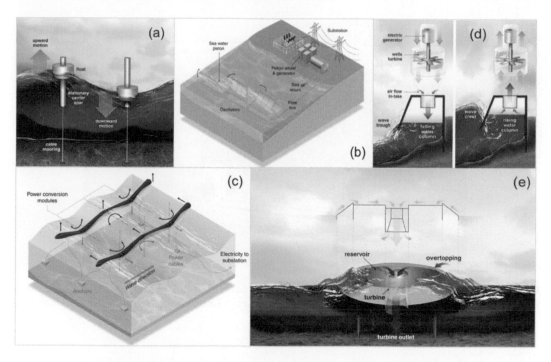

FIGURE 11.2 Wave generator: (a) buoy system (point absorber), (b) bottom-hinged flapper (terminator), (c) floating multibody pitching device (terminator), (d) OWC, and (e) overtopping device. (Courtesy of NREL, Denver, CO.)

FIGURE 11.3 Ocean Power Technologies' PowerBuoy point absorber in open-ocean testing [7]. (Photo from Ocean Power Technologies, NREL 22857, Pennington, NJ.)

FIGURE 11.4   A 1/7-scale point absorber developed by Columbia Power Technologies undergoing testing in the Puget Sound, Washington [22]. (Photo from Columbia Power Technologies, NREL 19381, Corvallis, OR.)

FIGURE 11.5   Ocean Energy's OWC device showing the air turbine that is used to generate energy [23]. (Photo from Ocean Energy Limited, NREL 17874, County Cork, Ireland.)

Several WEC prototypes have been built and deployed for testing. The examples of point absorber implementations are shown in Figures 11.3 and 11.4, and an OWC implementation is shown in Figure 11.5.

In the real world, the behavior of ocean waves is generally random in terms of amplitude, phase, and directionality. Because of the oscillatory nature of waves, the instantaneous power generated by a WEC oscillates from zero to a maximum twice per wave period, which is typically in the range of 5–20 s.

FIGURE 11.6 Verdant Power's axial-flow turbine CEC, similar in design to today's commercial-scale wind turbines, being deployed in the East River near New York City [25]. (Photo from Verdant Power, NREL 17209, New York, NY.)

11.2.2 CEC TECHNOLOGIES

Technologies that extract energy from ocean, tidal, and river currents are collectively called CECs. Tidal currents are generally driven by the Earth's rotation, the relative positions of the moon and the sun to the Earth, and local bathymetry, and they consist of multiple constituents with varying periods. In most places, the dominant constituent is the principal lunar semidiurnal, which has a period of 12 h and 25.2 min. The term *ocean current* refers to the wind-driven surface ocean current. The Gulf Stream (in the North Atlantic Ocean) and Kuroshio (in the North Pacific Ocean) are the two largest ocean currents.

Today, these technologies are significantly more mature than WECs because their designs draw on decades of research and development experience from the wind energy and shipbuilding industries. Accordingly, most CECs currently under development resemble wind turbines that have been adapted to operate in the ocean environment and marine propellers that have been modified to harness energy. The common types of CECs include (1) conventional axial-flow turbines, similar to today's commercial-scale wind turbines, and (2) cross-flow turbines, similar to many small-scale wind turbines. Figures 11.6 and 11.7 show the two prototype CEC designs currently being developed by the industry. The performance characteristics of CECs are well understood: it can be shown mathematically that the maximum theoretical efficiency of a CEC in an unrestricted flow is 59% [24]; and efficiencies of 50% have been achieved by prototype devices operating in real-world conditions.

11.3 ELECTRICAL GENERATION

In this section, the electrical generation of MHK is presented. The focus is on system integration and connecting the MHK generator to the grid. Different types of generators that can be used for MHK generators are listed.

FIGURE 11.7   A cross-flow turbine developed by Ocean Renewable Power Company being prepared for an open-water test [26]. (Photo from Ocean Renewable Power Company, NREL 24507, Portland, ME.)

11.3.1   MHK GENERATORS

Because the wind power industry has accumulated many years of experience, many types of generators used for wind power generation are adaptable to MHK generators.

A major difference between MHK generation and wind generation is the nature of the energy source. In addition to the salinity of the ocean, the specific mass density of water is 1000 times that of the air. The normal wind speed range for a wind turbine generator is between 4 and 25 m/s; on the other hand, the speed of the water in the current turbine is very low, between 0.5 and 3 m/s. Thus, the design and material used to build the hardware of the MHK generation must be carefully chosen to withstand corrosion, mechanical loads, and mechanical stresses. The electrical generator in wind generation mostly converts rotational motion into electrical energy, whereas in MHK generation, some linear motion is converted into electrical energy. However, from the perspective of electrical machine design and control, the linear generator and rotational generators have very similar principles—for example, the higher the rotational or translational speeds, the smaller the dimension of the generator.

Four types of generators are commonly used in wind generation and are being studied for MHK power generation.

11.3.1.1   Fixed-Speed Induction Generator: Type 1 MHK Generator

This type of generator is very simple, rugged, and requires no maintenance. It is directly connected to the grid, and it operates within a small range of slip (~1%) variation, so the rotational speed is practically constant. For 60 Hz grid frequency, this generator operates at 1800 revolutions per minute (rpm); thus, it requires a gearbox to reduce the speed to match the turbine speed for wind turbine, tidal turbine, and ocean current turbine applications (see Figure 11.8).

In addition, by its nature, it requires reactive power compensation (accomplished by installing parallel capacitor banks) to deliver unity power factor to the grid. An example application for this generator can be found in an installed pilot wave energy plant of the OWC type at Vizhinjam, on the west coast of India [27].

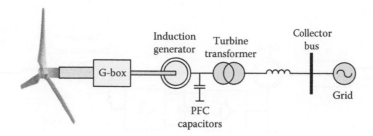

FIGURE 11.8   Type 1 MHK generator—induction generator. (Courtesy of NREL, Denver, CO.)

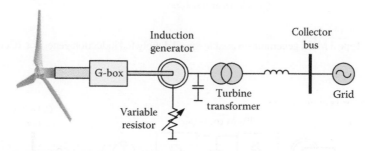

FIGURE 11.9   Type 2 MHK generator—variable-slip wound-rotor induction generator. (Courtesy of NREL, Denver, CO.)

11.3.1.2   Variable-Slip Wound-Rotor Induction Generator with Adjustable External Rotor Resistance: Type 2 MHK Generator

This type of generator is built out of a wound-rotor induction generator, and its rotor winding is connected to an adjustable external resistance. It was built to achieve up to 10% slip variation. The adjustable external resistance is controlled via simple power electronics and external resistors mounted on a rotating shaft. For 60 Hz grid frequency, this generator operates at 1200 rpm or 1800 rpm depending on the number of poles; thus, it requires a gearbox to reduce the speed to match the turbine speed for wind turbine, tidal turbine, and ocean current turbine applications. In addition, by its nature it requires reactive power compensation (accomplished by installing parallel capacitor banks) to deliver unity power factor to the grid. Although this type of generator is used in wind generation, it is less likely that it will be commercialized for MHK generation (see Figure 11.9).

11.3.1.3   Variable-Speed Doubly Fed Induction Generator–Partial-Size Power Converter: Type 3 MHK Generator

This type of generator is built out of a wound-rotor induction generator, and its rotor winding is connected to a three-phase slip ring. The slip rings are connected to a partial-size power converter to process the slip power. It was built to achieve up to +30% slip of the synchronous speed. For 60 Hz grid frequency and a four-pole generator, it operates between 1260 and 2340 rpm; thus, it requires a gearbox to reduce the speed to match the turbine speed for wind turbine, tidal turbine, and ocean current turbine applications. The reactive power compensation is accomplished by the power converter (see Figure 11.10).

This method of generation was analyzed, implemented, and tested in [28, 29] using the linear version intended for wave generator applications, and an application for the OWC is given in [30].

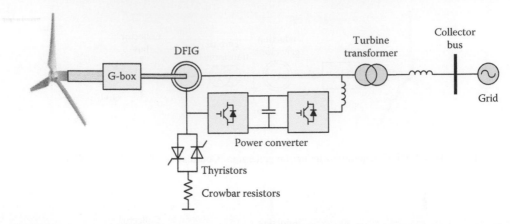

FIGURE 11.10   Type 3 MHK generator—variable-speed doubly fed induction generator. (Courtesy of NREL.)

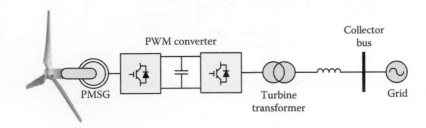

FIGURE 11.11   Type 4 MHK generator—variable-speed full-power converter with PMSG. (Courtesy of NREL.)

11.3.1.4   Variable-Speed Full-Power Converter: Type 4 MHK Generator

The unique feature of this generator is that the power converter connects the generator to the grid; thus, the power converter is sized to process the entire power generated. The power converter is the buffer between the grid and the generator, so the generator itself can be a direct-drive generator (shown as a permanent magnet synchronous generator [PMSG] in Figure 11.11), or it can be another variable-speed generator (switched-reluctance generator, simple AC induction generator, or reluctance generator) connected to the turbine blades through the gearbox to reduce the speed.

Examples of Type 4 MHK generator implementations for wave generation are given in [31] as a direct-drive linear wound-field generator, in [32] as a direct-drive linear permanent magnet generator, in [33, 34] as a linear switched-reluctance generator, and in [35] as a rotary permanent magnet generator. An example of its application to an ocean current is given in [36].

11.3.2   Prime Mover

The prime mover used to drive the generator is commonly called power "take-off" in MHK generation. Although the four types of generators illustrated earlier are known as rotary generators, the same types of generators can be designed to operate in linear motion.

Table 11.3 lists examples of different MHK generator applications for various types of WECs and CECs. A linear generator is suitable for a wave generator, whereas a rotary generator is commonly found for tidal and ocean current turbines. Thus, the four types of generators can be directly adaptable for CECs. WECs mostly use a linear version of a generator, but some use a rotary generator as well.

TABLE 11.3

Examples of Different Types of Marine and Hydrokinetic Generators

MHK Generator System	Example	Type of Energy Converter	Type of MHK Generator
Point absorber	Columbia Power Technologies	WEC	Rotary—Type 4 PMSG
Point absorber	Ocean Power Technologies	WEC	Rotary—Type 4 PMSG
Point absorber	Global Wedge, LLC	WEC	Linear—Type 4 reluctance generator
Terminator	Resolute Marine Energy	WEC	Rotary—Type 4 PMSG
OWC	Ocean Energy Buoy	WEC	Rotary—Type 1 induction generator
Attenuator	Pelamis Wave Power	WEC	Rotary—Type 4 PMSG
Overtopping device	Wave Dragon	WEC	Rotary—Type 4 synchronous generator
Axial-flow turbine	Verdant	CEC	Rotary—Type 4 PMSG
Cross-flow turbine	Ocean Renewable Power Company	CEC	Rotary—Type 4 PMSG

11.3.3 DIRECT DRIVE/GEARBOX

A gearbox is used to match the rotational speed of the turbine to the generator. Although the rotational speed of a small wind turbine can be very high, a larger turbine is usually designed to have a slow rotational speed. A wind turbine operates with much higher wind speeds. The normal cut-in wind speed is 4.5 m/s and the cutout wind speed is approximately 30 m/s. A water turbine operates at a much lower water speed; thus, the rotational speed of a water turbine is much lower than the rotational speed of a wind turbine. With a gearbox, it is possible to use a high-speed generator to match the rotational speed of the blades, so a much smaller size of the generator (often off the shelf) can be used. Unfortunately, a gearbox requires regular maintenance, so it is often the cause of generator's downtime and consequently the loss of productive hours.

11.3.4 CONTROL AND POWER CONVERSION

An MHK generator is usually controlled individually to control the real and reactive power. Because most modern MHK generators are equipped with power electronics, it is common to utilize a generator to control their real and reactive power independently and instantaneously. The real power is controlled for different purposes. Under normal conditions, the real power is usually controlled to maximize energy capture at the MHK generator level.

Like a wind turbine generator, in tidal generation, for example, it is common to adjust the output power of the generator to follow the rotational speed of the turbine to optimize the energy capture.

The reactive power of Type 3 and Type 4 MHK generators can be controlled to adjust a bus voltage, to adjust the power factor, or to control the output reactive power. In many places, the utility or grid operator does not allow the plant to adjust its voltage or reactive power in a normal situation. It is common for the generator to be controlled at unity power factor and for the grid operator to control the voltage using reactive power compensation (capacitor banks, static VAR compensators, etc.) or a tap-changing transformer or by adjusting the excitation of the conventional synchronous generator or synchronous condenser.

At the generator level, the power electronics will have its own protection to ensure that the voltage and current limits of its components (insulated gate bipolar transistors, capacitors, etc.) do not exceed the design values. Similarly, the grounding, bonding, and shielding (surge voltage suppressions) will be used to protect the circuit from possible interferences because of unwanted

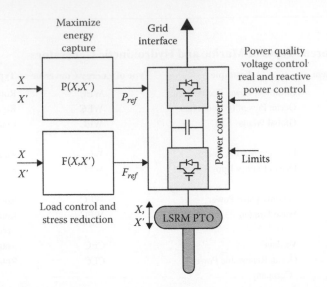

FIGURE 11.12 Control block diagram of a typical MHK power plant. (Courtesy of NREL, Denver, CO.)

induced signals (e.g., electromagnetic compatibility or radio frequency interferences during short circuits or lightning strikes).

Figure 11.12 shows a typical control block diagram of a WEC with linear switched-reluctance machines interfaced to the grid by the full-power conversion (Type 4 MHK generator). It also shows an additional control block that can be added to control and reduce the mechanical stresses and loads by controlling the electrical response to the mechanical excitations.

11.3.5 MHK POWER PLANT

As the MHK technology becomes more mature, future MHK generation can be predicted to follow the trend of wind generation, in which an MHK power plant may consist of hundreds of MHK generators. Similarly, in the future, large-scale MHK generation will become common practice, because the LCOE favors large-scale power generation.

It is easy to imagine that a future wave power plant or tidal power plant will consist of hundreds of generators connected to the point of interconnection (POI) to transmit the total output of the plant to the power system network. Similarly, the requirements for grid interconnection will adopt similar rules and grid codes for MHK generation as those used for wind power generation. As shown in Figure 11.13, a typical MHK power plant consists of many MHK generators interconnected in a daisy-chain fashion and eventually connected to the substation transformer.

Although at present the level of generation is very small, eventually, when it reaches a high level, MHK power plants will impact power system operations. System impact studies are commonly performed in the planning stage far before the actual implementation takes place [37].

11.3.5.1 Plant-Level Control

MHK generators within a power plant can also be controlled at the plant level via supervisory control. In an MHK power plant, the POI is the point to which all the generators are connected. To a power system network, the response of an individual turbine is less important than the effective response at the POI—or, in other words, the collective behavior of the generators in the power plant is more important than the behavior of an individual generator—thus, the plant controller is implemented to shape the characteristic of the entire power plant. Plant-level control is usually

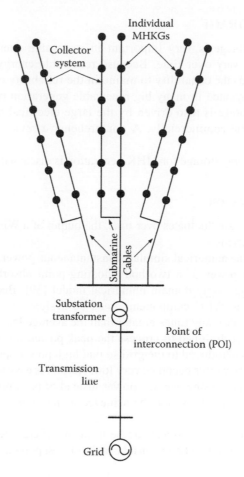

FIGURE 11.13 Layout of a typical MHK power plant. (Courtesy of NREL, Denver, CO.)

accomplished by supervisory control, in which the signals monitored at the POI are used to control the entire power plant.

During a grid disturbance, the real power is often used to support the grid by providing ancillary services to the grid. This kind of operation is usually controlled by the supervisory control to provide a collective response at the POI. One example is called "frequency response." Common practice in real power control is to provide inertial response, as is found in a conventional synchronous generators. Another common practice is to control the real power with the droop control as a function of the operating frequency.

11.3.5.2 Plant-Level Compensation

Another type of plant-level control can be accomplished by adding plant-level compensation. For example, centralized reactive power compensation can be installed at the POI. The plant-level reactive power compensation or energy storage is usually much larger in size than that of an individual turbine. Many older wind power plants that have Type 1 WTGs (induction generators) are enhanced by the installation of plant-level reactive compensation (e.g., DSTATCOM, DVAR, SVC, STATCOM). For MHK generation, this type of plant-level compensation is placed at the POI at the substation transformer onshore. This type of compensation is very significant, especially if the grid at the POI is very weak (e.g., short-circuit ratio <3).

11.4   ENERGY STORAGE

The topic of energy storage is very important in renewable energy generation because the resources continuously vary over time. Because renewable energy generation requires high-energy capture as well as the capability to maintain the reliability of the power system, energy storage has been implemented in many big renewable generation projects. On a positive note, research and energy storage is also driven by the large electrical transportation sector, which has been very successful commercially. A comprehensive review on energy storage can be found in [38].

In this section, the energy storage for MHK generation is discussed.

11.4.1   WEC GENERATOR

Because the nature of a wave fluctuates over time, the output of a WEC generator can be expected to follow the same characteristics.

Figure 11.14 shows the numerical simulated instantaneous power, maximum absorbed power, rated power, and mean power of a two-body floating-point absorber WEC system in irregular waves modeled using a time-domain numerical model [39]. Because of the oscillatory and irregular nature of waves, WEC components must be designed to handle loads (e.g., torques, forces, and powers) that are many times greater than the average load. Converting the peak power requires that the PTOS be designed to process the peak power, but WEC-generated power must be smoothed before it is introduced to the grid so that high power spikes are eliminated; thus, the utilization of the power converter becomes very low. Here, it is easy to see that energy storage can be very useful to allow processing average power instead of peak power. If designed correctly, a significant saving in the capital cost can be achieved along with the overall impact of lowering the cost of energy.

The implementation of energy storage can be in many different forms; however, to reduce the power converter size, the implementation must be at the closest point to the generator. For a hydraulic

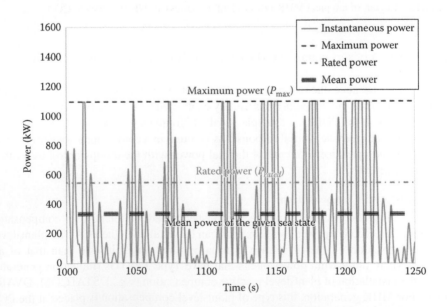

FIGURE 11.14   The instantaneous power, maximum absorbed power, rated power, and mean power for the RM3 floating-point absorber WEC system in irregular waves with a significant wave height of 4.25 m and a peak period of 10.7 s. (Courtesy of NREL, Denver, CO.)

PTOS, a pressure accumulator is often used to store energy [40]. In addition, most of the power converters presently used in MHK generators are AC–DC–AC systems, and placing the energy storage on the DC bus is a very viable option. Another example of energy storage is presented in [40, 41] as an indirect flywheel connected to the system, thus storing the energy as kinetic energy. In [42], energy storage is achieved using a superconducting magnetic energy storage system.

11.4.2 CEC GENERATOR

A CEC generator is usually smoother than a WEC generator. The ocean current is known to be very constant over time. Similarly, for a run-of-the river CEC generator, the flow is relatively constant. However, the tide changes direction every 6 h; thus, for a tidal current generator, the power generation varies within a longer time constant (6 h), and a longer-term energy storage may be beneficial for this type of generation.

For a CEC generator, the best place to install energy storage is at the POI at the substation transformer onshore. In addition, the size is much larger, the infrastructure cost is cheaper, and the environmental impact is much easier to manage if it is built onshore.

11.5 PROBLEMS AND EXERCISES

MHK renewable energy has a large diversity in its implementation. The purpose of the problems and exercise in this section is to help readers understand the basics of the typical issues faced by MHK engineers in understanding MHK generators, MHK generation, and supporting MHK infrastructure.

11.5.1 WEC GENERATOR

A useful ocean wave is affected by wind speed, the depth and surface characteristics of the ocean, and gravity. The ripple effect in the form of the ocean wave far from the exciting point is called the "swell." Although ocean waves may be excited by a tsunami or storm, the normal, day-to-day, somewhat predictable waves excited by normal wind speeds are of interest for conversion into electrical energy. Based on linear wave theory and assuming that the waves are in deep water, the power across each meter of wave front associated with a uniform wave with height H (m) and wavelength λ (m) is

$$P_w = 0.5\rho g H_s^2 \lambda \tag{11.1}$$

where
  P_w is the power density (kW/m) of the wave front
  ρ is the specific density of the water (kg/m³)
  H_s is the significant wave height (m)
  g is the specific gravity (m/s²)

Example Problem 11.1

At one site, the average significant wave height $H = 2.5$ m and the water density $\rho = 1$ kg/m³. Compute the power density across the wave front for this particular site if the specific gravity is given as 9.81 m/s² and the wavelength is $\lambda = 9$ m.

Solution

The power density is computed as

$$P_w = 0.5 * 9.81 * 2.5^2 * 9 = 275.9 \text{ kW} \tag{11.2}$$

Significant wave height, Hs [m]	Energy period, Te [sec]																	
	3.5	4.5	5.5	6.5	7.5	8.5	9.5	10.5	11.5	12.5	13.5	14.5	15.5	16.5	17.5	18.5	19.5	20.5
0.25					0.02	0.03												
0.75		0.02	0.46	1.49	2.68	1.91	1.10	0.53	0.17	0.02								
1.25		0.01	0.59	4.11	5.56	4.48	2.74	1.28	0.67	0.33	0.07	0.02	0.02					
1.75			0.12	3.27	5.14	4.62	3.93	2.11	1.24	0.76	0.31	0.10	0.03					
2.25				0.92	5.25	3.68	4.14	2.87	1.31	0.84	0.42	0.20	0.08	0.02				
2.75				0.14	2.43	2.60	2.82	2.85	1.57	0.80	0.32	0.14	0.06	0.02				
3.25					0.45	1.54	1.47	1.96	1.42	0.79	0.32	0.11	0.04	0.02	0.01	0.01		
3.75					0.05	0.49	0.63	1.08	1.01	0.63	0.29	0.10	0.05	0.02				
4.25						0.09	0.21	0.45	0.56	0.42	0.21	0.07	0.02	0.02				
4.75						0.02	0.08	0.12	0.26	0.27	0.19	0.07	0.02	0.01				
5.25							0.03	0.03	0.11	0.15	0.13	0.07	0.02					
5.75									0.02	0.07	0.05	0.05	0.02					
6.25										0.03	0.04	0.02	0.01					
6.75											0.02	0.02						
7.25																		
7.75																		
8.25																		
8.75																		
	4.1	5.2	6.4	7.5	8.7	9.9	11.0	12.2	13.3	14.5	15.7	16.8	18.0	19.1	20.3	21.5	22.6	23.8
	Peak period, Tp [sec]																	

FIGURE 11.15 Percentage occurrence of sea states at Humboldt Bay, California.

Because the nature of a wave fluctuates over time, the output of a WEC generator can be expected to follow a similar characteristic. The behavior of irregular waves at a reference site may be provided in terms of a joint probability distribution of possible binned sea states with each characterized by significant wave height, H_s, and energy period, T_e (or peak period, Tp); the percentage occurrence of each binned sea state at the reference site is shown in Figure 11.15.

Example Problem 11.2

A group of engineers are tasked to determine the optimal design for a point absorber WEC. The information is derived from Figure 11.15.

  a. What is the maximum percentage occurrence for wave height of 2.25 m at the site described in Figure 11.15?
  b. What is the wave period for the highest percentage of occurrence presented in Figure 11.15?

Solution

  a. The maximum percentage occurrence for a wave height of 2.25 m at the site described is 5.25%.
  b. The energy period for the highest percentage of occurrence presented in Figure 11.15 is 7.5 s.

An actual WEC device will convert only part of the theoretical wave energy into electricity. The WEC device capture efficiency is often represented by capture width, which is defined as

$$\text{Capture width} = \frac{P_m}{P_w} \tag{11.3}$$

where P_m is the device mechanical power without considering the power take-off losses. Generally, WECs are designed to have optimal capture efficiency at the frequency of the dominant waves at the deployment site. In addition, the power rating and the nature of resource determine the design of the WEC generator. The power generation performance of WECs in irregular seas is generally represented using a power matrix that is defined by a set of binned sea states. Figure 11.16 shows the

		Energy period, Te [sec]													
		4.5	5.5	6.5	7.5	8.5	9.5	10.5	11.5	12.5	13.5	14.5	15.5	16.5	17.5
Significant wave height, Hs [m]	0.75	5	8	10	11	13	14	15	14	13	12	10	10	8	7
	1.25	14	22	27	31	37	37	38	37	33	31	26	24	20	12
	1.75	28	42	52	61	70	71	71	68	61	57	48	44	36	21
	2.25	45	69	84	99	114	113	113	107	96	89	75	70	57	32
	2.75	66	102	124	146	167	164	162	152	137	127	108	100	82	46
	3.25	91	140	171	202	230	224	218	204	185	172	145	135	110	64
	3.75	119	183	226	266	301	292	282	264	239	222	188	175	142	85
	4.25	151	232	287	337	380	367	353	331	299	278	235	219	179	108
	4.75	186	286	355	416	468	451	432	404	365	339	287	267	218	135
	5.25	224	344	429	503	563	541	518	483	437	405	343	320	262	165
	5.75	265	408	510	597	666	639	610	569	515	477	404	377	309	199
		5.2	6.4	7.5	8.7	9.9	11.0	12.2	13.3	14.5	15.7	16.8	18.0	19.1	20.3
		Peak period, Tp [sec]													

FIGURE 11.16 Mechanical power matrix for a two-body floating-point absorber.

resulting mechanical power matrix for the two-body floating-point absorber system from numerical simulations.

To estimate the electrical power matrix, in which the maximum power output was also limited to the rated power (also referred to as capacity), the mechanical power matrix is multiplied with a mechanical-to-electrical conversion efficiency. The electrical power at each binned sea state is calculated from

$$P_e = \begin{cases} P_m \times \eta_1, \left| P_m \times \eta_1 < P_{rated} \right. \\ P_{rated}, \left| P_m \times \eta_1 \geq P_{rated} \right. \end{cases} \tag{11.4}$$

where
P_m and P_e are the predicted mechanical and electrical power that can be generated for each binned sea state
P_{rated} is the rated power
η_1 is the power take-off (PTO) efficiency that accounts for the losses between the absorbed mechanical power and the electrical power output

The rated power was calculated from

$$P_{rated} = \frac{P_{ae}}{C_f} \tag{11.5}$$

where
C_f is the capacity factor
P_{ae} is the annual averaged electrical power, which was obtained by summing the product of the electrical power matrix and the joint probability distribution for the reference site

Note that the annual averaged electrical power output increases and the capacity factor decreases as the rated power for the generator is increased, as shown in Figure 11.17. However, it costs more for a generator with larger capacity. Finally, the estimated annual energy production (AEP) (in megawatt-hours), which typically also includes the transmission losses and maintenance-related losses, can be calculated by

$$AEP = P_e \times 8776(h) \times \beta \tag{11.6}$$

where $\beta = \eta_2\eta_3$ is the parameter, accounting for the losses caused by device availability and the transmission efficiency.

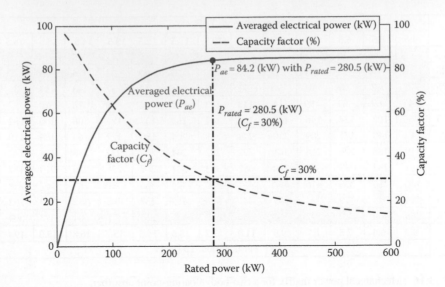

FIGURE 11.17 Annual averaged electrical power output (P_{ae}) and capacity factor for different rated powers (assuming a conversion efficiency of the PTOS of 80%).

Example Problem 11.3

Given the information in Figures 11.15 and 11.16 and assuming a PTO conversion efficiency of 80% and a P_{rated} of 280.5 kW,

   a. Compute the P_{ae} for the given WEC system
   b. Calculate the device capacity factor
   c. Calculate the AEP, assuming an availability of $\eta_2 = 0.95$ (95%) and a transmission efficiency of $\eta_3 = 0.98$ (98%)

Solution

   a. The electrical power matrix for the two-body floating-point absorber can be obtained following Equation 11.4. P_{ae} is then obtained by summing the product of the electrical power matrix and the joint probability distribution for the reference site (Figure 11.15) and $P_{ae} = 84.2$ kW.
   b. The capacity factor $= P_{ae}/P_{rated} = 30\%$.
   c. AEP = 84.2*8766*98%*95% = 687 MWh.

11.5.2 CEC GENERATOR

A CEC generator is usually designed very similarly to wind turbine generators, so these problems and exercises are derived in similar analogy to designing wind turbine generators. The available hydrokinetic power of a water turbine can be computed from the water flow and the turbine dimension as

$$P_{turbine} = 0.5\rho\pi R^2 C_p V^3 \qquad (11.7)$$

where
   R is the radius of the swept area of the turbine
   ρ is the water density
   V is the water flow
   C_p is the performance coefficient of the turbine

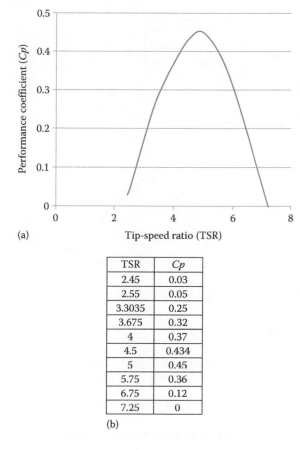

(a)

TSR	Cp
2.45	0.03
2.55	0.05
3.3035	0.25
3.675	0.32
4	0.37
4.5	0.434
5	0.45
5.75	0.36
6.75	0.12
7.25	0

(b)

FIGURE 11.18   The performance coefficient of a water turbine. (Image from Bahaj.)

The tentative performance coefficient of the turbine is shown in Figure 11.18. The performance coefficient of a hydroturbine is sharper than the performance coefficient of a wind turbine, and the tip-speed ratio (TSR) of the C_{pmax} occurs at a lower value (around 2–5 instead of 8 as in a wind turbine generator).

This type of C_p characteristics has a major advantage because the turbine can be stalled to control the rotational speed and thus prevent a runaway condition from occurring without the need for pitch control as in a wind turbine. The corresponding power versus speed characteristic for different flow speeds is given in Figure 11.19.

The performance coefficient can be approximated by a polynomial equation as follows:

$$C_p(\lambda) = a_6\lambda^6 + a_5\lambda^5 + a_4\lambda^4 + a_3\lambda^3 + a_2\lambda^2 + a_1\lambda + a_o \tag{11.8}$$

The TSR is defined as the ratio of the linear speed of the tip of the blade to the water speed:

$$\text{TSR} = \frac{\omega R}{V} \tag{11.9}$$

where the rotational speed, ω, is the rotational speed of the blade. As shown in Figure 11.19, there is a specific TSR that has a maximum operating C_p. This TSR corresponds to $\text{TSR}_{\text{cp\_max}} = 5$, corresponding to the $C_{pmax} = 0.45$.

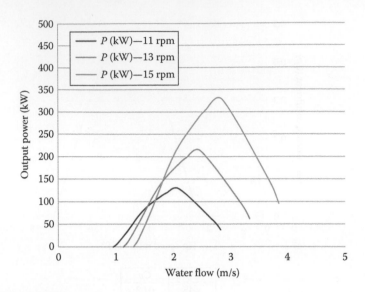

FIGURE 11.19   Tidal turbine operated at constant speeds.

Example Problem 11.4

A tidal turbine is connected to a generator operating at a constant rotational speed. The blade radius is given as 6 m, and the density of the flowing water is $\rho = 1$ kg/m³.

a. Compute the output power for different rotational speeds of 11, 13, and 15 rpm at the water flow of 2 m/s.
b. Compute and plot the output power as the water flow varies.

Solution

a. At 11 rpm, the rotational speed $\omega := 11 \dfrac{2\pi}{60 \text{ s}} = 1.152$ rad/s $\rho_{water} := 1000$ kg/m³

The corresponding TSR can be computed as follows:

$R_{blade} := 6$ m $V_{water} := 2$ m / s $A_{swept} := \pi \cdot$

$R_{blade}^2 = 113.097$ m² $\text{TSR} := \dfrac{\omega R_{blade}}{V_{water}} = 3.456$

$C_p := 0.44$

$P_{out} := 0.5 \cdot \rho_{water} \cdot A_{swept} \cdot C_p \cdot V_{water}^3 = 199.051$ kW

At 13 rpm, $\omega = 1.361$ rad/s, TSR = 4.085, $C_p = 0.35$, $P_{out} = 158.34$ kW.
At 15 rpm, $\omega = 1.361$ rad/s, TSR = 4.71, $C_p = 0.44$, $P_{out} = 199.1$ kW.

b. The output power as a function of the water flow can be plotted using the method described in Example Problem 11.4(a).

Example Problem 11.5

Because the speed of the water flow of the tides changes in magnitude and direction every 6 h, the operating performance coefficient, C_p, of the tidal turbine will not be constant if it is operated at the same speed

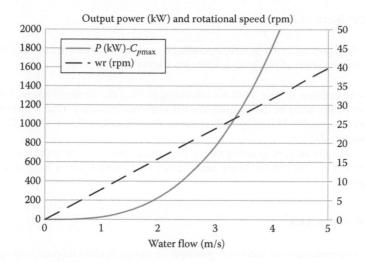

FIGURE 11.20 Output power and rotational speed as a function of the water flow.

all the time. From the operating characteristics of C_p-TSR, it is obvious that the rotational speed must follow the speed of the water flow.

a. Find the expression of the rotational speed for the turbine described in Problem 11.1 to maximize the performance coefficient, C_p, as the speed of the water flow varies.
b. Compute and plot the corresponding output power.

Solution

a. Operation at maximum C_p $C_{p\,\mathrm{max}} := 0.45$ $\mathrm{TSR}_{\mathrm{cpmax}} := 5$

$$\omega_{speed} := \frac{\mathrm{TSR}_{\mathrm{cpmax}} \cdot V_{water}}{R_{blade}} \quad \mathrm{RPM}_{speed} := \omega_{speed} \cdot \frac{60\ \mathrm{s}}{2\pi} \quad \mathrm{RPM}(V_{water}) := \frac{\mathrm{TSR}_{\mathrm{cpmax}} \cdot V_{water}}{R_{blade}} \cdot \frac{60\ \mathrm{s}}{2\pi}$$

$$P_{out} := 0.5 \cdot \rho_{water} \cdot A_{swept} \cdot C_{p\,\mathrm{max}} \cdot V_{water}^3$$

b. The output power as a function of the water flow can be plotted using the method described in (a) (Figure 11.20).

11.6 HOMEWORK PROBLEMS

Homework Problem 11.1

A two-body floating-point absorber WEC device is deployed in Humboldt Bay, California. Use the information given in Figures 11.15 and 11.16 and assume a conversion efficiency of the PTOS of 80% and a capacity factor of 25%. With this information,

a. Calculate the capture width for each binned sea, and plot the resulting matrix
b. Compute the P_{ae} and P_{rated} for the given WEC system
c. Plot the electrical power matrix
d. Calculate the AEP assuming an availability of $\eta_2 = 0.95$ (95%) and a transmission efficiency of $\eta_3 = 0.98$ (98%)

Homework Problem 11.2

A tidal turbine is connected to a generator operating in variable rotational speed. The blade radius is given as 6 m, and the density of the flowing water $\rho = 1$ kg/m³.

a. Suppose an engineer devises a turbine controlled to limit the rotational speed. Compute the minimum power rating of the generator/power converter (assume a Type 4 turbine) to limit the maximum speed to 13 rpm by stalling the turbine. (*Hint*: Use Figure 11.19. The answer is approximately 225 kW.)

b. Compute and plot the corresponding output power as a function of the rotational speed.

c. Compute and plot the corresponding performance coefficient, C_p, as a function of the rotational speed.

Homework Problem 11.3

A tidal turbine is connected to an induction generator operating in constant rotational speed. The stator winding of the induction generator can be connected in two different configurations such that the generator can be operated at 1800 rpm for high speeds of water flow and at 1200 rpm for low speeds of water flow. A gearbox is used to maximize the power coefficient, C_p, at the water flow of 2.5 m/s. The blade radius is given as 6 m, and the density of the flowing water is $\rho = 1$ kg/m³.

a. Choose the gear ratio of the gearbox so that the turbine will operate at maximum C_p at the generator speed of 1800 rpm.

b. Compute the performance coefficient, C_p, when the generator is operating at 1200 rpm at the same water flow (2 m/s).

c. Compute the water flow that will maximize the turbine when the generator is operating at 1200 rpm.

11.7 SUMMARY

The MHK renewable energy has great potential to provide a significant contribution to the electricity supply. However, MHK technologies are still emerging, and the cost of energy from MHK devices is not yet competitive with other forms of renewable energy. The cost of the PTOS represents approximately 10%–20% of the overall cost of energy and has a great influence on the power generation performance. MHK covers a broad spectrum of prime movers (i.e., PTOS) for electrical generators. This chapter presented many aspects of MHK generation to explore the potential of utilizing different types of electrical generators (linear or rotary), power converters, and energy storage systems and thus enable optimum energy capture, extend the life span of MHK generators, and enhance power system grid integration. The experience gained during the past few decades in the development of other forms of renewable energy generation (e.g., wind and solar) from technical, policy, market-driven, and power system integration points of view and the significant cost reduction of power electronics and energy storage will certainly help future developments in MHK generators.

In the near future, it is likely that MHK research and development will focus on the development of hardware for harsh ocean environments. Several concepts will compete to lower the capital and O&M costs. Eventually, the winning concepts will emerge with mature technologies that are economically viable for massive commercial deployment.

ACKNOWLEDGMENT

This work was supported by the U.S. Department of Energy under Contract No. DE-AC36-08-GO28308 with the National Renewable Energy Laboratory.

REFERENCES

1. B. Cross, ed., Wave energy paper, in *European Directory of Renewable Energy Suppliers and Services*, Institute of Mechanical Engineers, James and James Publishers, London, U.K., 1991.
2. U.S. Energy Information Administration, *Electric Power Monthly with Data for December 2013*, U.S. Department of Energy, Washington, DC, Tech. Rep., 2014.
3. K. Haas, Assessment of energy production potential from tidal streams in the United States: Final project report, Georgia Tech Research Corporation, Atlanta, GA, Tech. Rep., 2011.
4. K. Haas, Assessment of energy production potential from ocean currents along the United States coastline, Georgia Tech. Research Corporation, Atlanta, GA, Tech. Rep., 2013.
5. P. T. Jacobson, T. M. Ravens, K. W. Cunningham, and G. Scott, Assessment and mapping of the riverine hydrokinetic resource in the continental United States, Electric Power Research Institute, Palo Alto, CA, Tech. Rep., 2012.
6. P. Jacobson, G. Hagerman, and G. Scott, Mapping and assessment of the United States ocean wave energy resource, Electric Power Research Institute, Palo Alto, CA, Tech. Rep., 2011.
7. M. Previsic, J. Epler, M. Hand, D. Heimiller, W. Short, and K. Eurek, The future potential of wave power in the United States, RE Vision Consulting, Sacramento, CA, Tech. Rep., 2012.
8. Seabased [Online], Available: http://www.seabased.com/en/technology/seabased-wave-energy/, accessed December 5, 2016.
9. OPT/Ocean Power Technologies [Online], Available: http://www.oceanpowertechnologies.com/, accessed December 5, 2016.
10. Aquamarine [Online], Available: http://www.aquamarinepower.com/, accessed December 5, 2016.
11. NWEI/Northwest Energy Innovations [Online], Available: http://azurawave.com/, accessed December 5, 2016.
12. ORPC/Ocean Renewable Power Company [Online], Available: http://www.orpc.co/, accessed December 5, 2016.
13. Verdant Power [Online], Available: http://www.verdantpower.com/, accessed December 5, 2016.
14. MCT/Marine Current Turbines [Online], Available: http://www.marineturbines.com/, accessed December 5, 2016.
15. Carbon Trust, Accelerating marine energy, Carbon Trust, London, U.K., Tech. Rep., 2011.
16. V. S. Neary, M. Previsic, R. A. Jepsen, M. J. Lawson, Y.-H. Yu, A. E. Copping, A. A. Fontaine, K. C. Hallett, and D. K. Murray, Methodology for design and economic analysis of Marine Energy Conversion (MEC) Technologies, Sandia National Laboratories, Albuquerque, NM, Tech. Rep., 2014.
17. Marine_and_Hydrokinetic_Technology_Database [Online], Available: http://en.openei.org/wiki/Marine_and_Hydrokinetic_Technology_Database.
18. A. F. D. O. Falcão, Wave energy utilization: A review of the technologies, *Renew. Sustain. Energy Rev.*, 14(3), 899–918, 2010.
19. B. Drew, A. R. Plummer, and M. N. Sahinkaya, A review of wave energy converter technology, *Proc. Inst. Mech. Eng. A*, 223(8), 887–902, 2009.
20. Y. Li and Y.-H. Yu, A synthesis of numerical methods for modeling wave energy converter-point absorbers, *Renew. Sustain. Energy Rev.*, 16(6), 4352–4364, 2012.
21. J. Falnes, *Ocean Waves and Oscillating Systems*, Cambridge University Press, Cambridge, U.K., 2002.
22. CPT/Columbia Power Technologies [Online], Available: http://columbiapwr.com/, accessed December 5, 2016.
23. Ocean Energy [Online], Available: http://oceanenergy.ie/, accessed December 5, 2016.
24. J. F. Manwell, J. G. McGowan, and A. L. Rogers, *Wind Energy Explained: Theory, Design and Application*, John Wiley & Sons, New York, 2010.
25. Verdant Power [Online], Available: http://www.verdantpower.com/, accessed December 5, 2016.
26. ORPC/Ocean Renewable Power Company [Online], Available: http://www.orpc.co/, accessed December 5, 2016.
27. V. Jagadeesh Kumar, B. N. Biju, P. M. Koola, and M. Ravindran, Microcontroller-based instrumentation for control and PC based data acquisition system for a prototype wave energy plant, in *OCEANS 96, MTS/IEEE, Prospects for the 21st Century, Conference Proceedings*, vol. 3, pp. 1188–1192, Fort Lauderdale, Florida, September 23–26, 1996.
28. M. S. Lagoun, A. Benalia, and M. E. H. Benbouzid, A predictive power control of doubly-fed induction generator for wave energy converter in irregular waves, in *2014 First International Conference on Green Energy (ICGE)*, Sfax, Tunisia, March 25–27, 2014.

29. J. Vining, T. A. Lipo, and G. Venkataramanan, Experimental evaluation of a doubly-fed linear generator for ocean wave energy applications, in *2011 IEEE Energy Conversion Congress and Exposition*, Phoenix, AZ, September 17–22, 2011.

30. A. J. Garrido, I. Garrido, M. Alberdi, M. Amundarain, O. Barambones, and J. A. RomeroRobust, *Control of Oscillating Water Column (OWC) Devices: Power Generation Improvement, in OCEANs—Bergen, 2013 MTS/IEEE*, San Diego, CA, September 23–27, 2013.

31. J. Vining, T. A. Lipo, and G. Venkataramanan, Design and optimization of a novel hybrid transverse/ longitudinal flux, wound-field linear machine for ocean wave energy conversion, *2009 IEEE Energy Conversion Congress and Exposition*, San Jose, CA, September 20–24, 2009.

32. H. Polinder, M. E. C. Damen, and F. Gardner Linear, PM generator system for wave energy conversion in the AWS, *IEEE Trans. Energy Convers.*, 19(3), 583–589, September 2004.

33. R. P. G. Mendes, M. R. A. Calado, S. J. P. S. Mariano, and C. M. P. Cabrita, Design of a tubular switched reluctance linear generator for wave energy conversion based on ocean wave parameters, *2011 International Aegean Conference on Electrical Machines and Power Electronics and 2011 Electromotion Joint Conference (ACEMP)*, Istanbul, Turkey, September 8–10, 2011.

34. M. Santos, M. Lafoz, M. Blanco, L. García-Tabarés, F. García, A. Echeandía, and L. Gavela, Testing of a full-scale PTO based on a switched reluctance linear generator for wave energy conversion, *Presented at the Fourth International Conference on Ocean Energy*, Dublin, Ireland, October 17–19, 2012.

35. K. Rhinefrank, J. Prudell, and A. Schacher, Development and characterization of a novel direct drive rotary wave energy point absorber, in *OCEANS 2009, MTS/IEEE Biloxi—Marine Technology for Our Future: Global and Local Challenges*, October 26–29, 2009.

36. S. P. Bastien and R. B. Sepe, Jr., Optimal control of generators for water current energy harvesting, in *2012 IEEE Energy Conversion Congress and Exposition*, September 15–20, 2012.

37. H. Polinder and M. Scuotto, Wave energy converters and their impact on power systems, in *2005 International Conference on Future Power Systems*, November 18, 2005.

38. I. Gyuk, P. Kulkarni, J. H. Sayer, J. D. Boyes, G. P. Corey, and G. H. Peek, The United States of storage, *IEEE Power Energy Mag*, 3(2), 31–39, March/April 2005.

39. S. Muthukumar, S. Kakumanu, S. Sriram, R. Desai, A. A. S. Babar, and V. Jayashankar, On minimizing the fluctuations in the power generated from a wave energy plant, *2005 IEEE International Conference on Electric Machines and Drives*, May 15, 2005.

40. Y.-H. Yu, M. Lawson, K. Ruehl, and C. Michelen, Development and demonstration of the WEC-Sim wave energy converter simulation tool, Presented at the *Second Marine Energy Technology Symposium*, Seattle, WA, 2014.

41. S. H. Salter, J. R. M. Taylor, and N. J. Caldwell, Power conversion mechanisms for wave energy, *Proc. Inst. Mech. Eng. Part M: J. Eng. Marit. Environ.*, 216(1), 1–27, 2002.

42. S. Kakumanu, S. Sriram, and V. Jayashankar, Energy storage considerations for a stand-alone wave energy plant, *2005 IEEE International Conference on Electric Machines and Drives*, May 15, 2005.

12 Power Conversion and Control for Fuel Cell Systems in Transportation and Stationary Power Generation

Kaushik Rajashekara and Akshay K. Rathore

CONTENTS

Abstract 291
12.1 Fuel Cell Types and Operation 292
    12.1.1 PEMFC 292
    12.1.2 SOFC 293
    12.1.3 MCFC 293
    12.1.4 PAFC 294
    12.1.5 AFC 294
12.2 Reformers 294
12.3 Fuel Cell Characteristics and Properties 294
12.4 Modeling of Fuel Cells 297
12.5 Fuel Cells for Transportation 300
    12.5.1 Fuel Cell for Propulsion Systems 301
    12.5.2 Fuel Cell for APU and Plug-in-Fuel Cell Vehicles 304
12.6 Fuel Cells for Stationary Power Generation 305
12.7 Power Electronics for Fuel Cell Applications 309
12.8 Present Status and Future Strategies 311
12.9 Summary 312
Appendix 12A: Fuel Cell MPPT MATLAB® Example 313
    12A.1 Description 313
    12A.2 Explanation 313
    12A.3 Running the Simulink® File 315
References 316

ABSTRACT

Fuel cells are being explored for several applications such as transportation, microgrid, backup power, grid-connected and stationary power generation, and portable appliances. Fuel cells are preferred because of their secured and continuous output power, zero or low emission, and high efficiency. This chapter presents a review of the operation, power conversion, and control strategies of fuel cell systems in transportation and stationary power generation. The control characteristics and properties in fuel cells being used in these applications are discussed. Fuel cell modeling and thermodynamics to determine theoretical efficiency are explained. A few power conversion system architectures for propulsion and power generation are reviewed. The present status and future strategies for advancement of fuel cell–based systems are also reviewed.

12.1   FUEL CELL TYPES AND OPERATION

Fuel cells are modular, environmentally clean, sustainable, efficient, and of a versatile technology. Fuel cell systems have been demonstrated for clean energy, transportation, and various other applications. A fuel cell is a device that generates electricity by a chemical reaction. It has two electrodes, an anode and a cathode, with an electrolyte sandwiched between them. Hydrogen is the basic fuel for fuel cells, and exposing the anode to hydrogen and the cathode to oxygen derived from air results in electricity being produced without combustion of any form. Water and heat are the only by-products when pure hydrogen is used as the fuel source. Although hydrogen is considered the primary fuel source for fuel cells, the process of fuel reforming allows for the extraction of hydrogen from other fuels including methanol, natural gas, petroleum, or renewable hydrocarbon sources. The first fuel cell was developed in 1839 by Sir William Grove, a Welsh judge and a scientist [1]. Fuel cells are generally classified according to the nature of the electrolyte, each type requiring a particular type of electrolyte material. Also, each fuel cell type has its own unique characteristics and merits for a given application [1–3]. Table 12.1 lists different fuel cells, operating temperature, and electrical efficiency range.

12.1.1   PEMFC

The heart of the proton exchange membrane fuel cell (PEMFC) is the proton-conducting solid polymer electrolyte membrane in the form of a thin, permeable sheet. It is surrounded by two layers: a diffusion and a reaction layer. Under constant supply of hydrogen and oxygen, the hydrogen diffuses through the porous anode and diffusion layer up to the platinum catalyst. The reason for the diffusion current is the tendency of hydrogen–oxygen reaction. In the reaction layer supported by the catalyst and at a temperature of about 80 °C–100 °C, protons and electrons split as represented by $H_2 \rightarrow 2H^+ + 2e^-$.

The hydrogen ion passes through the polymer membrane on its way to the cathode, while the only possible way for the electrons is through an outer circuit. At the three-phase boundary between cathode and electrolyte, the hydrogen ions react with the oxygen, which has diffused through the porous cathode and the electrons from the outer electrical circuit, to form water. The resultant reaction is $2H^+ + \frac{1}{2}O_2 + 2e^- \rightarrow H_2O$. Excess airflow on the cathode side helps to remove the water resulting from the reaction. The overall fuel cell reaction is $2H_2 + O_2 \rightarrow 2H_2O + heat$.

TABLE 12.1
Fuel Cell Technologies

	Proton Exchange Membrane Fuel Cell	Solid Oxide Fuel Cell	Molten Carbonate Fuel Cell	Phosphoric Acid Fuel Cell	Alkaline Fuel Cell
Fuel	H_2	H_2, CO, CH_4, hydrocarbons	H_2, CO, CH_4, hydrocarbons	H_2	H_2
Electrolyte	Solid polymer (usually *Nafion*)	Solid oxide (yttria, zirconia)	Lithium and potassium carbonate	Phosphoric acid (H_3PO_4 solution)	Potassium hydroxide (KOH)
Charge carried in electrolyte	H^+	O^{2-}	CO_3^{2-}	H^+	OH^-
Operating temperature (°C)	80–100	800–1000	650	175–200	80–120
Electrical efficiency (%)	35–50	50–60	45–55	35–45	35–55

The efficiency of the PEMFC is about 40%–50%, and its operating temperature is about 80 °C–100 °C. Owing to low-temperature operation, faster response to load transients, and short start-up time (less than a minute), the PEMFC is suitable for propulsion applications. Several industries are developing this type of fuel cells for propulsion and other applications. However, the PEMFC requires pure hydrogen, and drastic performance degradation occurs even from a small amount of carbon monoxide, resulting in poisoning of the electrodes. A sophisticated water management system is also required because of the continuous water generation at the cathode and to keep the membrane at a certain humidity level. Although significant research has been done to reduce the level of platinum coating on the electrodes, this requirement makes the proton exchange membrane (PEM)-based systems relatively expensive.

PEMFCs operate at low temperatures (less than 100 °C), making them temperature compatible with many of today's automotive systems and also allowing for a faster start-up. However, due to a relatively small temperature gradient to the ambient atmosphere, the waste heat produced is low grade and requires large heat exchangers. Also, for the PEM electrolyte to operate properly it must be hydrated, resulting in the need for sophisticated water management and the humidification of the incoming fuel and oxygen flows. This hydration also causes issues because automobiles must operate below the freezing point of water, with typical specifications for today's automotive components being in the range of –30 °C to –40 °C. Finally, to achieve the high power density, the incoming reactant gases are pressurized to 2–3 atmospheres to increase power output, resulting in the need for high-pressure and high-flow air compressors. PEMFCs have been demonstrated in systems with the size ranging from 1 W to 250 kW.

12.1.2 SOFC

A solid oxide fuel cell (SOFC) usually uses a hard ceramic material of solid zirconium oxide and a small amount of yttria, instead of a liquid electrolyte, allowing operating temperatures to reach about 1000 °C. The solid electrolyte is coated on both sides with specialized porous electrode materials. The hydrogen is supplied at the anode, and oxygen, usually from air, at the cathode. At these high operating temperatures, oxygen ions (with a negative charge) migrate through the electrolyte to the anode. Electrons generated at the anode travel through an external load to the cathode, completing the circuit and supplying electric power along the way.

SOFC requires a simple reforming process and may not need an external reformer. SOFC operates at extremely high temperatures in the range of 700 °C–1000 °C and hence has less compelling requirements for reformate quality and uses carbon monoxide as fuel. High-grade heat exhaust of SOFC allows for smaller heat exchangers and the cogeneration option to produce extra power, thus increasing the overall system efficiency. Because the electrolyte is in a solid state and does not require hydration, water management is not a concern. The by-product is steam rather than liquid water, which must be drained out in a PEM system. In addition, fuel versatility makes SOFCs suitable for large to very large stationary power generation applications. SOFC has also been demonstrated as an auxiliary power unit (APU) in automotive systems [4, 5]. The SOFC is very suitable for combined heat and power (CHP) and for hybrid power generation (or cogeneration), where the exhaust of the SOFC can be used to run a turbine to generate additional electric power. The start-up time of the SOFC system is long because of its high-temperature operation and therefore is not suitable for propulsion applications.

12.1.3 MCFC

Molten carbonate fuel cell (MCFC) uses high-temperature compounds of salt (like a mixture of sodium and lithium or magnesium or lithium and potassium) carbonates (chemically CO_3) as the electrolyte and is the only fuel cell that requires CO_2 supply. Their nickel electrode catalysts are inexpensive compared to the platinum used in PEMFC. Carbonate ions from the electrolyte are used up in the reactions at the anode, making it necessary to compensate by injecting carbon dioxide at the cathode. The operating temperature of the MCFC is about 650 °C. Due to the high-temperature operation, MCFC is less prone to carbon monoxide "poisoning" than lower-temperature fuel cells, which makes coal-based fuels more attractive for this type of fuel cell. Like SOFC, it is best suited for stationary power generation applications and cogeneration. Units with output up to 2 megawatts (MW) have been constructed, and designs exist for units up to 100 MW. MCFC also has a long start-up time to reach the operating temperature and has a lower power density than SOFC and is not suitable for propulsion application.

12.1.4 PAFC

Phosphoric acid fuel cell (PAFC) uses phosphoric acid as the electrolyte and the operating temperature of these cells is about 200 °C. It uses platinum electrode catalysts to withstand the corrosive acid effects. Because of about 200 °C operation, PAFCs can tolerate a carbon monoxide concentration of about 1.5%, which broadens the choice of fuels they can use. However, the sulfur needs to be removed from the fuel. PAFC was the first fuel cell to cross the commercial threshold. The United Technology Corporation developed and placed a number of PAFC-based 50–200 kW range power units in operation for stationary power applications in the United States and overseas. Most fuel cell systems that were sold before 2001 for stationary power generation application used PAFC technology [6].

12.1.5 AFC

Alkaline fuel cells (AFCs) generally use a solution of potassium hydroxide in water as their electrolyte, and the operating temperature is 150 °C–200 °C. AFC uses platinum electrode catalysts. The hydroxyl ions (OH^-) migrate from the cathode to the anode. At the anode, hydrogen ions react with the OH^- ions to produce water and release electrons. Electrons generated at the anode supply electric power to an external circuit and then return to the cathode. At the cathode, the electrons react with oxygen and water to produce more hydroxyl ions that diffuse into the electrolyte. AFCs are highly reliable and efficient and are being used in space applications. NASA selected AFCs for their space shuttle fleet and Apollo program to provide both electricity and drinking water.

12.2 REFORMERS

A reformer is a device that produces hydrogen from fuels such as gasoline, methanol, ethanol, or naphtha. When using a fuel other than pure hydrogen for a fuel cell, a reformer or a fuel processor is required. The three basic reformer types are steam reforming, partial oxidation, and autothermal reforming. Steam reformers combine fuel with steam and heat to produce hydrogen. The heat required to operate the system is obtained by burning fuel or from excess hydrogen from the outlet of the fuel cell stack. Partial oxidation reformers combine fuel with oxygen to produce hydrogen and carbon monoxide. The carbon monoxide then reacts with steam to produce more hydrogen. Partial oxidation releases heat, which is captured and used elsewhere in the system. Autothermal reformers combine the fuel with both steam and oxygen so that the reaction is in heat balance. In general, both methanol and gasoline can be used in any of the three reformer types.

12.3 FUEL CELL CHARACTERISTICS AND PROPERTIES

In this section, the properties and control characteristics of the fuel cells are examined for designing the overall power conversion system to obtain the required voltage and power output for various applications.

A typical voltage–current characteristic of a fuel cell is shown in Figure 12.1. As can be seen, the actual voltage decreases as a function of the current drawn from the fuel cell. Several sources contribute to irreversible losses in a practical fuel cell. The losses, which are often called polarization losses, originate primarily from three sources: (1) losses due to the rate of reaction, (2) ohmic or resistance loss, and (3) concentration polarization or gas transport loss. These losses result in a cell voltage (V) for a fuel cell that is less than its ideal potential, E ($V = E -$ Losses). As shown, this characteristic is divided into three regions: R-1, R-2, and R-3. The point at the boundary of regions R-2 and R-3 is known as maximum power density point or knee/optimum point [1, 2, 8]. Loading the fuel cells above the maximum power point (MPP) current will shift the operating point right of the optimum point (region R-3) causing a sudden collapse of the fuel cell voltage to zero. Therefore, no power could be drawn from the cell. Extended operation in region R-3 may damage the fuel cell. Fuel cells are generally operated in the region R-2 of the characteristics shown in Figure 12.1. Various PEMFC characteristics as a function of typical values of current density are presented in Figure 12.2 [9]. Figure 12.2a is referred to as the performance curve showing the voltage as a function of the current density. The efficiency indicated on the secondary vertical axis is proportional to the cell voltage. The power density is shown in Figure 12.2b as a function of the current density. The efficiency as a function of power density is shown in Figure 12.2c. The efficiency increases at light load. The dotted line demonstrates above the maximum power regime. Figure 12.2d shows a typical efficiency characteristic of a complete fuel cell system in a vehicle, as a function of load power. It also compares with the thermal efficiency with a typical internal combustion engine (ICE) in a vehicle.

The fuel cell output current and output power are controlled by controlling the fuel flow input to the fuel cell. The MPP moves to higher current levels with an increase in fuel flow and therefore increases the ability of fuel cell to transfer higher power to the load as shown in Figure 12.3 [10]. Fuel flow is the main parameter that controls the power transfer capacity of the fuel cell. For efficient operation of the fuel cells, the operating point should be adjusted as a function of the electrical load. The flow rates of both fuel and oxidant are controlled to ensure that stoichiometric ratio remains in the design range to ensure a good balance between reactant supply, heat, water management, and pressure drop.

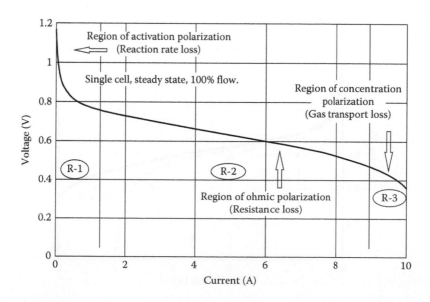

FIGURE 12.1 Typical fuel cell V–I characteristic.

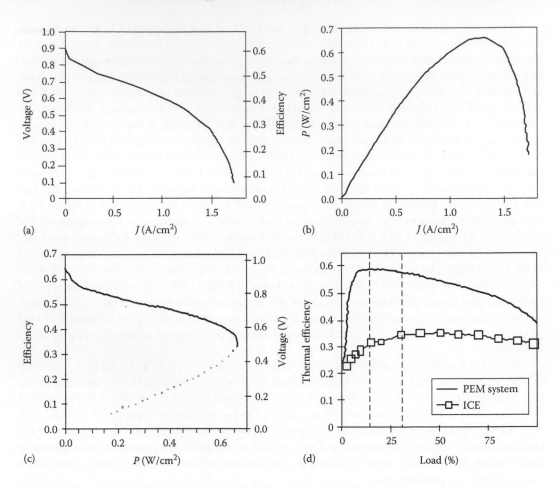

(a)   (b)   (c)   (d)

FIGURE 12.2   Typical characteristics of a PEMFC: (a) performance curve, (b) power density, (c) efficiency, and (d) system efficiency. (Based on Kartha and Grimes, *Phys. Today*, 11, 54, 1994.)

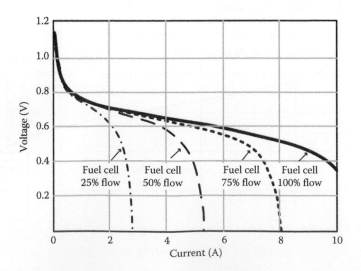

FIGURE 12.3   Variation of fuel cell output and MPP with fuel flow.

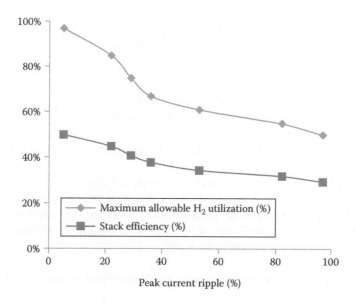

FIGURE 12.4 Effect of current ripple on fuel utilization in fuel cells.

Fuel cells are slow in responding to the fast electrical load transients because of slower internal electrochemical and mechanical dynamics. Load transients create a harmful low-reactant condition inside the fuel cells and shorten their life. The difference between the time constants of the fuel cell and electrical load calls for an energy storage unit that would supplement the peak power demand from the fuel cell during transients. Secondary energy source such as a battery or ultracapacitor: (1) compensates for the slow fuel cell dynamics, (2) responds to the fast-changing electrical load during transients, and (3) provides the power to the load until the fuel cell output is adjusted to match the new steady-state demand [11–16].

Fuel cells are very sensitive to low-frequency ripple current. While feeding the low-frequency alternating current to the utility or AC load from a fuel cell–based power plant, a second harmonic component of the line current may appear at the fuel cell stack. The low-frequency ripple current reaching the fuel cell may move the operating point from region R-2 to R-3 (Figure 12.3), thus leading to unstable operation of the fuel cell system. This may result in the maloperation of the fuel cell power unit, and hence, the system may shut down. Therefore, this low-frequency current should be absorbed.

Figure 12.4 shows how the peak-to-peak current ripple from a fuel cell stack affects the efficiency of a fuel cell and fuel utilization. Because of the ripple current, the fuel flow may have to be adjusted to peak value (instead of average) resulting in waste of fuel leading to higher cost of energy and poor efficiency and utilization [17, 18]. Fuel utilization is very important for a better overall system efficiency. In practice, the fuel utilization rate should be above 80%–85%.

12.4 MODELING OF FUEL CELLS

The modeling of the fuel cell enables to analyze the detailed operation of the fuel cell system for various operating conditions. In this section, the basic equations for developing a fuel model are described as follows. Thermodynamic equations to derive fuel cell efficiency are also given and explained. In general, for any arbitrary number of products and reactants, the ideal thermodynamic electrical potential voltage, E, is described by the Nernst equation given by [1, 2]

$$E = E_{cell}^o - \frac{RT}{nF} \ln \frac{\Pi a_{products}^{v_i}}{\Pi a_{reactants}^{v_i}}$$ (12.1)

where
  E^o is the open-circuit voltage of the fuel cell at no load
  R is the universal gas constant (R = 8.314472(15) J/K/mol)
  T is the absolute temperature
  F is the Faraday constant (the number of coulombs per mole of electrons is F = 9.64853399 × 10^4 C/mol)
  n is the number of moles of electrons transferred in the cell reaction
  a is the chemical activity for the relevant species (for ideal gas, $a_i = p_i/p_o$, where p_i and p_o are, respectively, the partial pressure of species i and the standard state pressure, that is, 1 atm)
  v_i denotes the stoichiometric coefficient of ith species involved in the chemical reaction (in this case the stoichiometric coefficient of O_2 is ½)

In case of a hydrogen–oxygen fuel cell system, $n = 2$, and therefore, Equation 12.1 reduces to

$$E = E_{cell}^o - \frac{RT}{2F} \ln \left(\frac{1}{p_{H_2} p_{O_2}^{1/2}} \right)$$ (12.2)

This expression gives the ideal thermodynamic potential or maximum theoretical voltage across the cell. It is also known as the open-circuit voltage at standard temperature and pressure with no current drawn from the cell.

When the activities of H_2 and O_2 are both unity, that is, the partial pressure of both H_2 and O_2 is 1 atm, the expression reduces to

$$E = E_{cell}^o$$ (12.3)

The E_{cell}^o decreases with increase in temperature. Therefore, at standard pressure conditions, $p_{H_2} = p_{O_2} = 1$ atm, the open-circuit voltage of an SOFC is lower than that of a PEMFC.

To calculate the fuel cell efficiency, some thermodynamic parameters, that is, Gibbs free energy G, enthalpy H, and entropy S, need to be defined and determined as a result of the chemical reaction.

Entropy is another thermodynamic property that is a measure of a system's thermal energy per unit temperature that is unavailable for doing useful work:

$$\Delta S = \frac{Q}{T}$$ (12.4)

where
  Q is the heat or thermal energy
  T is the absolute temperature of the system

In the case of a fuel cell (chemical reactions), entropy appears in the form of rejected or released heat in the process. Chemical reactions act as a source of enthalpy H and generate an amount of electrical energy W_e and reject an amount of thermal energy (heat) Q as shown in Figure 12.5.

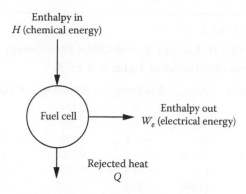

FIGURE 12.5 Fuel cell input–output in terms of thermodynamic parameters.

According to the law of thermodynamics, in a practical fuel cell, there will be a net increase in entropy. The entropy appearing in the form of rejected heat and water (liquid H_2O) must be greater than the entropy contained in the reactants (H_2 and O_2), that is, entropy gain $\geq$ entropy loss:

$$\frac{Q}{T} + \Sigma S_{products} \geq \Sigma S_{reactants} \tag{12.5}$$

$$Q \geq T \cdot \left(\Sigma S_{reactants} - \Sigma S_{products} \right) \tag{12.6}$$

where
$\Sigma S_{products}$ is the sum of the entropies of the products as a result of chemical reaction
$\Sigma S_{reactants}$ is the sum of the entropies of the reactants participating in the chemical reaction

The equation shows the minimum amount of heat generated in the fuel cell. The enthalpy supplied by the chemical reaction H equals the electrical energy W_e produced plus the heat rejected Q:

$$H = W_e + Q \tag{12.7}$$

The fuel cell's efficiency η is given by

$$\eta = \frac{W_e}{H} = \frac{H - Q}{H} = 1 - \frac{Q}{H} \tag{12.8}$$

The Gibbs free energy ΔG refers to the maximum possible, entropy-free, electrical or mechanical output from a chemical reaction. It can be calculated by taking the difference between the sum of the Gibbs free energies of the reactants and the products:

$$\Delta G = \Sigma G_{products} - \Sigma G_{reactants} \tag{12.9}$$

where
$\Sigma G_{products}$ is the sum of the Gibbs free energies of the products as a result of the chemical reaction
$\Sigma G_{reactants}$ is the sum of the Gibbs free energies of the reactants participating in the chemical reaction

---

TABLE 12.2

Enthalpy H, Entropy S, and Gibbs Free Energy G of Various Substances at 1 atm and 25 °C

Substance	State	H (kJ/mol)	S (kJ/mol K)	G (kJ/mol)
H_2	Gas	0	0.13	0
O_2	Gas	0	0.205	0
CH_4	Gas	−74.9	0.186	−50.8
CH_3OH	Liquid	−238.7	0.1268	−166.4
H_2O	Gas	−241.8	0.1888	−228.6
H_2O	Liquid	−285.8	0.0699	−237.2
CO_2	Gas	−393.5	0.213	−394.4

Fuel cell efficiency η can be calculated by evaluating the ratio of the Gibbs free energy ΔG to the enthalpy change ΔH in the chemical reaction:

$$\eta = \frac{\Delta G}{\Delta H} \tag{12.10}$$

The change in enthalpy in a chemical reaction can be calculated by

$$\Delta H = \Sigma H_{reactants} - \Sigma H_{products} \tag{12.11}$$

where

$\Sigma H_{products}$ is the sum of the enthalpies of the products as a result of chemical reaction
$\Sigma H_{reactants}$ is the sum of the enthalpies of the reactants participating in the chemical reaction

Table 12.2 shows the enthalpy, entropy, and Gibbs free energy at 1 atm and 25 °C for a few substances. The values in Table 12.2 can be utilized based on the state of the reactants and the products to calculate Gibbs free energy or electrical energy output ΔG, enthalpy change due to chemical reaction ΔH, net entropy ΔS, heat liberated Q at an absolute temperature T, and theoretical efficiency η of the fuel cells.

12.5 FUEL CELLS FOR TRANSPORTATION

Transportation electrification is growing to meet the challenges of reducing emissions and meeting the increasing power demand of electrical loads in automobiles. Fuel cell vehicles (FCVs) are promising owing to their merits of high efficiency and low or zero emissions as compared to conventional internal combustion engine vehicles. FCVs offer the benefits of zero emission without the charging time and driving range limitations of electric vehicles (EVs). Automotive industries like Honda, Toyota, GM, Ford, and Kia are all developing their FCVs. Fuel cells are being considered for providing the propulsion power to the vehicle and also for onboard power generation to power the electrical loads in the vehicle.

Because of a short start-up time, relatively low operating temperature, and faster response, PEMFCs are considered mainly for vehicle propulsion applications. In the past, fuel cells were mainly considered for the propulsion applications. But recently, fuel cells are being investigated as APU for onboard power generation to supply power to the accessory loads for either engine-on or engine-off conditions. For APU, either the PEMFC or the high-temperature SOFC could be used.

12.5.1 FUEL CELL FOR PROPULSION SYSTEMS

A fuel cell system designed for propulsion applications should match the weight, volume, power density, start-up, and transient response with present-day ICE-based gasoline vehicles. Other requirements are high performance over a short start-up time, better fuel economy, fast acceleration, and the meeting of all the safety requirements. Expected lifetime and cost are also major considerations [7]. A FCV drivetrain system usually consists of a fuel cell stack, power conditioner (or DC–DC converter) to convert the variable output voltage of fuel cell stack to a fixed DC voltage, propulsion inverter to obtain variable voltage and variable frequency AC power, propulsion motor, and a transmission to transmit the power from the electric motor to the wheels as shown in Figure 12.6 [5, 7]. The fuel cell stack for propulsion application in the majority of vehicles is based on PEMFC. The hydrogen fuel input to the fuel cell can be obtained from (1) electrolysis of water and storage of hydrogen in pressurized cylinders, (2) metal hydrides such as sodium aluminum hydride and lithium aluminum hydride, and (3) reformation of liquid fuels such as gasoline, methanol, and other hydrocarbon-based fuels. Most of the early fuel cell demonstration vehicles were based on the onboard reformer for producing hydrogen to fuel the fuel cell stack. Because of the reformer's additional weight and volume, these vehicles were not efficient and had a poor performance. Automotive manufacturers are presently focusing on direct hydrogen-based FCVs instead of onboard reformer-based vehicles.

A battery is generally connected in parallel with the fuel cell system to enable the most efficient usage of high power density of the battery and the inherently high energy density of the fuel cell. During high power demand such as acceleration, batteries will supply the required power. During normal driving operation such as cruising, the fuel cell supplies the required power. During low power demand, batteries will be recharged. Therefore, based on the energy and power requirements, the fuel cell and battery could be designed to supply cruising power and peak power, respectively. The battery also assists in the rapid start-up of the fuel cell and protects it against cell reversal during this operation. In addition, the battery supplies the peak power and enables the vehicle's system to respond faster for load changes and to capture the regeneration energy. Several other benefits of using the battery are that the vehicle can be started without the preheating requirement of the fuel cell and can be operated in pure electric mode until the fuel cell reaches its nominal output voltage level. The battery technology that has been found in most of the present-day FCVs is based on lithium ion because of its high energy density, high power density, life cycle, etc.

Propulsion motor determines the propulsion system characteristics of the vehicle, operation of the motor controller, and the ratings of the semiconductor devices of the power converters. The main requirements for a propulsion motor are ruggedness, high torque-to-inertia ratio, high torque density, wide speed range, low noise, little or no maintenance, small size, ease of control, and low cost. Most of the present-day electric, hybrid, and fuel cell vehicles are based on the permanent

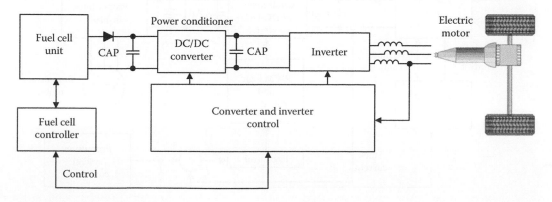

FIGURE 12.6 A fuel cell propulsion system for an automobile.

magnet motors with Honda focusing on the concentrated winding–type and Toyota on distributed winding–type motors [24]. The permanent magnet AC motor offers merits of higher power density and thus lower weight compared to other types of motors of similar rating. Tesla is the only major EV manufacturer using induction motors for propulsion. Induction motors were employed for most of the early EVs including GM EV1 and FCVs [24]. Induction motors are rugged, have high-speed capability, and are of a smaller size compared to a separately excited DC motor.

A power electronics system forms a major portion of the FCV propulsion system. It consists of a power conditioner or a DC–DC converter to match the output voltage of the fuel cell stack to the DC input voltage of the inverter. The inverter converts the DC voltage to the required voltage and frequency to drive the propulsion motor. Also, additional power converters are required for powering the balance of plant electrical loads of the fuel cell system and all the accessory loads of the vehicle. Several requirements related to power converter and propulsion motor control strategies are similar among the electric, hybrid, and fuel cell vehicles.

A fuel cell propulsion system with a battery pack and a power conditioner is shown in Figure 12.7 [7]. Both the fuel cell stack and the battery provide the power required for propulsion. The power conditioner must be sized depending on the maximum power capacity of the fuel cell stack. To prevent the negative current from going into the stack, a diode is connected in series with the fuel cell stack. Negative current may cause cell reversal and damage the fuel cell stack. The power for the accessory loads of the fuel cell system is derived from the battery side of the converter to make sure that accessory loads are always powered even when the fuel cell stack is not producing any power. This would help to start the system faster. The battery power could also be used to initially warm up the system and bring the stack output voltage to a nominal level. Thus, in this kind of layout, starting the fuel cell system will not be a concern. The battery voltage has to be selected to be equal to the required DC input voltage to the power inverter. If the propulsion system is designed to be operated at a DC voltage above the fuel cell voltage, the power conditioner boosts the fuel cell stack voltage up to the battery voltage and also charges the propulsion batteries. If the fuel cell stack voltage is higher than the battery voltage, the power conditioner acts as a buck converter.

In the system shown in Figure 12.8, the battery voltage is independent of the DC input voltage. Hence, a lower battery voltage can be selected and a bidirectional DC–DC converter is used to match the battery voltage to the DC-link voltage [7–19]. The fuel cell stack voltage determines

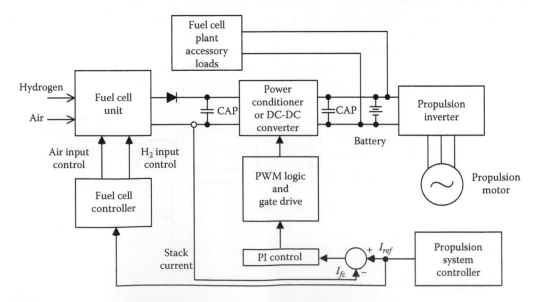

FIGURE 12.7   A fuel cell propulsion system with a power conditioner and battery.

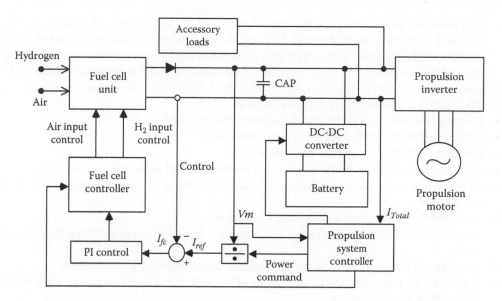

FIGURE 12.8 A fuel cell propulsion system with different voltages for the battery and the DC input to inverter.

the input voltage to the propulsion inverter. Generally, the battery voltage is selected lower than the DC-link voltage. The DC–DC converter acts as a boost converter to provide the propulsion power. During charging of the battery unit from the fuel cell, the DC–DC converter is operated in a buck mode. When the vehicle is started, the battery voltage is boosted to supply power to the propulsion motor. Both the fuel cells and the battery supply power during rapid acceleration. Once the vehicle attains its steady speed, only the fuel cell supplies the propulsion power and also charges the battery. In this case, the DC–DC converter operates in the buck mode. During regeneration, the converter acts as a buck converter and charges the battery. The power drawn from the fuel cell is controlled by controlling the output current of the fuel cell stack for a given output voltage. The stack current is proportional to the hydrogen fuel flow to the stack input. Thus, the reference current from a vehicle system controller controls the fuel flow rate into the stack. This configuration has been used in Toyota fuel cell demonstration vehicles [20, 21].

If the load is suddenly applied across the fuel cell stack, the fuel cell may not respond instantly because of the additional fuel flow requirement and also the change in fuel flow rate. Hence, the rate of increase of the output power of the fuel cell stack has to be limited. If the amount of hydrogen flow into the stack is higher than that required by the electrical load, then energy is wasted in the exhaust. If the fuel flow is lower than that required by the electrical load, then the stack impedance increases causing overheating of the stack. Therefore, it is essential to match the hydrogen flow rate to the stack to meet the expected electrical load output. Another important item is the coordination of the power delivered from the battery and from the fuel cell stack for optimum energy management from the energy sources. In FCVs, it is important to select the optimum DC voltage for the stack and for the propulsion drive. A large number of cells in series affect the efficiency of the system because of higher series impedance, but a higher voltage results in a lower current and hence lower losses in the power electronics and motor.

The issues to be considered in the design of the fuel cell propulsion system are the following:

- Optimum DC voltage for the stack and for the propulsion drive. This determines the number of cells to be connected for the stack.
- Rate of increase of the output power of the fuel cell stack. Due to the sudden application of the load, the fuel cell may not respond instantly because of the requirement of additional

fuel flow and also a change in the rate of fuel flow. If the amount of hydrogen flow to the stack is higher than that required by the electrical load, then energy is wasted in the exhaust. If the fuel flow is less than that required by the electrical load, then the impedance of the stack increases, thus overheating the stack. Hence, it is necessary to match the amount of hydrogen flow to the stack to meet the desired electrical load at the output.

- The sequence of shutting off the entire system.
- Effect on the fuel input if the load is suddenly disconnected.
- Isolation of the fuel cell stack from the drive system.
- Connecting the battery and the fuel cell stack.
- Charging the capacitors of the inverter and the limitation of capacitor inrush current.
- Coordination of the battery charging simultaneously using regenerative energy and from the fuel cell.
- Charging the battery from the fuel cell stack alone.
- Coordination of the power delivered for propulsion from the battery and from the fuel cell stack.
- Limiting the battery current during regeneration and charging at the same time.
- Supplying the power to the accessory loads of the vehicle and to the accessory loads of the fuel cell stack.
- Matching the fuel cell output characteristics with those of the battery and the drive system.
- Advanced sensors for fuel flow measurement, temperature/pressure regulation, and sensing the various chemical reactions in the fuel cell unit.

12.5.2 FUEL CELL FOR APU AND PLUG-IN-FUEL CELL VEHICLES

Start–stop technology for automobiles based on 48 V DC is rapidly gaining momentum particularly in Europe. This technology will automatically switch off the engine every time the vehicle stops or when the engine is idling and restart it instantly as and when needed. However, instead of the engine-driven alternator, a fuel cell can produce electricity when the engine is running as well as not running and thus could eliminate the need for an alternator. Also, it has higher efficiency as compared to an engine-driven generator. The fuel cell APU can be used in conventional or hybrid vehicle systems and is not linked to a fully electric drivetrain [4, 22]. It could be used to generate power at 42 or 48 V and meet the challenges of providing higher power to the increasing electrical loads such as electrical air-conditioning, drive-by-wire, and steer-by-wire systems. Instead of the PEMFC, high-temperature SOFC could also be used for automotive APU because of the potential for internal reforming of more conventional petroleum fuels.

An electrical architecture with a typical fuel cell APU is shown in Figure 12.9. In this system, the fuel cell unit generates 48 V and powers the 48 and 12 V vehicle loads [22, 23]. A 48 V/12 V buck–boost converter is employed for charging the 12 V battery in the vehicle and for powering all the 12 V accessories. The system is started by boosting the 12–48 V to power the accessories of the fuel cell unit. The function of the power conditioner is to obtain controlled 48 V to power the loads. The control system operation is similar to the systems in Figures 12.7 and 12.8.

The fuel cells could be used as range extenders instead of the ICE-driven generators in series hybrid vehicles. A range extender–type fuel cell with a lower rating could be used only to charge the batteries. The battery should be designed to supply the full propulsion power to the vehicle. For such applications, it is possible to use the PEMFC or SOFC. The battery unit of this type of vehicle can be charged at night similar to a plug-in hybrid electric vehicle. This plug-in hybrid fuel cell vehicle (PFCV) consisting of a smaller fuel cell and a larger battery (battery dominant) may be the future direction for automobiles [24]. The PEM technology is already available and combining this with a compressed hydrogen cylinder for onboard charging would lead to a zero-emission vehicle with a range greater than 500 miles (800 km). If SOFC is used as shown in Figure 12.10, it would also provide electric power to the house as vehicle-to-grid operation and sustained heat. The hydrogen

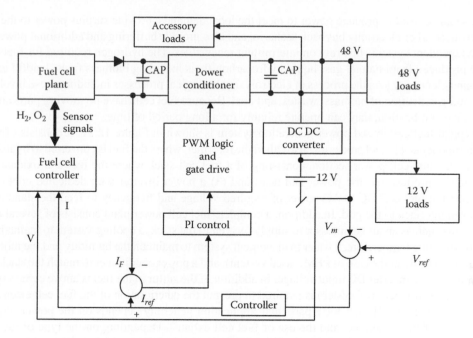

FIGURE 12.9 Fuel cell–based APU powering 48 and 12 V automotive loads.

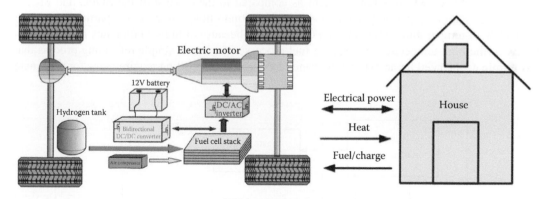

FIGURE 12.10 Plug-in fuel cell hybrid vehicle that can be committed to a home.

infrastructure is not stringent for the PFCVs. If hydrogen is generated using renewable energy and lithium-ion battery is charged using the power generated from the renewable energy sources, the PFCVs could be dominant in the automotive market. In addition to using fuel cell for propulsion and as APUs in automobiles, there is an increasing interest in using fuel cell as an APU for onboard power generation in airplanes and ships [25].

12.6 FUEL CELLS FOR STATIONARY POWER GENERATION

Fuel cells are also used for commercial, industrial, residential, and grid-tied (or microgrid) power generation applications. Fuel cells have also been used as power sources in remote weather stations, communication centers, rural locations, and military applications. The stationary fuel cell power systems can be used to produce power to meet the local demands, provide surplus power to the grid in a net-metered or electricity buyback scenario, and also provide buffering and additional power for grid-independent systems that rely on intermittent renewables. The hydrogen required for fuel cells can be produced from natural gas; liquid hydrocarbon fuels including biomass fuels, landfill gases, water, and electricity (via the process of electrolysis); biological processes including those involving algae; and gasification of biomass, wastes, and coal. Because fuel cells have no moving parts and do not involve combustion, they can operate reliably for a long period of time.

A typical fuel cell–based power generation system is shown in Figure 12.11. It consists of three main components: (1) fuel processor, also called the reformer, where the fuel is converted to hydrogen-rich gas; (2) fuel cell power section, consisting of a fuel cell stack where the fuel and oxidants are combined to produce electric power and heat; and (3) a power inverter with dedicated controllers to convert the fuel cell DC to AC power of required voltage and frequency to feed the stand-alone loads or to interface to the grid. In addition, a complete fuel cell power plant consists of several other subsystems such as an air compressor to supply oxygen to the stack, a cooling system to maintain the desired operating temperature, a water management system to maintain the humidity and the moisture in the system (only in the case of PEM-based system), and a power conditioner to match the stack output voltage to the inverter DC input voltage. In addition, if the sulfur in the fuel is above certain specified limits, a sulfur removal system is required to prevent the deterioration of the fuel cell electrodes.

The type of fuel cell used for a particular application generally depends on the power requirements, type of fuel available, and the use of fuel cell exhaust. Depending on the type of fuel cell system, waste heat is available at different temperatures and can be used for various heating and cooling applications. Generally, for low power and residential applications, AFCs or PEMFCs are used. Although PEMFC has been used at high power (>200 kW), the waste heat available is only about 80 °C–90 °C, which is not as useful as compared to the exhaust of the SOFC and MCFC systems. For stationary power generation and for interconnection to microgrids, particularly above 200 kW, high-temperature SOFC or MCFC is suitable because of higher efficiency and fuel flexibility. In the high-temperature SOFC, the fuel processor requires a simple reforming process that may eliminate the need for an external reformer and has less compelling requirements for reformate

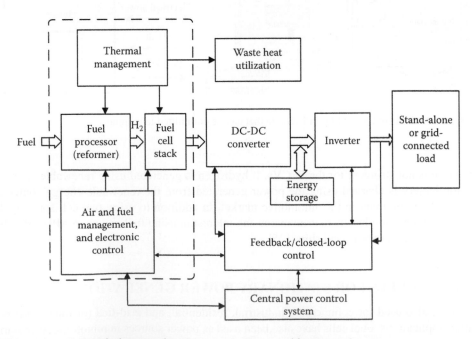

FIGURE 12.11 Typical fuel power plant for a stand-alone or grid-connected system.

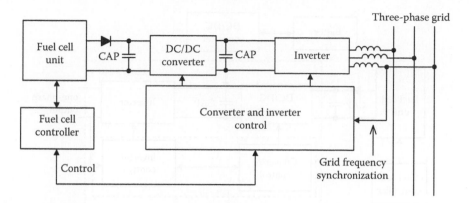

FIGURE 12.12 A grid-connected fuel cell–based power conversion system.

quality and uses carbon monoxide directly as fuel. The high-quality waste heat can be used for heating buildings and for cogeneration. Also, water management is not an issue because the electrolyte is a solid-state material that does not require hydration, and the exhaust is steam rather than liquid water. MCFC is suitable for stationary power generation because of high efficiency, flexibility of fuels, and high-temperature operation close to 650 °C. Again, the high-quality heat generated can be utilized for CHP resulting in higher efficiency. As already mentioned, PAFCs have been used in stationary power generation, but presently most manufacturers are developing SOFC and MCFC based systems.

The power conversion system for a grid-connected fuel cell generation system is shown in Figure 12.12. The DC–DC converter converts the fuel cell voltage to the required DC voltage input to the inverter. It is also possible to operate the system without a DC–DC converter if the nominal fuel cell voltage is compatible with the required input voltage to the converter [8, 11, 13, 14, 26]. In this case, the inverter devices have to be selected to withstand the open-circuit voltage of the fuel cell.

The inverter is the interface unit between the fuel cell and the electrical network in a fuel cell power plant for electrical utility applications. It adjusts the voltage and frequency according to the electrical load. The interface conditions require that the inverter unit has the ability to synchronize with the network and regulates the output voltage to 208 or 480 V or (as specified) within ±2% regulation. The inverter also needs to supply the required reactive power to the network within the inverter capabilities, adjustable between 0.8 lagging and unity power factor based on the inverter used and without impacting maximum kW output. It should also suppress the ripple voltage fed back to the fuel cells and the output voltage harmonics so that the power quality is within the IEEE 519 harmonic limit requirement. When the inverter is connected to the grid, it is operated in current control mode. When the fuel cell plant is operating as a stand-alone mode, the inverter is operated in voltage control mode. The response of the fuel cell unit to system disturbances or load swings also must be considered whether it is connected to a dedicated load or utility's grid. A few of the demonstrated fuel cell power conditioners have no transient overload capability beyond the kW rating of the fuel cell, a load ramp rate of 80 kW/s when operated independently of the utility grid, a load ramp rate of 0%–100% in one cycle when operated independently of the utility grid, and a load ramp rate of 10 kW/s when connected to the utility grid and following initial ramp up to full power.

In residential and other low power applications, a battery is generally used to provide the power during starting, peak demand, and transients. A typical system is shown in Figure 12.13. Battery voltage is different and is generally lower than fuel cell stack. The power conditioner is a DC–DC converter that boosts and regulates the fuel cell stack voltage at a DC voltage above the peak of the AC voltage at the input of the inverter. DC–DC converter connected to the battery translates the battery voltage up to common DC bus of inverter input. This same DC–DC converter is also used for charging the battery.

Fuel cells could be made modular and can be connected in series and parallel to obtain the desired voltage and power [27]. A series connection of fuel cell stacks to increase the voltage level

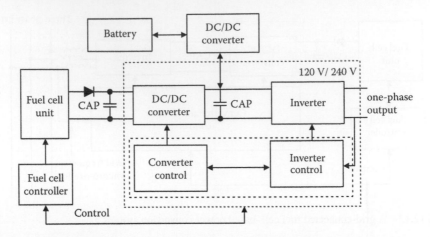

FIGURE 12.13   A stand-alone low power fuel cell power generation system.

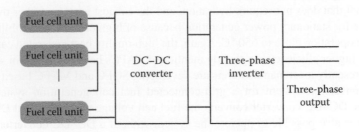

FIGURE 12.14   Modular series-connected fuel cell units for high power application.

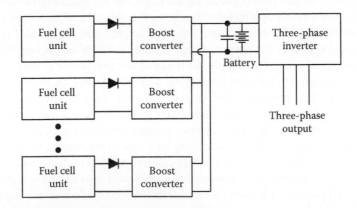

FIGURE 12.15   A modular fuel power system with independent boost converters for high power application.

is shown in Figure 12.14. The fuel flow in all fuel cell units should be the same to produce the same amount of current to achieve optimal operation. A DC–DC converter is used to meet the input voltage requirement of a three-phase inverter. This configuration needs only a single DC–DC converter and one inverter. In Figure 12.15, the fuel cell stacks are connected in parallel using individual boost converters to obtain the required DC input voltage to the inverter.

High-temperature fuel cell–based systems could be combined with another power generation technology such as a turbine or a microturbine resulting in higher power generation efficiency than

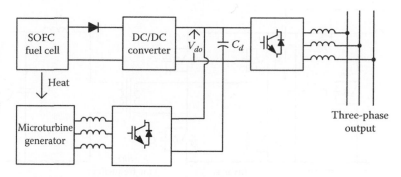

FIGURE 12.16   Power conversion systems for combining SOFC and turbine generator outputs.

could be provided by either system operating alone [1, 26]. In this type of hybrid power generation system, the high-temperature exhaust of SOFC unit can be used to run the turbine generator. The output of the turbine generator and the fuel cell output can be combined as shown in Figure 12.16 [28]. Fuel cell DC–DC converter and the AC to DC PWM rectifier at the output of the turbine generator must be controlled to share the output power according to their VA ratings and regulate the DC-link voltage of the inverter. In a hybrid system, the microturbine generator functions as a turbo charger for the SOFC, resulting in system efficiencies greater than 65%. Fuel cell power units and also hybrid power generation systems can be combined with other power generation sources in a microgrid to feed power to DC- and AC-type loads [29].

12.7   POWER ELECTRONICS FOR FUEL CELL APPLICATIONS

Power electronics is an essential component in a fuel cell power conversion system to convert variable fuel cell stack DC voltage into regulated output DC voltage or desired AC voltage and frequency. In fuel cell inverter applications, the fuel cell stack voltage must first be boosted to at least the peak of the utility line AC voltage at the DC link of the inverter. This DC–AC power conversion is possible using single-stage or multistage power conversion. Depending on fuel cell stack voltage, particularly at lower voltages, a transformer-isolated DC–DC power converter is necessary to be used as a boost converter. Also, the transformer isolates fuel cell stack from the utility line in case of fault and also ensures the safety of personnel.

A single-stage DC–AC power conversion system using a transformer-isolated inverter is shown in Figure 12.17 that uses line frequency transformer to scale the voltage before connecting to the utility line. Figure 12.18 shows a two-stage power conversion system using a front-stage nonisolated

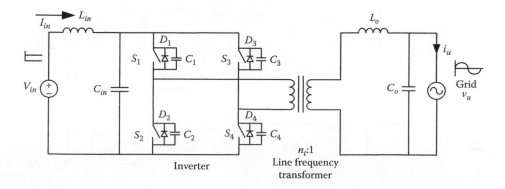

FIGURE 12.17   Single-stage DC–AC inversion using line frequency transformer isolation.

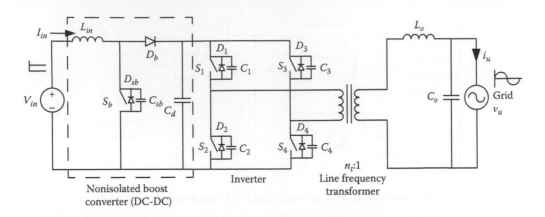

FIGURE 12.18 Two-stage DC–AC conversion using line frequency transformer isolation.

DC–DC boost converter to boost the input voltage partially, followed by an inverter connected to the utility line using a line frequency transformer to offer the rest of the voltage gain. This configuration needs a lower transformer turns ratio than the former due to an additional stage of boost converter. Similarly, several other options are also available using line frequency transformer isolation. Line frequency transformer scales the fuel cell stack voltage up to the utility line peak voltage. Such configurations show high efficiency but require the line frequency transformer, which is heavy, bulky, and costly. Therefore, multistage power conversion with high-frequency (HF) transformer isolation as shown in Figure 12.19 is desirable to realize a small-sized, economical, and lightweight system design. Voltage gain is obtained by a HF transformer-isolated DC–DC converter. Input DC voltage is first inverted to HF AC, which is elevated by HF transformer and then converted into DC voltage by a rectifier. The first two HF stages form a DC–DC converter to form an intermediate high-voltage DC bus. The elevated voltage of the DC bus is inverted to the utility line or load voltage through an inverter at desired frequency. However, increasing the number of stages of power conversion increases the number of components and reduces the efficiency of the system. Voltage control, usually, sine pulse width modulation, is implemented for stand-alone applications and current control is adopted for grid-tied applications. Voltage-fed or current-fed and resonant or PWM converter topologies have been reported in the literature for power conditioning of a fuel cell system [29]. In all configurations, a front-end converter is expected to be used to track MPP based on current fuel flow rate and temperature conditions. A maximum power point tracking (MPPT) algorithm needs to

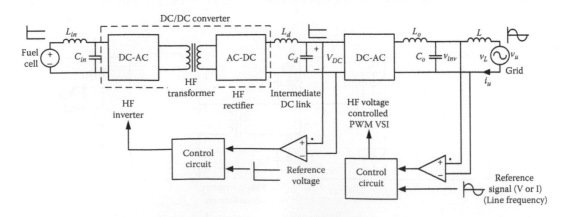

FIGURE 12.19 Multistage DC–AC conversion using HF transformer.

be implemented to operate the front-end converter to operate the fuel cell at MPP under all operating conditions for the best utilization of the fuel. An example of perturb and observe (P&O) MPPT through a boost converter implemented using MATLAB® is explained in Appendix 12A.

12.8 PRESENT STATUS AND FUTURE STRATEGIES

Despite the challenges in developing fuel cell systems, significant progress has occurred, especially over the last decade, due to government funding and private sector investment. In 2014, the fuel cell industry grew by almost $1 billion, reaching $2.2 billion in sales, up from $1.3 billion in 2013. Major increases were spurred by fuel cells for material handling (the United States) and large-scale stationary power unit sales by U.S. companies and residential fuel cells in Japan [30]. In addition, large commercial and industrial buildings as well as data centers are using fuel cells for reliable power or CHP. Although fuel cell technology has shown great promise, the FCVs continue to remain merely as demonstration vehicles or limited use vehicles. This is because of the issues related to cost, manufacturing, robustness of the technology, hydrogen production, and hydrogen infrastructure. However, automakers have renewed their interest in FCVs. In 2015, at least five auto companies including Honda, Toyota, and Hyundai launched their FCVs. The Honda FCV will seat five persons and will have a range greater than 300 miles (480 km). Honda developed a fuel cell concept in 2006 that became available on a lease-limited basis in 2008 as the Honda FCX Clarity, a four-seat sedan. About 200 of these vehicles are on monthly lease in southern California, where there are hydrogen refueling stations. Nissan signed an agreement in January 2013 with Daimler and Ford to build a common fuel cell system and hopes to release an FCV in 2017 [31]. Toyota Mirai FCV is already available in California. Its range is about 300 miles (480 km), refueling will take about 5 min, and fuel is included for the first 3 years of ownership. The power train has an eight-year/100,000 mile (162,000 km) warranty to allay early-adopter concerns.

The future of FCVs depends on the advancement of the battery technologies for EVs. If the lithium ion or lithium air batteries advance sufficiently to provide a range of about 500 miles, and also the charging infrastructure receives continued interest and investment, the EVs could become dominant and there would be less interest in FCVs. Also, if the EV batteries are charged using renewable energy sources, there will not be any emissions due to the charging of EVs. However, with the advancement of PEMFC and SOFC technologies, the fuel cell systems could be used as range extenders instead of the ICE-driven generators in series hybrid vehicles. The batteries in these vehicles could also be charged by connecting to an external power source. These PFCVs, consisting of a smaller fuel cell and a larger battery (battery dominant), may be the future direction for FCVs. Using a compressed hydrogen cylinder for onboard charging would lead to a zero-emission vehicle with a range greater than 500 miles (800 km). The fuel cell unit in these vehicles could also provide backup or sustained heat and power to a residence. The hydrogen infrastructure is not stringent for the PFCVs. If the hydrogen is generated using renewable energy and the battery is charged by the power generated by the renewable energy sources, in addition to the pure EVs, the PFCVs could take up a significant share of the auto market. The pure fuel cell (or fuel cell–dominant) vehicle could still play a role in certain geographic locations where abundant amount of hydrogen is produced using renewable energy sources and for specialized vehicles, buses, etc.

In spite of the advancement in research and technology over several years, high manufacturing cost, long-term durability, weight, and volume are still the major hurdles to commercializing the fuel cell–based systems. By the year 2020, technical targets set by the U.S. Department of Energy for integrated PEMFC power systems and fuel cell stacks operating on direct hydrogen for transportation applications for an 80 kW system are power density of 850 W/L, specific power of 650 W/kg, and a cost of $30/kW. These targets exclude hydrogen storage, power electronics, and electrical drive [32]. The U.S. Department of Energy has also set goals for residential and commercial fuel cells. A 1–10 kW residential CHP-distributed generation fuel cell system operating on natural gas needs to have an efficiency greater than 45%, an energy efficiency of 90%, and the 10 kW system should be costing less than $1700/kW by the year 2020. A 100 kW to 3 MW CHP-distributed generation fuel cell system operating on natural gas needs to have an efficiency greater than 50%, an energy efficiency of 90%, and the installed cost should be less than $1500/kW by the year 2020.

PEMFC technology continues to be the most popular type of fuel cell being used in various applications and also with regard to unit shipments. This can be attributed to its suitability of use for both small- and large-scale power generation. In high power stationary applications greater than 500 kW, the picture is distributed more evenly between PEMFC, MCFC, and SOFC. Stationary fuel cells are emerging as alternative to combustion heat engines for the production of electric power and for the cogeneration. Although PEMFCs have been used in many stationary power generation applications, high-temperature fuel cells show tremendous promise. By combining with gas turbine as a hybrid system, efficiencies greater than 65% could be achieved.

Fuel cells for prime power applications continue to be dominated by three companies: FuelCell Energy (MCFC, 300 kW+), Bloom Energy (SOFC, 200 kW+), and Doosan Fuel Cell America (PAFC, 400 kW+). Ballard Power Systems, in the 1990s, pioneered the advancement of PEMFCs for transportation applications. Since the overall technology of fuel cells did not advance as predicted, Ballard diversified its market to other applications such as distributed power generation, material handling, and backup power. Ballard continues to sell a small number of its PEMFC ClearGen™ units and the company commissioned a 1 MW system at the headquarters of Toyota USA at the end of 2012. In early 2013, it also announced the development of 175 kW systems to run using hydrogen produced from biomass gasification. The technology of high-temperature MCFC has been evolving over the past 30 years. A leading U.S. manufacturer, FuelCell Energy, pioneered the development of 250 kW MCFC–based system in partnership with the U.S. Department of Energy. FuelCell Energy delivered its first commercial unit in 2003, and several units are now operating at more than 50 facilities worldwide. Presently, FuelCell Energy produces 300 kW, 1.4 MW, and 2.8 MW fuel cell power plants. These systems offer CHP capabilities for use in industrial processes or facility heating and absorption chilling [33]. There is a significant interest by the academia and industry to advance the technology of SOFC. Initially, the focus was on tubular type cells, but presently the focus is mainly on high-efficiency planar-type SOFC. Bloom Energy is one of the major suppliers of SOFC-based power generation system and has installed these systems at several industrial and commercial locations. Bloom Energy is producing 262 kW, 210 kW, and 110 kW systems [34]. There are other manufacturers such as Delphi, Halder Topsoe, and Ceramic Fuel Cells, all working on lower power systems.

12.9 SUMMARY

In this chapter, types of fuel cells, the control characteristics, and power conversion strategies for transportation and stationary power generation are discussed. PEMFCs are being used for automotive propulsion applications and are better suited as an automotive APU if pure hydrogen is the chosen fuel source, without onboard reforming. SOFC could also be used as an APU, which does not need pure hydrogen. SOFC and MCFC are mainly used for stationary power generation. The high efficiency of fuel cells along with high-quality heat has found ways for CHP generation as well as heating options, thus increasing the overall efficiency of the system [35].

Power electronics is an enabling technology for the advancement of fuel cell–based propulsion systems and fuel cell power generation systems. The power converters match the fuel cell voltage to the traction inverter input voltage to produce the required output voltage and frequency for propulsion and power generation applications. Progress has been made in the area of power electronics to reduce the cost and volume and improve the efficiency of the system. In addition to the fuel cell stack itself, the cost of power electronics has also to be considered. But the technology has significantly advanced to such an extent that the power electronics system costs considerably less than the similarly rated fuel cell stack and its balance of plant. In spite of all the challenges associated with fuel cells, there is intense competition around the world to develop FCVs and fuel cell–based power generation systems.

APPENDIX 12A: FUEL CELL MPPT MATLAB® EXAMPLE

12A.1 DESCRIPTION

MATLAB® simulation of maximum power extraction of fuel cells using boost converter with per-turb and optimize (P&O) MPPT control algorithm

12A.2 EXPLANATION

The MATLAB® Simulink® diagram of the power circuit shown in Figure 12.17 consists of an input fuel cell model, a boost converter, an output DC bus with load, and an MPPT controller. Explanation of each block is given below. The simulation diagram is shown in Figure 12.20.

1. *Fuel cell model*: The inputs to the fuel cell block are temperature and fuel flow, and the output is generated electric power.

   All the constant parameters are defined inside the function block. Open-circuit voltage of 33.5 V and short-circuit current of 8.21 A have been updated inside the model to produce 240 W. Series and parallel combinations of cells are represented as Ns (9) and Np (6), respectively. These parameters can be changed to replicate any module.

2. *Boost converter*: Boost converter with input and output capacitor has been developed in the MATLAB® Simulink® with SimPowersystem tools. The converter is designed for 500 W with the following parameters. These design parameters are defined inside the m-file *paramBoostMPPT.m*:

$Fs = 100e3;$	*% Switching frequency 100 kHz*
$gV_o = 48;$	*% Output Voltage*
$Tsamp = 1/(Fs*10);$	*% Sampling time of simulation 100 times of switching frequency*
$Prated = 500;$	*% Output Power 500 W*
$gV_{in} = 30;$	*% Input Voltage*
$CurRIP = 5/100;$	*% Input Current Ripple*
$gVolRIP = 5/100;$	*% Output Voltage Ripple*
$Po = Prated/100 * 100;$	*% 100 means full load*
$MinLoad = 1/100 * 10;$	*% 10% load is the minimum*
$C_{in} = 50e-6;$	*% 50 uF*

3. *DC bus*: In order to maintain constant voltage at the output side, a voltage source of 48 V is connected in parallel with load.

4. *MPPT control*: The purpose of the MPPT control is to track the maximum power by vary-ing the duty cycle of the boost converter. The concept of P&O method has been utilized in the MPPT control. The algorithm works based on modifying (increasing or decreasing) the operating voltage until maximum power is reached. A simple flowchart of P&O is shown in Figure 12.21.

Under steady-state operation, the boost converter output V_o is given by $V_o = V_{in}/(1-D)$ or $V_{in} = V_o(1-D)$. The relationship between change in duty cycle with respect to change in voltage for boost converter can be derived as $delD = -delV_{in}/V_o$. A tolerance limit of $\pm delP$ (change in power) is set in order to stop at a certain value of duty cycle. A sampling time tmppt should be decided for MPPT algorithm such that the boost converter duty cycle is modified only during steady-state operation of the converter.

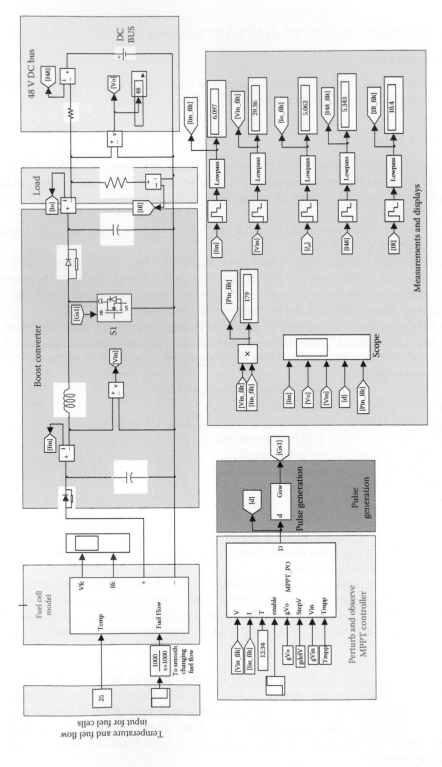

FIGURE 12.20 MATLAB® Simulink® circuit model of boost converter with MPPT control.

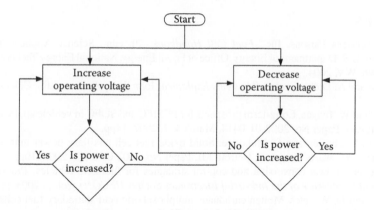

FIGURE 12.21 MPPT P&O algorithm flowchart for the fuel cell system used in Figure 12.20.

The algorithm has been coded in the MATLAB® S-function block to implement the MPPT control. A sampling time of 2.5 ms has been set with delVin of 0.3 V. In addition to this, the lower and upper bound of the D has been set to 0.1 as minimum and 0.9 as maximum. These parameters have been set in the *paramBoostMPPT.m*.

12A.3 RUNNING THE SIMULINK® FILE

- First, run the *paramBoostMPPT.m* m-file to load the initial parameters.
- Then, check inside the Simulink file *BoostMPPTV3_onlyD.mdl* if the constant values of fuel cell model are as per design. In addition to this, set the initial voltage values of input and output capacitors of boost converter as 30 and 48 V.
- Make sure the simulation parameters are set as shown in Figure 12.22.
- Run the Simulink model to check the MPPT performance. Please note that there is a fuel flow step change at 0.04 s.

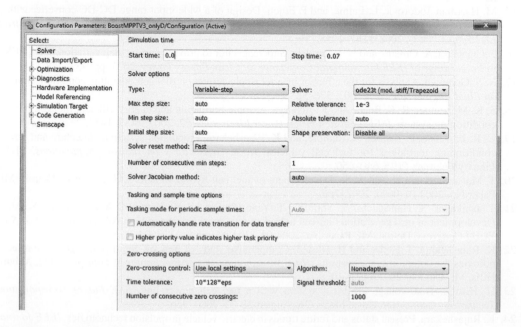

FIGURE 12.22 Simulation set up parameters running the fuel cell system.

REFERENCES

1. EG & G Services Parsons, Inc., *Fuel Cell Handbook*, 7th edn., Science Applications International Corporation, U.S. Department of Energy, Office of Fossil Energy, National Energy Technology Laboratory, Morgantown, WV, 2004.
2. J. Larminie and A. Dicks, *Fuel Cell Systems Explained*, John Wiley & Sons, Ltd., Chicester, U.K., 2000, pp. 63–65.
3. J. Thijssen and W. Teagan, Long-term prospects for PEMFC and SOFC in vehicle applications, *SAE 2002 World Congress*, Paper No. 2002-01-0414, March 4–7, 2002, 14pp.
4. J. Zizelman, S. Shaffer, and S. Mukerjee, Solid oxide fuel cell auxiliary power unit—A development update, *SAE 2002 World Congress*, Detroit, MI, Paper No. 2002-01-0411, 2002, 10pp.
5. K. Rajashekara, Power conversion and control strategies for fuel cell vehicles, *Proceedings of IEEE International Conference of the Industrial Electronics Society (IECON)*, Vol. 3, 2003, pp. 2865–2870.
6. R. Remick and D. Wheeler, Molten carbonate and phosphoric acid stationary fuel cells: Overview and gap analysis, Technical report, NREL/TP-560-49072, September 2010.
7. K. Rajashekara, Propulsion system strategies for fuel cell vehicles, *SAE 2000 World Congress*, Detroit, MI, March 6–9, 2000, 11pp.
8. E. Santi, D. Franzoni, A. Monti, D. Patterson, F. Ponci, and N. Barry, A fuel cell based domestic uninterruptible power supply, in *Proceedings of IEEE Applied Power Electronics Conference*, 2002, pp. 605–613.
9. Kartha and Grimes, *Physics Today*, 11, 1994, p. 54.
10. P. T. Krein, R. S. Balog, and X. Geng, High-frequency link inverter for fuel cells based on multiple-carrier PWM, *IEEE Transactions on Power Electronics*, 19(5), September 2004, 1279–1288.
11. T. A. Nergaard, J. F. Ferrell, L. G. Leslie, and J. S. Lai, Design considerations for a 48 V fuel cell to split single phase inverter system with ultracapacitor energy storage, in *Proceedings of IEEE Power Electronics Specialists Conference*, 2002, pp. 2007–2012.
12. A. Drolia, P. Jose, and N. Mohan, An approach to connect ultracapacitor to fuel cell powered electric vehicle and emulating fuel cell characteristics using switched mode converter, in *Proceedings of IEEE Conference of Industrial Electronics Society (IECON)* 2003, pp. 897–901.
13. M. Uzunoglu and M. S. Alam, Dynamic modeling, design and simulation of a combined PEM fuel cell and ultracapacitor system for stand-alone residential applications, *IEEE Transactions on Energy Conversion*, 21(3), September 2006, 767–775.
14. C. Wang and M. H. Nehrir, Load transient mitigation for stand-alone fuel cell power generation systems, *IEEE Transactions on Energy Conversion*, 22(4), December 2007, 864–872.
15. M. Harfman Todorovic, L. Palma, and P. Enjeti, Design of a wide input range DC-DC converter with a robust power control scheme suitable for fuel cell power conversion, *IEEE Transactions on Industrial Electronics*, 55(3), 2008, 1247–1255.
16. F. Ciancetta, A. Ometto, and N. Rotondale, Supercapacitor to provide current step variation in FC PEM, in *Proceedings of IEEE International Conference on Clean Energy Power*, 2007, pp. 439–443.
17. R. S. Gemmen, Analysis for the effect of inverter ripple current on fuel cell operating condition, *Journal on Fluid Engineering*, 125, 2003, 576–585.
18. S. K. Mazumder, R. K. Burra, and K. Acharya, A ripple-mitigating and energy-efficient fuel cell power-conditioning system, *IEEE Transactions on Power Electronics*, 22, 2007, 1437–1452.
19. U. R. Prasanna, P. Xuewei, A. Rathore, and K. Rajashekara, Propulsion system architecture and power conditioning topologies for fuel cell vehicles, *IEEE Transactions on Industry Applications*, 51(1), January/February 2015, 640–650.
20. T. Matsumoto and N. Watanabe, Development of fuel cell hybrid vehicle, *SAE Congress*, Detroit, MI, Paper No. 2002-01-0096, 2002, 7pp.
21. T. Ishikawa, S. Hamaguchi, T. Shimizu, T. Yano, S. Sasaki, K. Kato, M. Ando, and H. Yoshida, Development of next generation fuel-cell hybrid system-consideration of high voltage system, *2004 SAE World Congress*, Detroit, MI, Paper No. 2004-01-1304, March 8–11, 2004, 8pp.
22. K. Rajashekara, J. Fattic, and H. Husted, Comparative study of new on-board power generation technologies for automotive applications, *IEEE Workshop on Power Electronics in Transportation*, Auburn Hills, MI, October 2002, pp. 3–10.
23. H. Husted, Dual-voltage electrical system with a fuel cell power unit, *SAE Future Transportation Technology Conference*, Costa Mesa, CA, August 21–23, 2000, 9pp.
24. K. Rajashekara, Present status and future trends in electric vehicle propulsion technologies, *IEEE Journal of Emerging and Selected Topics in Power Electronics*, 1(1), March 2013, 3–10.

25. K. Rajashekara, J. Grieve, and D. Daggett, Solid oxide fuel cell/gas turbine hybrid APU system for aerospace applications, in *Proceedings of IEEE Industry Applications Conference—IAS Annual Meeting*, 2006, pp. 2185–2192.
26. K. Rajashekara, Hybrid fuel cell strategies for clean power generation, *IEEE Transactions on Industry Applications*, 41(3), 2005, 682–689.
27. B. Ozpineci, Z. Du, L. M. Tolbert, D. J. Adams, and D. Collins, Integrating multiple solid oxide fuel cell modules, *Proceedings of IEEE International Conference of the Industrial Electronics Society (IECON)*, 2003, Vol. 2, pp. 1568–1573.
28. E. Cengelci and P. Enjeti, Modular PM generator/converter topologies, suitable for utility interface of wind/micro turbine and flywheel type electromechanical energy conversion systems, *Proceedings of IEEE Industry Applications Conference*, 2000, Vol. 4, pp. 2269–2276.
29. A. K. Rathore, A. K. S. Bhat, and R. Oruganti, A comparison of soft switched DC-DC converters for fuel cell to utility interface application, *IEEJ Transactions on Industry Applications*, 2008, 128(4), 450–458.
30. Fuel cell technologies market report 2014, 2014, U.S. Department of Energy.
31. http://oilprice.com/Energy/Energy-General/Are-Hydrogen-Fuel-Cell-Vehicles-Dead-On-Arrival.html.
32. Fuel cell technologies office multi-year research, development, and demonstration plan—3.4 Fuel Cells, pp. 3.4-1–3.4-49, http://energy.gov/eere/fuelcells/downloads/fuel-cell-technologies-office-multi-year-research-development-and-16.
33. MCFC and PAFC R&D workshop summary report held November 16, 2009, Palm Springs, CA.
34. http://www.bloomenergy.com/fuel-cell/energy-server/.
35. N. H. Behling, Chapter 6: History of solid oxide fuel cells, *Fuel Cells: Current Technology Challenges and Future Research Needs*, Elsevier, 2013, pp. 223–421.

13 Batteries and Ultracapacitors for Electric Power Systems with Renewable Energy Sources

Seyed Ahmad Hamidi, Dan M. Ionel, and Adel Nasiri

CONTENTS

Abstract 320
13.1 Introduction 320
13.2 Battery Energy Storage System: Types, Characteristics, and Modeling 323
    13.2.1 Lead-Acid 323
    13.2.2 Lithium-Ion 323
    13.2.3 Sodium Sulfur 323
    13.2.4 Other Types of Batteries and Energy Storage Systems 324
    13.2.5 BES Modeling and Test Setups 324
13.3 Ultracapacitor Energy Storage: Types, Characteristics, and Modeling 327
    13.3.1 Ultracapacitor Types and Characteristics 327
    13.3.2 Ultracapacitor Modeling 328
13.4 Energy Storage Management Systems (ESMS) 330
    13.4.1 Main Concepts 330
    13.4.2 State of Charge (SOC) 330
    13.4.3 State of Health (SOH) 332
    13.4.4 State of Life (SOL) 332
    13.4.5 Cell Balancing Systems 332
13.5 Energy Storage Interface Systems 333
    13.5.1 AC/DC Converters 333
    13.5.2 Non-Isolated DC/DC Converters 334
    13.5.3 Isolated DC/DC Converters 337
13.6 Utility-Level Storage Systems 339
    13.6.1 Ancillary Services 339
    13.6.2 Energy Time Shifting 340
    13.6.3 Capacity Credit 340
    13.6.4 Renewable Energy Integration 340
    13.6.5 Microgrids and Islanded Grids 341
    13.6.6 Examples of ESS Field Installation 342
13.7 Simulation Examples 343
    13.7.1 EV Charging Station 343
    13.7.2 Load Leveling/Energy Time Shifting in a System with Integrated BES 343
    13.7.3 Multiphysics and Multidomain Simulations of BES 343
13.8 Summary 348
References 349

ABSTRACT

Energy storage devices and systems are now playing a major role in electrical systems from small electronics to automotive applications and to the utility power grid. This chapter reviews energy storage devices, management, control, interface, and demonstrations for electrical power systems. Various types of energy storage systems (ESSs) are discussed with a main focus on batteries and ultracapacitors. Different types of batteries and their electrical models are explained. Three major types of ultracapacitors are also discussed. The main concepts of battery management systems, including functions, controls, and hardware, are introduced. Various power electronics–based interface systems for battery and ultracapacitor charging and discharging are presented. Finally, applications of ESSs for the utility power grid, including renewable firming, power shifting, and ancillary services, are discussed.

13.1 INTRODUCTION

There is a shift in the decades-old paradigm of energy generation, distribution, and consumption. Several technical and nontechnical factors are driving this change including concerns on impacts of fossil-based fuels, advancement of alternative energy technologies, increasing penetrations of distributed generations (DG), and demand for higher energy efficiency and reliability [1]. Energy storage systems (ESSs) are acting as enabler to support this paradigm shift. They have been employed in a wide range of electrical systems from consumer electronics to automotive industry and to utility-level transmissions and generations. The higher energy density and smaller size of newer ESS technologies have significantly improved mobile and portable consumer electronic devices. In the automotive industry, advanced ESSs have enabled efficient hybrid electric vehicles (HEVs) and electric vehicles (EV). Advancements in energy storage and power electronics technologies have also transformed industrial energy conversion systems such as uninterruptible power supplies (UPS) and systems with pulse loads/sources.

In electrical power systems, the emerging paradigm includes new elements such as deregulations, distributed generations (DG), energy storage, DC systems, and power electronics–based systems at different power scales. With increasing concerns over energy reliability and security, energy cost, and environmental concerns over fossil-based sources, alternative energy systems have also experienced a large growth in recent years. The majority of growth has been happening in renewable energy systems, especially in solar photovoltaic (PV) and wind energy. When the penetration of these sources increases in the grid, they are required to participate in the grid support functions. The intermittent nature of these sources needs support from ESSs to make them dispatchable and more predictable in the grid operation.

Support for renewable energy systems can be realized at several time lengths from short term (less than 5 s) to long term (several hours). Short-term support is needed during solar PV clouding, wind power gusting, and power ramp rate limitation. The appropriate energy storage for this application does not need to have high energy capability but must have a large power capability. Long-term support can include power shifting and spinning reserve function. The daytime solar PV power can be moved to early evening peak demand hours and energy storage can supplement when forecasted renewable energy does not occur. For these applications, the ESS must have appropriate energy and power capabilities.

Electrical utility systems can also take advantages of ESSs at distribution, transmission, and generation levels. ESSs can help to stabilize the electric power system by providing voltage and frequency support, load leveling and peak power shaving, spinning reserve, and other ancillary services. They also count as extra energy capacity for the generation and transmission systems in order to defer the upgrade cost. Time shifting of the energy delivery is another major benefit of integrating ESSs into grid.

For various applications, different types of ESSs with respective strengths and weaknesses have been developed. Electrical energy can be converted to several other types of energy and can then be stored. The energy can be released by converting back to the electrical form when needed. There are several types of energy storage technologies, including electrochemical, mechanical, and electrical/ magnetic fields. Figure 13.1 shows the classification of electrical ESSs and examples for each category. Electrochemical energy storages are batteries in which electrical energy is stored as electrochemical reactions and then released by changing back to electrical energy.

Lead-acid (LAB), lithium-ion (Li-ion), nickel–metal hydride (NiMH), sodium sulfur (NaS), and reduction–oxidation (redox) flow batteries are the most common electrochemical ESSs. Electrical energy is also changed to kinetic energy and saved in mechanical ESSs such as flywheel, compressed air, and pumped hydro. In capacitors, electrical energy is saved in the electrical field established between the capacitor plates.

Mechanical energy storage technologies have primarily been adapted for large-sized storage systems, while batteries and capacitors cover a wide range of applications from costumer electronics to industrial and utility-level applications. Table 13.1 reports an estimation of total electrical energy storage capacity that is currently installed worldwide, according to the Electric Power Research Institute (EPRI) [1].

Currently, the majority of bulk storage capacity worldwide is in the forms of pumped hydro and compressed air, while the rest of storage technologies account for less than 1% of the total capacity. However, due to geographical limitations, distance from the demand, and also initial capital cost, other storage technologies, which are accessible everywhere, are gaining market share. Batteries are receiving significant attention for industrial and grid applications, due to their portable characteristics, and they are considered to be the most promising energy storage type of future. Various characteristics of the energy storage technologies, which have been extracted from [1–4], are summarized in Figure 13.2 and Table 13.2.

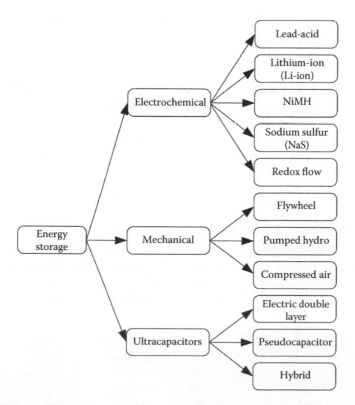

FIGURE 13.1 Different types of energy storage systems (ESSs).

TABLE 13.1

Worldwide Installed Electrical Energy Storage Capacity Based on Fraunhofer Institute Data and Rastler, EPRI, Technical Update, 2010

Energy Storage Technology	Installed Capacity (MW)
Pumped hydro	127,000
Compressed air energy storage (CAES)	440
Sodium sulfur battery (NaS)	316
Lead-acid battery	35
Nickel-cadmium battery	27
Flywheels	25
Lithium-ion battery	20
Redox flow battery	3

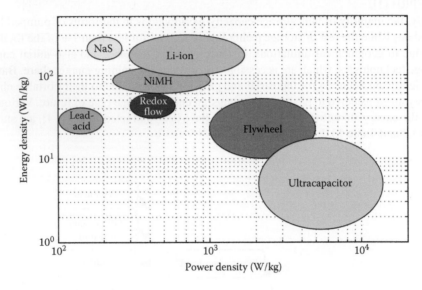

FIGURE 13.2 Ragone plot for energy and power density for ultracapacitors (UCap), FES, and batteries of the nickel–metal hydride (NiMH), zinc–bromide (ZnBr), lead-acid (LAB), lithium-ion (Li-ion), and sodium sulfur (NaS) types.

TABLE 13.2

Characteristics of Common Energy Storage Systems

Characteristic/ Energy Storage Type	Specific Power (W/kg)	Specific Energy (Wh/kg)	Cycle Life (cycles)	Self-Discharge at 25 °C (%) per Month	Efficiency (%)
Ultracapacitor	2,000–14,000	1.5–15	10^5–10^6	Very low	>90
Lead-acid battery	100–200	20–40	200–2500	Medium	70–80
Li-ion battery	300–1,500	100–300	2000–5000	Low	80–90
NiMH battery	220–1,000	60–120	500–2000	High	50–80
NaS battery	150–230	150–240	2000–4500	Very low	75–90
ZnBr flow battery	300–600	30–60	2000–3000	Very low	70–80
Flywheel	1,000–5,000	10–50	10^5–10^7	Very high	80–90
Pumped hydro	N/A	0.3–30	>20 years	Very low	65–80
CAES	N/A	10–50	>20 years	High	50–70

This chapter is organized in eight sections. The main types and characteristics of batteries and ultracapacitors along with their models are discussed in Sections 13.2 and 13.3. Energy storage management systems (ESMSs) are covered in Section 13.4. In Section 13.5, interface systems for ESSs are presented. Utility-level storage systems are discussed in Section 13.6 and the last section of the chapter before the final summary includes simulation examples.

13.2 BATTERY ENERGY STORAGE SYSTEM: TYPES, CHARACTERISTICS, AND MODELING

13.2.1 LEAD-ACID

The oldest rechargeable battery technology, which was invented some 150 years ago, is based on the use of lead-acid. The modern version of the technology is able to deliver relatively large power for relatively low cost, making such batteries strong candidates for applications in which a surge power support with low depth of discharge (DOD) is required, including backup power supply like UPS, emergency power, and power quality management.

A short life cycle and very low energy density are two main disadvantages [5]. Deep cycling and high discharging rate have a serious impact on the life span of the battery. The latest developments, including advanced materials, resulted in lead-acid batteries to have better performance and longer life cycle and include low-maintenance versions, such as GEL cells and Absorbed Glass Mat collectively known as valve-regulated lead-acid batteries [6].

13.2.2 LITHIUM-ION

One of the most popular types of batteries commercially available is based on Li-ion, which provides comparatively very good performance, with high power density and satisfactory energy density. A long life cycle without memory effect, together with high columbic efficiency and low self-discharge characteristics, makes this type of battery the preferred energy storage choice for a wide variety of applications, spanning from customer electronic devices and mobile products, all the way to the latest generation of plug-in HEV and systems for frequency regulation at the utility level [7].

The electrode material greatly influences the battery specifications in terms of power and energy density, voltage characteristics, lifetime, and safety. A typical cathode, that is the positive active electrode, is made of a lithium metal oxide and common materials such as cobalt ($LiCoO_2$ or LCO) and manganese ($LiMn_2O_4$ or LMO). Combined chemistries including nickel, cobalt, aluminum (NCA); nickel, manganese, cobalt (NMC); and iron phosphate (LFP) are also employed for the cathode. Graphite and lithium titanate ($Li_4Ti_5O_{12}$ or LTO) are the typical choices for the anode, that is the negative active electrode.

A comparison of the battery chemistries, clearly illustrating the advantage of different battery types, is presented in Table 13.3.

13.2.3 SODIUM SULFUR

Sodium sulfur (NaS) rechargeable batteries are mostly developed for large-scale applications, as they operate at a high operating temperature of 300 °C–350 °C. Such batteries are made with inexpensive materials, are known as high-power and energy storage devices with high columbic efficiency up to 90%, good thermal behavior, and are made with long life cycle. The primary applications are large-scale power and energy support, such as load leveling, renewable energy integration, and UPS systems. The battery contains hazardous materials like sodium, which can burn spontaneously in contact with air and moisture or sodium polysulfide that is highly corrosive [8].

TABLE 13.3

Characteristic Comparison between Batteries of Li-Ion Family Chemistry, Lead-Acid, and Ultracapacitor

Energy Storage Type	Power Density	Energy Density	Safety	Cycle Life	Cost
LFP	4	4	4	4	4
LTO	4	4	4	4	2
NCA	5	6	2	4	3
NMC	4	6	3	3	4
LCO	3	3	2	2	4
LMO	4	5	3	2	4
Lead-acid	3	2	3	2	6
Ultracapacitor	6	1	3	6	2

Note: The highest figure of merit, which is associated with best performance, is equal to six.

13.2.4 OTHER TYPES OF BATTERIES AND ENERGY STORAGE SYSTEMS

Some of the other common energy storage technologies include NiMH batteries, which can be recharged and have higher energy density and shorter life cycle compared to nickel–cadmium (NiCd) chemistries, but still suffer from strict maintenance requirements due to the memory effect. The high rate of self-discharging is the main disadvantage of NiMH batteries [9].

Flow batteries, also known under the redox (reduction–oxidation) name, employ for storage chemical compounds, dissolved in the liquid electrolyte and separated by a membrane. Such batteries have been developed using zinc–bromide (ZnBr), sodium bromide (NaBr), vanadium bromide (VBr), or polysulfide bromide (PSB). A unique advantage of flow batteries is that their energy capacity is completely separated from their power, and therefore, the design can be scaled with more flexibility [10]. Redox batteries can be matched very well for the integration of renewable energy to the grid and for frequency regulation [11, 12]. A zinc–bromide (ZnBr) flow battery system, shown in Figure 13.3, is used in the Power Electronics Laboratory at the University of Wisconsin–Milwaukee (UWM) for demonstrating techniques of mitigating wind power fluctuations [12].

Energy can also be stored using electromechanical systems employing high-speed high-inertia flywheels. The absorption and the release of electrical energy will result in an increase or decrease of the flywheel speed, respectively. A main advantage is represented by the rapid response time, recommending the technology especially for applications such as transportation, backup power, UPS, and power quality improvement [13].

Other forms of energy storage suitable for large-scale grid applications employ pumped hydro and compressed air. In the first case, water is pumped uphill in a natural or man-made reservoir, for example, during off-peak hours, and released downhill to turn a turbine and produce electricity when needed, for example, during peak hours. In the second case, air is typically stored underground and then used as needed to generate electricity from a generator coupled to a turbine. High capital investment and installation costs, coupled with geological availability, environmental concerns, and restrictions, represent challenges for these types of storage and may generate opportunities for developments for electrical batteries.

13.2.5 BES MODELING AND TEST SETUPS

In order to design, analyze, and optimize the ESSs, suitable battery models, which can address the main characteristics and the behavior for the application specifics, are a vital requirement. The battery model should be able to satisfactorily predict the dynamics of the system with a reasonable low

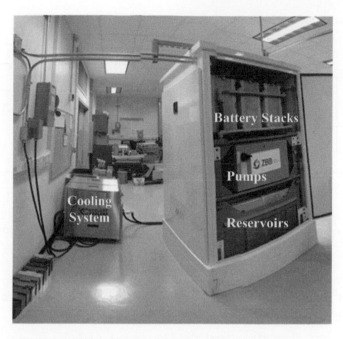

FIGURE 13.3 Zinc–bromide (ZnBr) flow battery used in the University of Wisconsin-Milwaukee (UWM) lab for experimentally demonstrating the mitigation of power variability from renewable energy sources. The battery is rated at 50 kW and 675 Ah when fully charged.

computational complexity. Reduced-order models that neglect the phenomena of less significance may provide a suitable trade-off between accuracy and simplicity.

Battery models may be classified into three major groups: physical or electrochemical, mathematical, and electrical. A physical model is based on the electrochemical reactions and thermodynamic phenomena that take place inside the battery cell. Such models involve high-order differential equations, and they are complex and time-consuming, but provide, in principle, the basis for the most accurate results [14, 15]. In order to reduce complexity, reduced-order simplified electrochemical models have been proposed [16–18].

Mathematical battery models, without any electrical properties, are limited to the prediction of system-level performance indices, such as energy efficiency, runtime, and capacity. In this type of models, the result accuracy is highly dependent on the experimental data employed for model identification, the models are typically applicable only to a reduced range of devices and ratings, and they do not include terminal voltage and current characteristic, which are essential for circuit analysis and system simulation [19, 20].

The electrical models for batteries employ lumped equivalent circuit parameters with sources and passive elements, that is, resistances and capacitance. Such models are the most familiar to electrical engineers and can be successfully employed for system simulation. A comprehensive model that combines the transient capability of a Thevenin-based model, the AC features of an impedance-based model, and the information specific to a runtime-based model has been proposed and validated for lead-acid, NiMH, and Li-ion batteries [21–26]. The model, which is schematically represented in Figure 13.4, includes two equivalent circuits: for battery lifetime, capacity, state of charge (SOC) and runtime of the battery (Figure 13.4a) and for the voltage–current characteristics of the battery (Figure 13.4b).

Battery lifetime has been modeled through three elements, a resistance, R_{self}, which quantifies the self-discharge energy loss during storage operation; a current-dependent source for charging and discharging, I_{bat}; and a capacitance, C_{cap}, which provides the SOC for the battery as a scaled voltage

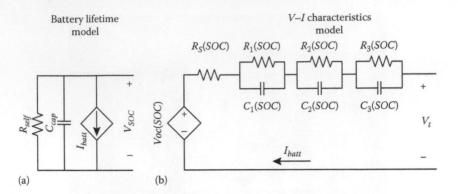

(a) (b)

FIGURE 13.4 Combined detailed equivalent circuit models for batteries: (a) battery lifetime model and (b) V–I characteristics model. (Based on the concept proposed by Chen, M. and Rincon-Mora, G.A., *IEEE Trans. Energy Conv.*, 21(2), 504, 2006.)

drop, V_{SOC}, with a per unit value between 0 and 1. The capacitance, C_{cap}, accounts for the entire charge stored in the battery and can be calculated as

$$C_{cap} = 3600 \cdot C_n \cdot f_1(T) \cdot f_2(n) \cdot f_3(i) \tag{13.1}$$

where

C_n is the nominal battery capacity in Ah

$f_1(T)$, $f_2(n)$, and $f_3(i)$ are correction factors dependent of temperature, number of cycles, and current, respectively.

The battery voltage–current characteristics are modeled through the equivalent circuit depicted in Figure 13.4b. In this case, all equivalent circuit elements are dependent on the SOC. The voltage–current nonlinearity is incorporated through a dependent voltage source, V_{oc}, and a resistor, R_s, is responsible for immediate voltage change in step response. A number of RC parallel networks, that is R_i and C_i, are connected in series to provide multiple time-transient constants. Typically, three such time-constant RC networks are considered satisfactory for most practical purposes. The parameters identified in Figure 13.4b are a function of SOC, as shown in the following equations, and they are also affected by other operational characteristics, such as temperature.

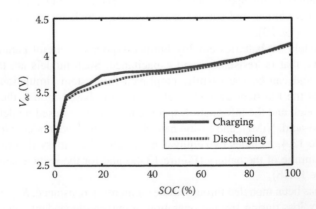

FIGURE 13.5 Open-circuit voltage (V_{oc}) versus *SOC*, for example, lithium-ion battery of 2.6 Ah.

The SOC and the terminal voltage, V_t, are calculated as

$$SOC(t) = SOC_0 - \frac{1}{c_{cap}} \int_0^t i(t) \cdot d(t) \tag{13.2}$$

$$V_t = V_{OC}(SOC) - (V_{Rs} + V_{C1} + V_{C2} + V_{C3}) \tag{13.3}$$

where
V_{C1}, V_{C2}, and V_{C3} are the voltages across capacitors
V_{Rs} is the voltage drop across the internal resistor (R_s)
An open-circuit voltage, V_{oc}, versus SOC characteristic is exemplified in Figure 13.5.

13.3 ULTRACAPACITOR ENERGY STORAGE: TYPES, CHARACTERISTICS, AND MODELING

13.3.1 ULTRACAPACITOR TYPES AND CHARACTERISTICS

Ultracapacitors, which are also referred to as super capacitors, provide energy storage, and they are able to fast charge/discharge and delivering high power for a very short period of time, in the order of fraction of seconds. The significant improvements in capacity and energy density over conventional capacitors, while maintaining the same high power density values, are possible through the use of a much larger surface area for the electrodes and thinner dielectrics. In comparison with other energy storage devices, ultracapacitors have a very high power density while their energy density is substantially lower than that of electric batteries. Ultracapacitors are especially suitable for applications that require high-rate and short deep cycles, such as backup power supplies, HEVs, automotive start–stop applications, DC link voltage support in converter, and power quality correction in utility applications.

Based on their electrode design, ultracapacitors may be classified into three main groups: electrical double-layer capacitors, pseudocapacitors, and hybrid capacitors (see Figure 13.6). The double-layer capacitors rely on the electrostatic field between two plates, while pseudocapacitors employ electrochemical reactions in order to store the electric charge. The hybrid capacitors combine the two phenomena and include the popular Li-ion ultracapacitors [27, 28].

Ultracapacitors are superior to batteries both in terms of life cycling, with more than 10^5 cycles being possible, as well as in energy efficiency. Deep cycling does not have a significant influence on ultracapacitors' life span, while this is definitely not the case for the lead-acid or Li-ion batteries. Examples of Li-ion ultracapacitors of 2200 F, 2300 F, and 3300 F Li-ion ultracapacitors are shown in Figure 13.7. In order to reach higher voltages, currents, or capacities, ultracapacitors are connected in series or parallel in banks as the one shown in Figure 13.8, which have been used in the UWM lab for demonstrating grid integration methods for intermittent renewable energy generation [29, 30].

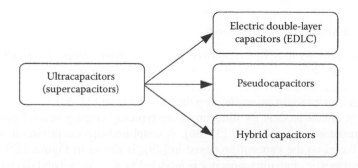

FIGURE 13.6 Classification of ultracapacitors based on the electrode design.

FIGURE 13.7   Examples of Li-ion ultracapacitors of 2200 F, 2300 F, and 3300 F.

FIGURE 13.8   Ultracapacitor bank in the UWM lab. The approximate overall dimensions are $1.03 \times 0.93 \times 1.17$ m. Rated at 720 V and 0.5 Ah.

13.3.2   ULTRACAPACITOR MODELING

Models for ultracapacitors may be categorized into three main groups: physical/electrochemical, equivalent circuit, and behavioral neural network types. Physical models rely on electrochemical reactions and corresponding high-order differential equations, and therefore, they are the most phenomenological, relevant and, in principle, also the most accurate [31, 32].

The equivalent circuit models for ultracapacitors typically employ several passive circuit elements, such as resistors and capacitors [33–36]. A simplified equivalent circuit model for Li-ion ultracapacitors, based on the concept proposed in [29], is shown in Figure 13.9, where the self-discharging characteristic of an ultracapacitor is modeled by R_{sd}, the cell and the junction resistance are totally quantized by series resistance R_s, and the RC elements, R_{ss} and C_{ss}, model the transient

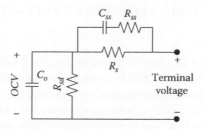

FIGURE 13.9 Electric equivalent circuit model for an ultracapacitor. (Based on the concept proposed in Manla, E. et al., *Proceedings of Energy Conversion Congress and Exposition (ECCE)*, pp. 2957–2962, 2011.)

response of an ultracapacitor. The parameter identification requires multiple AC and DC tests. The equivalent incremental internal capacitance is calculated as

$$C_0\left(OCV_i\right) = \frac{\Delta Q_i}{\Delta OCV_i} = \frac{\sum_i I_i(t)\Delta t}{\Delta OCV_i}$$

(13.4)

where

OCV is the open-circuit voltage
I is the current
Q is the columbic charge for the data point *i*

It should be noted that a columbic counting technique has been employed to estimate the SOC, which varies approximately linearly with OCV.

An 1100 F Li-ion ultracapacitor was charged during 16 cycles in which each cycle consists of 10 s charging and then rest (disconnected from the source) for 20 s. Figure 13.10 verifies the accuracy of the equivalent circuit model of Figure 13.9 by comparing the terminal voltage of the ultracapacitor and the simulation model.

Some models that combine equivalent circuits and electrochemistry fundamentals have also been proposed [31, 32]. Yet in another approach, training data represented by the voltage, current, and temperature values measured during charging and discharging experiments have been used to train a neural network [37].

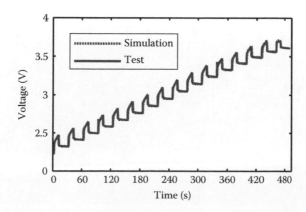

FIGURE 13.10 Experimental data and simulation results for a Li-ion ultracapacitor, 10 A pulsed charging DC test at 25 °C. (Based on the concept proposed in Manla, E. et al., *Proceedings of Energy Conversion Congress and Exposition (ECCE)*, pp. 2957–2962, 2011.)

13.4   ENERGY STORAGE MANAGEMENT SYSTEMS (ESMS)

13.4.1   MAIN CONCEPTS

An ESMS is typically employed in order to ensure optimal and safe operation of devices, such as batteries and ultracapacitors. A typical ESMS configuration includes an effective cell balancing mechanism; cooling and ventilation; data acquisition and controls; communications and interfaces between subsystems and with the power system; protections; for example, to overvoltage and short circuit; condition monitoring for SOC and state of health (SOH); and temperature.

The management systems for both batteries and ultracapacitors share the same basic principles, but batteries require additional care as their lifetime and safe performance are highly sensitive to parameters like high temperatures, DOD, and current rate. Consequently, the focus in this section is on battery management systems (BMS) and includes topics related to the SOC, SOH, and state of life (SOL).

In terms of functionality, BMSs may be divided into three categories: centralized, modular or master–slave, and distributed. In a centralized BMS, parameters such as voltage, current, and temperature are measured for individual cells and sent to the main BMS board. This topology is compact, cost efficient, and well suited for troubleshooting. In a modular BMS, slave cards collect the data from each cell and send them to a master card, which coordinates the management of the entire system (Figure 13.11). This topology enables a modular expansion for larger-sized packs. In a distributed BMS, as shown in Figure 13.12, each cell has its own individual electronic board and a main controller, which is responsible for communications and necessary computations [38].

Xing et al. [39] have proposed a generic BMS structure in which various sensors are installed in the battery pack and gather real-time data for system safety and battery state calculation [39]. The data are employed for cell balancing and thermal management, protection, and state determination, which in turn are used for the electrical control, as shown in Figure 13.13.

13.4.2   STATE OF CHARGE (SOC)

The SOC is an indicator of the amount of remaining energy or charge available in a battery as a fraction of the nominal value, that is rated value of capacity. Since the SOC is not measurable directly, it needs to be evaluated based on other parameters that significantly affect the SOC like battery current, temperature, and the number of lifetime cycles. It should be noted that the maximum

FIGURE 13.11   A modular BMS topology where several slave boards collect cells data and send it to the master board.

FIGURE 13.12 A distributed BMS: each cell sends the data to the main controller.

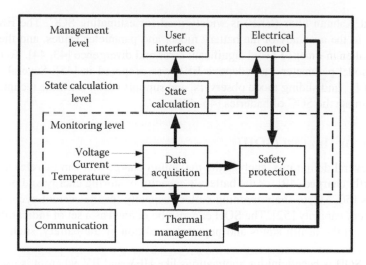

FIGURE 13.13 Block diagram of a typical BMS. (Based on the concept proposed in Xing, Y. et al., *Energies*, 4, 1840, 2011.)

capacity of a battery gradually and nonlinearly degrades over time, making it very challenging to extract and estimate the exact value of the SOC. Extensive research has been conducted in recent years in this respect.

The most common approach for estimating SOC is Coulomb counting, according to which the capacity of the battery is calculated by integrating the battery current over time. This method is well suited for Li-ion batteries, which have high columbic efficiency [7]. The accuracy of this method is highly dependent on the initial value of the SOC and the nominal capacity of the battery, which is decreasing as a battery ages. In order to reduce the possible initial SOC error and also to compensate for the possible accumulated error due to integration, the SOC estimation based on the open-circuit voltage (OCV) versus SOC table, a table in which each OCV value is associated with an SOC value, has been proposed [40]. Although the online (real-time) measurement of OCV has its own challenges, computationally intelligent methods, such as neural and fuzzy algorithms, have been developed in this respect to estimate SOC [41, 42]. Because these methods are very sensitive to the model error and disturbance, the estimated results may fluctuate widely. Furthermore, the OCV versus SOC or DOD curves for some battery chemistries are almost flat for most of the operating ranges, due to their cathode chemistry (Figure 13.14). Some types, such as LFP, also have a very long voltage relaxation time, limiting the practical application of this technique.

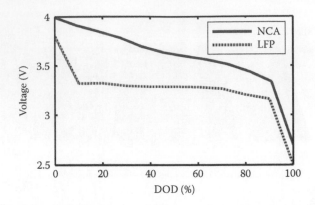

FIGURE 13.14 Open-circuit voltage (OCV) versus DOD for NCA and LFP types of Li-ion batteries.

The extended Kalman filter (EKF) is widely used for estimating SOC. The EKF approach is highly sensitive to the accuracy of the battery model and parameter values, and therefore, special care should be taken in order to avoid significant error and divergence [43, 44]. To reduce the sensitivity to the model parameters, an adaptive EKF was proposed in [45]. Several other methods, including robust H_∞ and sliding mode observers, and support vector machine techniques have been employed to estimate the SOC of batteries [46–48].

13.4.3 State of Health (SOH)

The SOH has several definitions, such as the maximum charge that can be released after the battery has been fully charged [49], or the battery's capacity of storing energy and preserving charge for long periods [50, 51], or the remaining battery capacity for the current cycle as compared to the original battery capacity [52]. The SOH can also be defined as a set of indicators or diagnostic flags, which reflect the health status and physical condition of the battery, such as loss of rated capacity [53].

The value of SOH is beneficial for applications like HEV and EV, where it is used as an indication of specified power or to estimate the driving range. Similar to the SOC problem, several techniques have been developed for SOH estimation, including EKF [53], adaptive observer [47], and probabilistic neural networks [49]. Measuring the internal equivalent DC resistance of a cell, which increases with capacity degradation, is another characterization tool for SOH [52].

13.4.4 State of Life (SOL)

The SOL is defined as the remaining useful life (RUL) of a battery or as the time when a battery should be replaced [39]. This indicator is considered from the design stage in order to plan ahead maintenance and replacement schedules, prevent failures during operation, and increase the reliability and availability of an ESS. Several methods have been published for RUL estimation [54–56].

13.4.5 Cell Balancing Systems

In order to achieve higher voltage and current, a battery pack consists of several cells, which are connected in series and parallel layouts, respectively. The cells in a string could have different SOC levels due to several internal and external sources of unbalancing, which may result in different capacity fading rates between cells. Internal imbalances include different self-discharging resistance and impedance, and external causes may include thermal variation across the string [57, 58]. During charging or discharging, imbalances in between cells may lead to extreme voltages and hence severe overcharging and over discharging that could seriously damage cells, reduce useful lifetime, and even cause fires and explosions. Therefore, an effective battery cell balancing system, which maintains the SOC of the cells at the same level, is an important feature of any BMS.

Cell balancing systems are either passive or active. According to the typical passive balancing methods, the extra energy of an imbalanced cell is released by increasing the cell body temperature, a technique that is useful especially for small battery packs with low voltage [59]. This technique is relatively straightforward and inexpensive to implement, but its applicability is mostly limited to cells that do not damage severely due to overcharging [60].

The active balancing technique utilizes an active circuit to distribute as evenly as possible the energy among the cells [57]. Active balancing techniques are employed in several different ways including shunting and shuttling and energy converting method. From the energy flow point of view, active balancing methods comprise dissipative and non-dissipative methods [61]. The extra energy in dissipative methods is wasted as heat across a resistor, while in nondissipative techniques, the excess energy is distributed among the string cells, leading to a higher system efficiency.

13.5 ENERGY STORAGE INTERFACE SYSTEMS

Power electronics converters are required in order to ensure the connection between the AC grid and electric energy storage devices such as batteries and ultracapacitors. These converters, which are the basis of energy storage interface systems, operate bidirectionally and they are of the AC/DC and DC/DC types.

13.5.1 AC/DC CONVERTERS

AC/DC power converters may employ one or multiple phases, with the three-phase versions being the most popular for AC grid connection. Based on the circuit topology, bidirectional AC/DC converters may be categorized into boost, buck, buck–boost, multilevel, and multipulse types.

Bidirectional boost-type converters, which may have different topologies (see Figure 13.15), have been typically employed in applications like energy storage and line interactive UPS systems. A conventional configuration is based on a three-leg, six-switch (controllable switches, each connected in parallel with a freewheeling diode) H-bridge voltage source converter for which PWM-based closed-loop control is the most common technique for DC link voltage regulation at close to unity power factor. Alternative topologies for bidirectional boost converters were also proposed in a four-switch configuration, which is typically employed for variable speed induction motor drives [62], and in a four-wire arrangement that is used to reduce the DC link voltage ripple and balance the supply currents [63–65].

Some of the most common bidirectional buck-type converters are shown in Figure 13.16 and include a six-switch configuration with gate turn-off thyristors (GTOs) for higher power and insulated gate bipolar transistors (IGBTs) for lower power ratings. The four-leg topology is advantageous for reducing the DC link voltage ripple and for balancing the currents in case of an unbalanced supply voltage [66, 67].

Bidirectional buck–boost converters are employed in cases where the DC link voltage is substantially variable around an average value. Configurations reported include cascades of buck and boost converters or matrix converters, as shown in Figure 13.17. From a performance point of view, matrix converters are advantageous because they are capable of operation as bidirectional buck or boost converters and because the higher switching frequency reduces the size of the input and output filters [68, 69].

For higher voltages and higher power, bidirectional multilevel power converters have been proposed, including multilevel diode-clamped [70, 71] (Figure 13.18), flying capacitor [71], and cascaded topologies. The multilevel and multipulse types of high-power converters provide better AC quality with lower total harmonic distortion (THD) and electromagnetic inference (EMI) noise and a higher power factor [72].

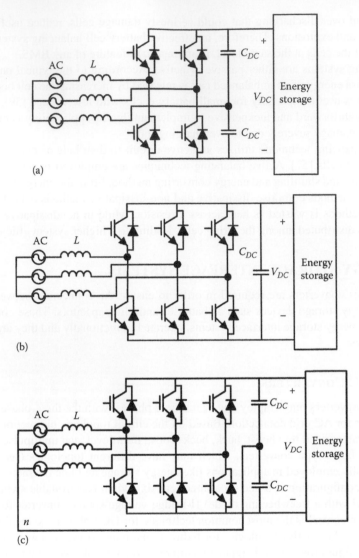

FIGURE 13.15 Three-phase bidirectional boost converter topologies: four-switch (a), six-switch H-bridge voltage source converter (b), and four-wire (c).

13.5.2 NON-ISOLATED DC/DC CONVERTERS

Bidirectional DC/DC converters can be classified into two main groups as isolated and non-isolated. Non-isolated versions are simpler in design, control, and implementation and may have lower cost and better efficiency. Nevertheless, isolated converters are absolutely required in certain applications. The applications of bidirectional DC/DC converters include renewable energy systems integrated with ESS (as an energy buffer to smooth and mitigate output power fluctuations), line interactive UPS (interface between ESS and DC link), PHEV, and EV (to connect the ESS to the electric traction drive system).

An important requirement for DC/DC converters used in ESS applications is a high voltage ratio with low input current ripple, such that the ESS lifetime is not negatively affected due to the ripple itself [73, 74]. Several configurations have been published [75–78], including basic converter topologies of the Cuk, SEPIC/Luo, half-bridge (HB) type, and more complex configurations, such as multiphase interleaved, cascaded HB, and, more recently, multilevel bidirectional DC/DC converters [79–82].

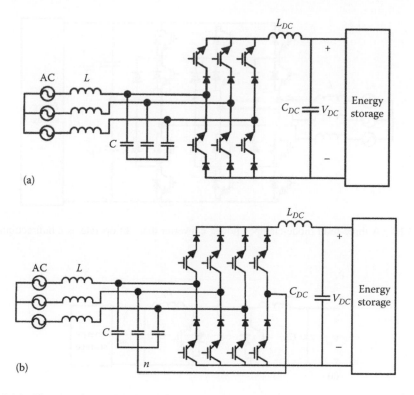

FIGURE 13.16 Three-phase bidirectional buck converter topologies: six-switch with IGBTs (a) and a four-leg (b).

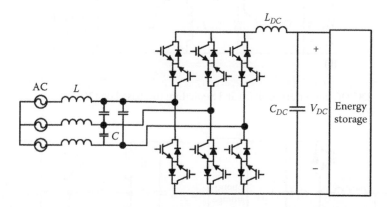

FIGURE 13.17 A matrix converter that can perform the function of a bidirectional buck–boost converter.

Cuk and Sepic/Luo converters have the capability of bidirectional power conversion by utilizing only two active switches, as depicted in Figure 13.19.

The circuit schematic for a HB converter, which is also a bidirectional power converter widely used in different applications, is shown in Figure 13.20. In step-down, buck, or charging mode, the upper switch is on and the power flow is from the higher DC link voltage to the lower voltage for the energy storage device. The power flow is reversed in the step-up, boost, or discharging mode, when the lower switch conducts. HB converters use a smaller number of passive elements and smaller inductors and experience a lower current stress for active switches and diodes as compared with Cuk and Sepic/Luo converters operating at the same voltage and power. Therefore, HB converters are expected to achieve higher efficiency.

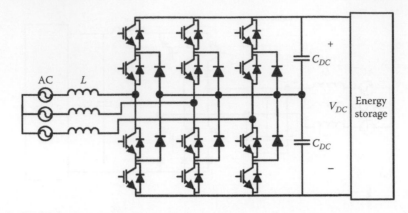

FIGURE 13.18 A three-level diode-clamped power converter that can operate as a bidirectional multilevel converter.

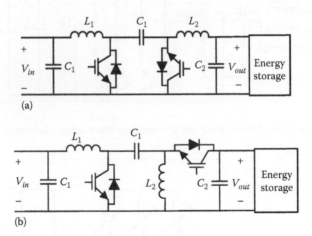

FIGURE 13.19 DC/DC converters: Cuk (a) and Sepic/Luo (b), based on [75–78].

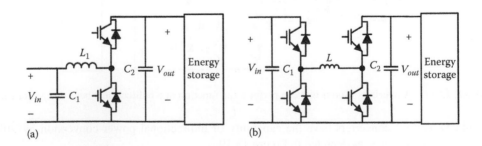

FIGURE 13.20 DC/DC converters: Half-bridge (HB) (a) and cascaded HB (b).

In order to further reduce the current ripple through the inductor, reduce the size and weight of the inductor, and minimize the current stress on the active switches, cascaded HB and multi-phase interleaved converters have been developed [83, 84]. These two types of converters have similar principles of operation. A two-phase interleaved bidirectional DC/DC converter is shown in Figure 13.21.

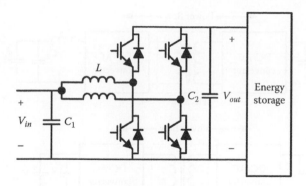

FIGURE 13.21 Two-phase interleaved bidirectional converter.

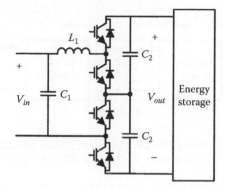

FIGURE 13.22 Three-level neutral point clamped bidirectional DC/DC converter.

Bidirectional three-level DC/DC converters, as illustrated in Figure 13.22, have been developed for high DC link voltage applications, in order to lower the switch voltage stress on the switches and minimize the passive elements, inductors and capacitors, in comparison with HB converters. Due to the reduced switch voltage stress, switching can be performed with a lower voltage rate, yielding lower on-state voltage and switching losses [84].

13.5.3 ISOLATED DC/DC CONVERTERS

Isolated converters typically comprise three main blocks: two high-frequency AC/DC converters (inverter/rectifier) and a high-frequency transformer (HF Xfm), which provides the galvanic isolation between the two sources input/output (Figure 13.23). Configurations that have been employed for the two high-frequency converters include the HB, full bridge (FB), push–pull, and L-type HB. FB and HB converters experience comparable voltage stress on the active switches, but FB

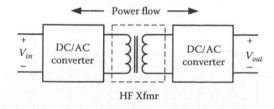

FIGURE 13.23 An isolated bidirectional DC/DC converter typically comprises a DC/AC converter, an HF Xfm, and an AC/DC converter.

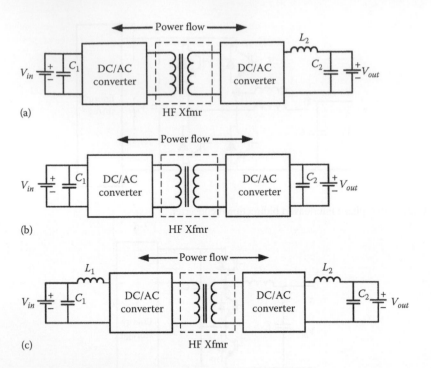

(a)

(b)

(c)

FIGURE 13.24 Main layouts for isolated DC/DC converters, voltage-fed and current-fed converter (a), voltage-fed converter (b), and current-fed converter (c). Note the different use of capacitors and inductors in the input and output, based on [86–91].

converters have lower current switch stress, making this type a strong candidate for higher power applications. The FB converter topology is also superior to the push–pull configuration, in which lower transformer utilization, the requirement for additional snubber circuitry, and increased conduction losses are the main drawbacks [85].

Based on the type of converters employed on each side, isolated DC/DC converters are classified into three main categories including current fed, voltage fed, and combination of current and voltage fed. The current-fed converter has an inductor at its terminals, which acts like a current source, while the voltage-fed converter has a capacitor at its terminals, which acts as a voltage source (Figure 13.24). Voltage-fed and current-fed FBs are widely used in practice and publications can be found in [86–91].

On one side, the FB, shown in Figure 13.25, is voltage fed and connected to a DC link capacitor and on the other side is current fed and connected to an inductor. In order to reduce the voltage switch stress caused by the energy stored in the leakage inductance of the transformer and DC choke, an active clamp snubber circuit is added into the current-fed FB to ensure zero voltage switching (ZVS) for all the switches in the current-fed side and zero current switching or ZVS for all the switches in voltage-fed side. The absence of a circulating current results in higher efficiency.

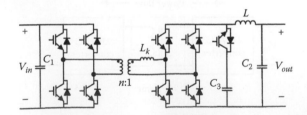

FIGURE 13.25 Voltage-fed and current-fed FB bidirectional DC/DC converter with active clamp.

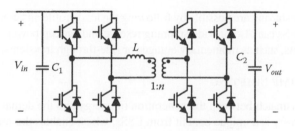

FIGURE 13.26 A bidirectional DC/DC converter with DAB topology with voltage-fed FB.

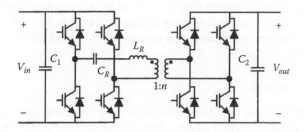

FIGURE 13.27 Isolated and bidirectional DC/DC SRC topology.

The dual active bridge (DAB) topology, which consists of two voltage-fed FBs as illustrated in Figure 13.26, is also widely used for bidirectional isolated converters. This solution is particularly attractive for ESS because it has the capability to connect two DC link sources. The configuration is advantageous because the number of switches is lower than that for voltage-fed and current-fed FB topology and ZVS is achievable for all switches at both sides. The voltage switch stress is lower than that for current-fed FBs and a high efficiency is expected [92, 93]. The narrow voltage range for optimal operation is a disadvantage of the DAB topology. Overall, voltage-fed topologies are preferred over current-fed ones due to their reliability and converter cost [85]. Other converter topologies that have been derived from the DAB topology are presented in [85–93].

Another topology for a bidirectional isolated converter can be obtained by adding a capacitor in series to the leakage inductance of the transformer, which corresponds to a series resonant converter (SRC) as shown in Figure 13.27 [94, 95]. In this case, ZVS is achievable in a wide range of voltages and loads with the exception of light load conditions [87].

13.6 UTILITY-LEVEL STORAGE SYSTEMS

ESSs are evolving as a viable technical and economical solution for improving both short-term and long-term utility operations. Increased peak power demand, limited generation capacity, variability of renewable energy generation, and transmission and distribution issues can be addressed, in principle, with ESSs. Different types of ESSs can be employed for various utility-level applications, such as voltage and current regulation, power quality management, load leveling, and peak power shaving, as explained in the following.

13.6.1 ANCILLARY SERVICES

ESSs can improve the operation of ancillary services. Due to the bidirectional power flow capability, ESSs can absorb or inject active and reactive power from and into the grid, respectively, in order to regulate the voltage and frequency and provide higher quality and smoother power. The stored energy can be used to manage the power quality in the short term and to provide increased reliability for grid in the long term.

Load leveling and peak shaving are possible by following the load profile and controlling EESs to match the required energy. EESs can also provide spinning reserve and backup power supply, which could be utilized in sudden events, such as momentary outages or ride-through transients.

13.6.2 Energy Time Shifting

Typically, there is a mismatch between the generation of energy and the demand in peak hours. Time shifting of energy delivery is a unique benefit from ESSs, which enables the utilities to absorb excess energy in off-peak hours by charging the ESS and release it during peak hours of high demand by discharging the storage. Functional and economical benefits, related to different higher on-peak and lower off-peak tariffs, can be achieved in this manner. Without ESSs, the traditional approach is to provide additional and costly generation capacity able to cover up to the maximum possible load demand.

The arbitrage over time or the time-shifting technique can also be used for renewable energy generation plants, in which case the harvested energy can be stored during off-peak demand and made available during peak time.

13.6.3 Capacity Credit

Capacity credit for ESSs allows electrical utilities to defer the installation and upgrade for existent generation or transmission infrastructures. Since expansions for power plants, power lines, and transformers, which may be required in order to meet expected demand, are very time-consuming and costly, electrical utilities can take the advantage of the extra capacity provided by ESSs in order to satisfy the demand and reduce the load shedding caused by an inability of the electrical utility to provide adequate capacity.

13.6.4 Renewable Energy Integration

Due to the intermittent nature of renewable energy sources, which are highly affected by geological, time, and weather conditions, their output power may fluctuate unpredictably and may not meet the quality required in order to directly connect to the electrical grid. In this case, ESSs can provide smoothening services, such that the intermittent output power from renewable energy sources is adjusted, and fluctuations mitigated, as schematically illustrated in Figures 13.28 and 13.29 [96, 97].

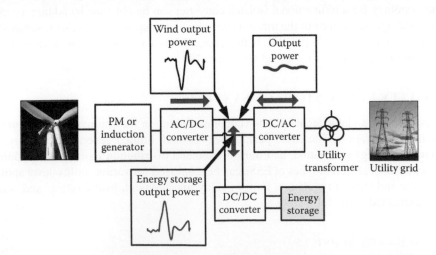

FIGURE 13.28 Block diagram of a wind turbine with battery integrated at turbine level. (Based on the concept proposed in Esmaili, A. and Nasiri, A., *Proceedings of IEEE Energy Conversion Congress and Exposition (ECCE)*, pp. 3735–3740, 2012.)

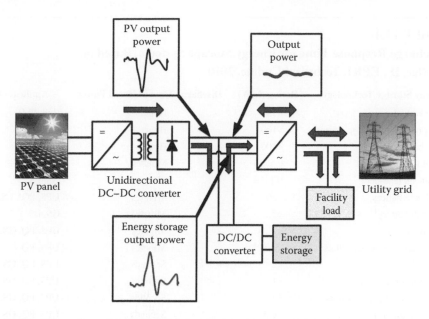

FIGURE 13.29 Illustration of a PV system integrated with energy storage (Based on the concept proposed in Esmaili, A. and Nasiri, A., *Proceedings of IEEE Annual Conference of Industrial Electronics (IECON)*, pp. 3957–3962, Portugal, 2009.)

13.6.5 Microgrids and Islanded Grids

One of the main challenges for microgrids is to provide a stable and reliable power system for grid-connected and islanded operation. Voltage and frequency stability are major concerns, especially if generation includes renewables, such as solar and wind, and the load demand is variable. EESs can provide voltage and frequency stability for short-term and long-term applications. Practical large installations have been demonstrated in combination with wind turbines and PV systems, and ongoing research is conducted for optimizing the location and size of microgrid EESs in order to minimize the system cost and energy loss, while ensuring improved system reliability.

In microgrids, EESs can be employed, distributed, or centralized ESSs. In a distributed installation, EESs are capable of actively managing and supporting local loads. A centralized configuration for ESSs is typically employed in smaller microgrids with critical loads, where a main backup power supply is required in case of power outages [98–100].

Energy storage technologies have been developed also in order to meet utility-level requirements. Some applications, such as load shifting for duration of hours, require an ESS with high energy, while voltage and frequency regulation need high power density. Lifetime, response time, DOD, and efficiency are other significant factors for the selection of energy storage type for a given application.

Typical electrochemical batteries have a lifetime in the range of 5,000–10,000 cycles, while pumped hydro, compressed air, flywheels, and capacitors are rated at 10,000–100,000 cycles (see Table 13.2). On the other hand, pumped hydro and compressed air are classified as having a rather slow response time in order of minutes, while batteries, capacitors, and flywheels are characterized as fast responders in order of fractions of seconds (Table 13.4). In terms of technology maturity and commercial availability, pumped hydro, lead-acid batteries, and sodium sulfur batteries have the highest rankings.

TABLE 13.4

Discharge Response Time for Energy Storage Systems, Based on Rastler, D., EPRI, Technical Update, 2010

Energy Storage Technology	Rating (MW)	Discharge Time at Rated Power	Applications
Pumped hydro	>100	Hours	Bulk PM[a]
Compressed air energy storage	>100	Hours	Bulk PM
Flow batteries	<100	Hours	GS, LS[a]
Sodium sulfur (NaS)	0.5–50	Minutes	GS, LS
Advanced lead-acid	0.1–10	Minutes	GS, LS
Lead-acid battery	<50	Minutes	UPS, PQ[a], LS, GS[a]
NaNiCl$_2$ battery	<5	Minutes	GS, LS
Li-Ion battery	<2	Minutes	UPS, PQ, GS, LS
High-energy supercapacitors	<0.5	Minutes	UPS, PQ
NiMH	<2	Seconds	UPS, PQ, GS, LS
High-power flywheels	<1	Seconds	UPS, PQ, GS, LS
High-power supercapacitors	<1	Seconds	UPS, PQ, GS, LS
Nickel-cadmium	<0.5	Seconds	UPS, PQ, GS

[a] PM, power management; GS, grid support; LS, load shifting; PQ, power quality.

13.6.6 EXAMPLES OF ESS FIELD INSTALLATION

1. LMO-based Li-ion batteries—A version employing NMC/graphite for the active material of the electrodes has been installed by Mitsubishi Heavy Industries. The battery is rated for 1 MW and 400 kWh, is installed in a container, and is used for load leveling and peak shaving in the grid-connected operation of the wind farm and mega solar power generation [101].

2. LTO-based Li-ion batteries—Two 1 MW units based on LTO chemistry have been developed for frequency regulation application by Altairnano. It is estimated that this battery will have a lifetime in the range of 20 years by performing 10%–30% DOD with several cycles per day for grid stabilization applications, while for applications like peak shifting with 80%–100% DOD and one cycle per day, it is expected to reach a lifetime of 50 years [102].

3. Lead-acid batteries—These are the most mature and commercialized electrochemical battery technology with more than 35 MW installed worldwide for utility-level applications, such as grid stabilizing, short and long duration power quality improvement [1].

4. Flywheel energy storage (FES)—The largest installed system of the kind is rated for 20 MW and 5 MWh and consists of 200 spinning mass units. This FES is developed and installed by Beacon Power in Stephentown, New York, and is employed for frequency regulation [103].

5. Sodium sulfur (NaS) batteries—A 2 MW, 14 MWh system was installed by the American Electric Power at Churubusco, Indiana, and is used for smoothening wind farm intermittency. The largest NaS battery, rated at 80 MW, will be installed at the Noshiro thermal power plant in northern Japan [104].

6. Pumped hydro energy storage—This is the most mature energy storage technology available and has a worldwide installed power capacity of 127 GW. This technology is well matched for load leveling and peak shaving applications for electrical utilities as well as long-term storage, even at the transmission level [1].

13.7 SIMULATION EXAMPLES

13.7.1 EV CHARGING STATION

An electric vehicle (EV) charging station is schematically represented in Figure 13.30 and the corresponding MATLAB®/Simulink® model is shown in Figure 13.31. The system is supplied from the grid through a diode bridge. Two DC/DC converters are employed on the grid side and on the battery energy storage (BES) side, respectively. The BES is charged by the AC grid and discharged on the load represented by the EV.

In the simulation example with the main results depicted in Figure 13.32, the BES is charging at a low rate from the grid. At a time of 0.015 s into the simulation, the BES is considered charged and the grid is disconnected. From there on, the BES provides power to a 1 kW load, and at 0.020 s, the 5 kW load, which models a scale-down of EV, is connected to the system for charging. System operation is controlled at constant DC link voltage.

13.7.2 LOAD LEVELING/ENERGY TIME SHIFTING IN A SYSTEM WITH INTEGRATED BES

The example system shown in Figure 13.33 is supplied from the AC grid and includes an AC/DC converter and an integrated BES system in order to perform several ancillary services, such as load leveling and energy time shifting. The system performs energy time shifting by storing in the BSS low-cost energy during off-peak demand and delivering energy to the load during peak hours demand, when the electricity grid price is high. In the load-leveling service, the BES supports the grid power input by providing a portion of the power to the load at peak time and in this way smooths the grid input power.

In the simulation example with the main results depicted in Figure 13.34, during the first 0.02 s, the grid is charging the BES while providing 1 kW power for the load. At 0.02 s, the load is increased to 10 kW, corresponding to a peak demand, while the grid current and hence power are maintained constant and the excess load demand is covered by the BES.

13.7.3 MULTIPHYSICS AND MULTIDOMAIN SIMULATIONS OF BES

As previously discussed in this chapter, circuit and control simulations with BES typically employ equivalent circuit models (ECMs), such as those exemplified in Figure 13.4. Experimentation on a physical device, when available as a lab prototype or manufactured product, or numerical simulations with high-fidelity multiphysics models, at the research and design stages, can be used to determine terminal characteristics that are then employed for the identification of model parameters.

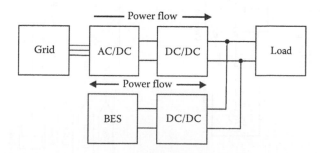

FIGURE 13.30 Schematic of an EV charging system including a BES.

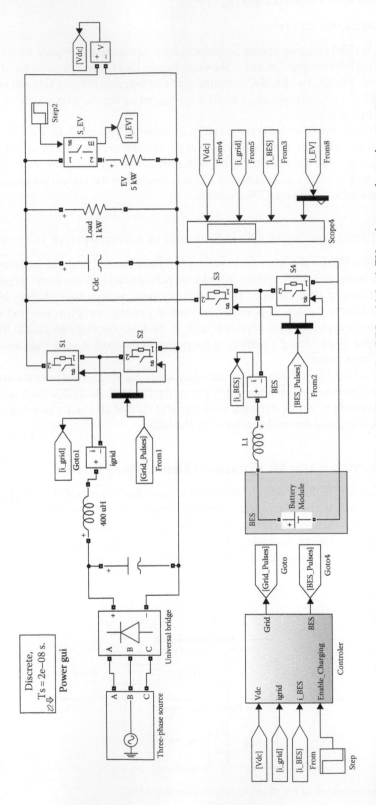

FIGURE 13.31   MATLAB®/Simulink® schematic of the system from Figure 13.30, including BES, grid, load, EV, and power electronic converters.

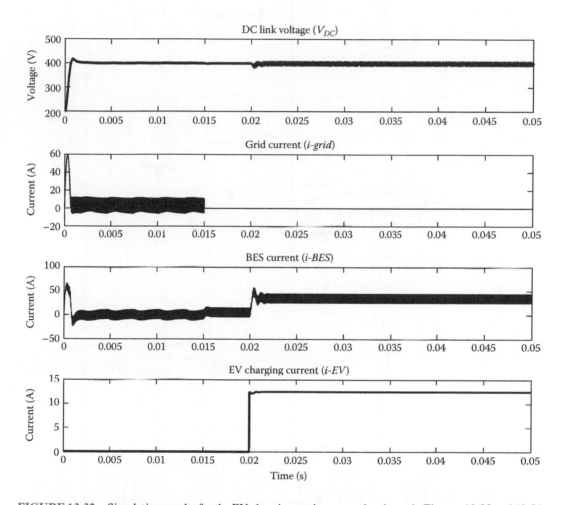

FIGURE 13.32   Simulation results for the EV charging station example, shown in Figures 13.30 and 13.31.

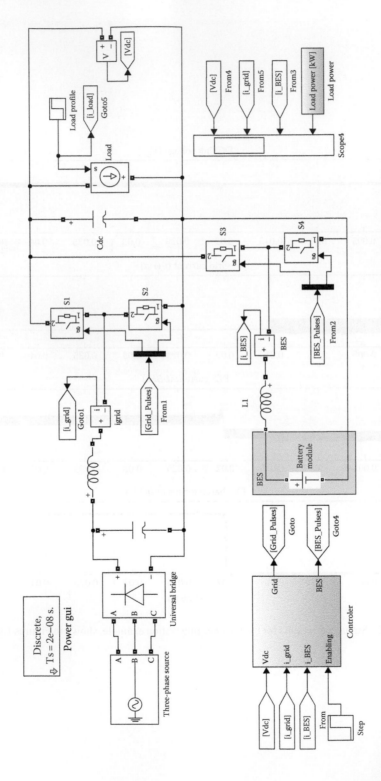

FIGURE 13.33   Simulink® model illustrating the load-leveling and time-shifting features of a power system integrated with BES.

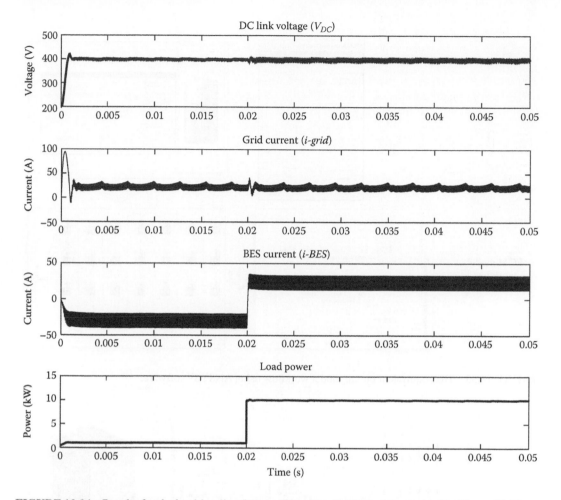

FIGURE 13.34 Results for the load-leveling feature of the Simulink® model presented in Figure 13.33.

The ECM workflow and the system integration illustrated in Figures 13.35 and 13.36 are part of a collection of examples developed for ANSYS®, Inc. by Xiao Hu et al. In this case, the ECM parameter values are obtained based on a combination of data for open-circuit voltage versus SOC and transient voltage under pulse discharge, respectively. Multiple cell models are connected to create a battery or a pack circuit model, which can be used for the prediction of electrical performance. For the example shown in Figure 13.35, this technique was validated with less than 0.2% error between measurements and simulations for terminal current and voltage.

The simulation of large systems incorporating many BES has to consider multiple aspects related to different physical phenomena and multiple domains. As exemplified in Figure 13.36, the behavior of intercell bus bar electromagnetic parasitics can be captured through a frequency-dependent model. The influence of temperature can be studied through computational fluid dynamics (CFD) and, in order to minimize the computational effort, can be used together with linear time-invariant (LTI) techniques and reduced-order models, which can then be employed in order to simulate control blocks, including temperature rise control on individual modules with closed feedback from the electric circuit. More details regarding such methods are described, for example, by Xu and Stanton in [105].

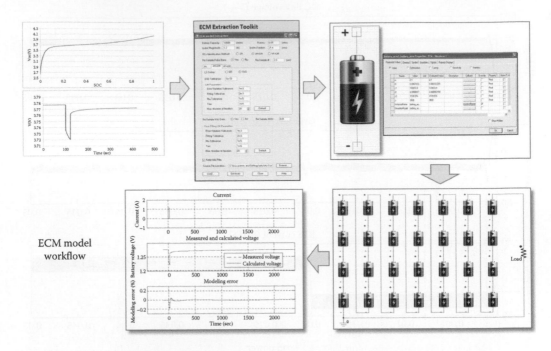

ECM model
workflow

FIGURE 13.35   BES equivalent circuit model workflow using ANSYS® software.

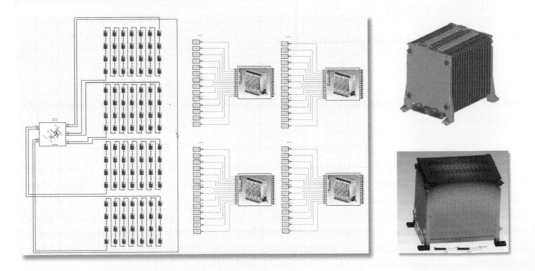

FIGURE 13.36   Comprehensive multiphysics and multidomain simulation of battery packs, including the effects of electromagnetic parasitics, thermal, and mechanical stress, using the ANSYS® software.

13.8   SUMMARY

This chapter presents a comprehensive review of ESSs and their applications with emphasis on batteries and ultracapacitors. Different types of battery and ultracapacitor ESSs have been discussed along with their modeling techniques. Management system principles and the main functions of the BMS have been reviewed. Different power converter topologies employed in ESSs have been presented, and the utility-level applications of ESSs have been described. Despite all the research and studies that have been conducted on ESSs in recent decades, electrical energy systems are still an active field of research. ESSs are the main key to many concepts and tasks, including smart grids, microgrids, and energy-efficient systems. Active research areas include investigation and improvement in lifetime, energy and power density, and thermal behavior. Although the majority of the installed electrical energy storage capacity is currently pumped hydro and compressed air energy storage (CAES) systems, batteries and ultracapacitors are very promising ESSs for future systems.

REFERENCES

1. D. Rastler, Electricity Energy Storage Technology Options; a white paper primer on applications, costs and benefits, Electric Power Research Institute (EPRI), Technical Update, December 2010.
2. S. Vazquez, S. M. Lukic, E. Galvan, L. G. Franquelo, and J. M. Carrasco, Energy storage systems for transport and grid applications, *IEEE Transactions on Industrial Electronics*, 57(12), 3881–3895, 2010.
3. A. Khaligh and Z. Li, Battery, ultracapacitor, fuel cell, and hybrid energy storage systems for electric, hybrid electric, fuel cell, and plug-in hybrid electric vehicles: State of the art, *IEEE Transactions on Vehicular Technology*, 59(6), 2806–2814, 2010.
4. A. R. Sparacino, G. F. Reed, R. J. Kerestes, B. M. Grainger, and Z. T. Smith, Survey of battery energy storage systems and modeling techniques, *IEEE Power and Energy Society General Meeting*, pp. 1–8, 2012.
5. A. Esmaili and A. Nasiri, Energy storage for short-term and long-term wind energy support, *Annual Conference on IEEE Industrial Electronics Society, IECON'2010*, pp. 3281–3286, 2010.
6. B. McKeon, J. Furukawa, and S. Fenstermacher, Advanced lead–acid batteries and the development of grid-scale energy storage systems, *Proceedings of the IEEE*, 102(6), 951–963, 2014.
7. T. Horiba, Lithium-ion battery systems, *Proceedings of the IEEE*, 102(6), 939–950, 2014.
8. Z. Wen, Study on energy storage technology of sodium sulfur battery and its application in power system, *Proceedings of International Conference on Power System Technology*, pp. 1–4, 2006.
9. Q. Fu, A. Hamidi, A. Nasiri, V. Bhavaraju, S. Krstic, and P. Theisen, The role of energy storage in a microgrid concept, examining the opportunities and promise of microgrids, *IEEE Electrification Magazine*, 1(2), 21–29, 2013.
10. H. Ibrahim, A. Ilincaa, and J. Perronb, Energy storage systems—Characteristics and comparisons, *Renewable and Sustainable Energy Reviews*, 12(5), 1221–1250, 2008.
11. D. Banham-Hall, G. Taylor, C. Smith, and M. Irving, Flow batteries for enhancing wind power integration, *IEEE Transactions on Power System*, 27(3), 1690–1697, 2012.
12. E. Manla, A. Nasiri, and M. Hughes, Modeling of zinc energy storage system for integration with renewable energy, *Proceedings of IEEE Industrial Electronics Conference, IECON'2009*, pp. 3987–3992, 2009.
13. H. Liu and J. Jiang, Flywheel energy storage—An upswing technology for energy sustainability, *Energy and Buildings*, 39(5), 599–604, 2007.
14. D. W. Dennis, V. S. Battaglia, and A. Belanger, Electrochemical modeling of lithium polymer batteries, *Journal of Power Sources*, 110(2), 310–320, 2002.
15. J. Newan, K. E. Thomas, H. Hafezi, and D. R. Wheeler, Modeling of lithium-ion batteries, *Journal of Power Sources*, 119(3), 838–843, 2003.
16. K. Smith, C. Rahn, and C. Wang, Control oriented 1D electrochemical model of lithium ion battery, *Energy Conversion Management*, 48(9), 2565–2578, 2007.
17. K. Smith, C. Rahn, and C. Wang, Model-based electrochemical estimation and constraint management for pulse operation of lithium ion batteries, *IEEE Transaction on Control Systems Technology*, 18(3), 654–663, 2010.
18. K. Smith, Electrochemical control of lithium-ion batteries, *IEEE Control Systems*, 38(2), 18–25, 2010.
19. D. Rakhmatov, S. Vrudhula, and D. A. Wallach, A model for battery lifetime analysis for organizing applications on a pocket computer, *IEEE Transactions on VLSI Systems*, 11(6), 1019–1030, 2003.
20. P. E. Pascoe and A. H. Anbuky, VRLA battery discharge reserve time estimation, *IEEE Transaction on Power Electronics*, 19(6), 1515–1522, 2004.
21. M. Chen and G. A. Rincon-Mora, Accurate electrical battery model capable of predicting runtime and I-V performance, *IEEE Transactions on Energy Conversion*, 21(2), 504–511, 2006.
22. R. C. Kroeze and P. T. Krein, Electrical battery model for use in dynamic electric vehicle simulations, *Proceedings of IEEE Power Electronics Specialists Conference, PESC'2008*, pp. 1336–1342, 2008.
23. B. Schweighofer, K. M. Raab, and G. Brasseur, Modeling of high power automotive batteries by the use of an automated test system, *IEEE Transactions on Instrumentation and Measurement*, 52(4), 1087–1091, 2003.

24. S. Abu–Sharkh and D. Doerffel, Rapid test and non–linear model characterization of solid–state lithium–ion batteries, *Journal of Power Sources*, 130(1–2), 266–274, 2003.
25. L. Gao and S. Liu, Dynamic lithium–ion battery model for system simulation, *IEEE Transactions on Components and Packaging Technologies*, 25(3), 495–505, 2002.
26. A. Hamidi, L. Weber, and A. Nasiri, EV charging station integrating renewable energy and second–life battery, *Proceedings of International Conference on Renewable Energy Research and Applications (ICRERA)*, pp. 1217–1221, Spain, 2013.
27. L. Zubieta and R. Bonert, Characterization of double-layer capacitor (DLCs) for power electronics application, *IEEE Transactions on Industrial Applications*, 36(1), 199–205, 2000.
28. S. Sivakkumar and A. Pandolfo, Evaluation of lithium-ion capacitors assembled with pre-lithiated graphite anode and activated carbon cathode, *Electrochimica Acta*, 65, 280–287, 2012.
29. E. Manla, G. Mandic, and A. Nasiri, Testing and modeling of lithium-ion ultracapacitors, *Proceedings of Energy Conversion Congress and Exposition (ECCE)*, pp. 2957–2962, 2011.
30. G. Mandic and A. Nasiri, Modeling and simulation of a wind turbine system with ultracapacitors for short-term power smoothing, *Proceedings of IEEE International Symposium on Industrial Electronics (ISIE)*, pp. 2431–2436, Italy, 2010.
31. N. Bertrand, O. Briat, J.-M. Vinassa, J. Sabatier, and H. El Brouji, Porous electrode theory for ultracapacitor modelling and experimental validation, *Proceedings of IEEE Vehicle Power and Propulsion Conference (VPPC)*, pp. 1–6, 2008.
32. N. Bertrand, J. Sabatier, O. Briat, and J. M. Vinassa, Embedded fractional nonlinear supercapacitor model and its parametric estimation method, *IEEE Transactions on Industrial Electronics*, 57(12), 3991–4000, 2010.
33. L. Shi and M. L. Crow, Comparison of ultracapacitor electric circuit models, *Proceedings of IEEE PES General Meeting*, pp. 1–6, 2008.
34. S. Buller, E. Karden, D. Kok, and R. W. De Doncker, Modeling the dynamic behavior of supercapacitors using impedance spectroscopy, *IEEE Transactions on Industry Applications*, 38(6), 1622–1626, 2002.
35. W. Yang, J. E. Carletta, T. T. Hartley, and R. J. Veillette, An ultracapacitor model derived using time-dependent current profiles, *Proceedings of Midwest Circuit and Systems, MWSCAS*, pp. 726–729, 2008.
36. A. Grama, L. Grama, D. Petreus, and C. Rusu, Supercapacitor modelling using experimental measurements, *International Signals, Circuits and Systems, ISSCS*, pp. 1–4, 2009.
37. J. N. Marie-Francoise, H. Gualous, and A. Berthon, Supercapacitor thermal and electrical behavior modeling using ANN, *IEE Proceedings, Electric Power Applications*, 153(2), 255–261, 2006.
38. D. Andrea, *Battery Management Systems for Large Lithium Ion Battery Packs*, 1st edn., Artech House, U.K., 2010.
39. Y. Xing, E. Ma, K. Tsui, and M. Pecht, Battery management systems in electric and hybrid vehicles, *Energies*, 4, 1840–1857, 2011.
40. K. Ng, C. Moo, Y. Chen, and Y. Hsieh, Enhanced coulomb counting method for estimating state-of-charge and state-of-health of lithium-ion batteries, *Applied Energy*, 86(9), 1506–1511, 2009.
41. J. Kozlowski, Electrochemical cell prognostics using online impedance measurements and model-based data fusion techniques, *Proceedings of IEEE Aerospace Conference*, 7, pp. 3257–3270, 2003.
42. A. Salkind, C. Fennie, and P. Singh, Determination of state-of-charge and state-of-health of batteries by fuzzy logic methodology, *Journal of Power Sources*, 80(1–2), 293–300, 1999.
43. N. Windarko, J. Choi, and G. Chung, SOC estimation of LiPB batteries using extended Kalman filter based on high accuracy electrical model, *Proceedings of Power Electronics and ECCE Asia (ICPE & ECCE)*, pp. 2015–2022, Korea, 2011.
44. G. Plett, Extended kalman filtering for battery management systems of LiPB-based HEV battery packs part 2. State and parameter estimation, *Journal of Power Sources*, 134, 277–292, 2004.
45. R. Xiong, H. He, F. Sun, and K. Zhao, Evaluation on state of charge estimation of batteries with adaptive extended Kalman filter by experiment approach, *IEEE Transactions on Vehicular Technology*, 62(1), 108–117, 2013.
46. J. Yan, G. Xu, Y. Xu, and B. Xie, Battery state-of-charge estimation based on H∞ filter for hybrid electric vehicle, *Proceedings of International Conference on Control, Automation, Robotics and Vision (ICARCV 2008)*, pp. 464–469, Vietnam, 2008.
47. M. Gholizadeh and F. Salmasi, Estimation of state of charge, unknown nonlinearities, and state of health of a lithium-ion battery based on a comprehensive unobservable model, *IEEE Transactions on Industrial Electronics*, 61(3), 1335–1344, 2014.
48. T. Hansen and C. Wang, Support vector based battery state of charge estimator, *Journal of Power Sources*, 141(2), 351–358, 2004.

49. H. Lin, T. Liang, and S. Chen, Estimation of battery state of health using probabilistic neural network, *IEEE Transactions on Industrial Informatics*, 9(2), 679–685, 2013.
50. N. Watrin, B. Blunier, and A. Miraoui, Review of adaptive systems for lithium batteries state-of-charge and state-of-health estimation, *Proceeding of IEEE Transactions on Electrification Conference*, pp. 1–6, 2012.
51. M. Shahriari and M. Farrokhi, Online state-of-health estimation of VRLA batteries using state of charge, *IEEE Transactions on Industrial Electronics*, 60(1), 191–202, 2013.
52. T. L. Matthew, B. Suthar, P. W. C. Northrop, Sumitava De, C. Michael Hoff, O. Leitermann, M. L. Crow, S. Santhanagopalan, and V. R. Subramanian, Battery Energy Storage System (BESS) and Battery Management System (BMS) for grid-scale applications, *Proceedings of IEEE*, 102(6), 1014–1030, 2014.
53. B. S. Bhangu, P. Bentley, D. A. Stone, and C. M. Bingham, Nonlinear observers for predicting state-of-charge and state-of-health of lead-acid batteries for hybrid-electric vehicles, *IEEE Transactions on Vehicular Technology*, 54(3), 783–794, 2005.
54. A. Widodo, M. Shim, W. Caesarendra, and B. Yang, Intelligent prognostics for battery health monitoring based on sample entropy, *Expert System Applications*, 38,(9), 11763–11769, 2011.
55. B. Saha, K. Goebel, S. Poll, and J. Christophersen, Prognostics methods for battery health monitoring using a bayesian framework, *IEEE Transactions on Instrument and Measurement*, 58(2), 291–296, 2009.
56. D. Stroe, M. Swierczynski, A. Stan, R. Teodorescu, and S. Andreasen, Accelerated lifetime testing methodology for lifetime estimation of lithium-ion batteries used in augmented wind power plants, *IEEE Transactions on Industry Applications*, 50(6), 4006–4017, 2014.
57. J. Cao, N. Schofield, and A. Emadi, Battery balancing methods: A comprehensive review, *Proceedings of IEEE Vehicle Power and Propulsion Conference (VPPC)*, pp. 1–6, China, 2008.
58. W. Bentley, Cell balancing considerations for lithium-ion battery systems, *Proceeding of Annual Battery Conference on Applications and Advances*, pp. 223–226, 1997.
59. N. H. Kutkut, H. L. N. Wiegman, D. M. Divan, and D. W. Novotny, Charge equalization for an electric vehicle battery system, *IEEE Transaction on Aerospace and Electronics Systems*, 34(1), 235–246, 1998.
60. S. Moore and P. Schneider, A review of cell equalization methods for lithium ion and lithium polymer battery systems, *Proceeding of Society of Automotive Engineers, SAE*, pp. 1–5, 2001.
61. M. Uno and K. Tanaka, Influence of high-frequency charge–discharge cycling induced by cell voltage equalizers on the life performance of lithium-ion cells, *IEEE Transaction on Vehicular Technologies*, 60(4), 1505–1515, 2011.
62. G. T. Kim and T. A. Lipo, VSI-PWM rectifier/inverter system with a reduced switch count, *IEEE Transactions on Industrial Applications*, 32(6), 1331–1337, 1996.
63. B. T. Ooi, J. W. Dixon, A. B. Kulkarni, and M. Nishimoto, An integrated ac drive system using a controlled current PWM rectifier/inverter link, *IEEE Transactions on Power Electronics*, 3(1), 64–71, 1988.
64. P. Verdelho and G. D. Marques, Four-wire current-regulated PWM voltage converter, *IEEE Transactions on Industrial Electronics*, 45(5), 761–770, 1998.
65. R. Zhang, F. C. Lee, and D. Boroyevich, Four-legged three-phase PFC rectifier with fault tolerant capability, *Proceedings of IEEE Power Electronics Specialists Conference, PESC*, pp. 359–364, 2000.
66. M. Hombu, S. Ueda, and A. Ueda, A current source GTO inverter with sinusoidal inputs and outputs, *IEEE Transactions on Industrial Applications*, 23(2), 247–255, 1987.
67. R. J. Hill and F. L. Luo, Current source optimization in AC-DC GTO thyristor converters, *IEEE Transactions on Industrial Electronics*, 34(4), 475–482, 1987.
68. D. G. Homes and T. A. Lipo, Implementation of a controlled rectifier using AC-AC matrix converter theory, *IEEE Transactions on Power Electronics*, 7(1), 240–250, 1992.
69. J. B. Ejea, E. Sanchis, A. Ferreres, J. A. Carrasco, and R. D. L. Calle, High-frequency bi-directional three-phase rectifier based on a matrix converter topology with power factor correction, *Proceedings of IEEE Applied Power Electronics Conference and Exposition, APEC*, pp. 828–834, 2001.
70. D. Carlton and W. G. Dunford, Multilevel, unidirectional AC-DC converters, a cost effective alternative to bi-directional converters, *Proceedings of IEEE Power Applied Electronics Specialists Conference*, pp. 1911–1916, 2001.
71. J. S. Lai and F. Z. Peng, Multilevel converters—A new breed of power converters, *Transactions on Industrial Applications*, 32(3), 509–517, 1996.
72. B. Singh, B. N. Singh, A. Chandra, K. Al-Haddad, A. Pandey, and D. Kothari, A review of three-phase improved power quality AC–DC converters, *IEEE Transactions on Industrial Electronics*, 51(3), 641–660, 2004.

73. G. Fontes, C. Turpin,S. Astier, and T. Meynard, Interactions between fuel cells and power converters: influence of current harmonics on a fuel cell stack, *IEEE Transactions on Power Electronics*, 22(2), 670–678, 2007.

74. M. Kabalo, D. Paire, B. Blunier, D. Bouquain, M. Simões, and A. Miraoui, Experimental evaluation of four-phase floating interleaved boost converter design and control for fuel cell applications, *IET Power Electronics*, 6(2), 1–12, 2012.

75. M. Ortúzar, J. Dixon, and J. Moreno, Ultracapacitor-based auxiliary energy system for an electric vehicle: Implementation and evaluation, *IEEE Transactions on Industrial Electronics*, 54(4), 2147–2156, 2007.

76. R. M. Schupbachj and C. Bald, Comparing DC-DC converters for power management in hybrid electric vehicles, *IEEE International Electric Machines and Drives Conference*, pp. 1369–1374, 2003.

77. J. Czogalla, J. Li, and C. R. Sullivan, Automotive application of multi-phase coupled-inductor DC-DC converter, *Proceedings of IEEE Industry Applications Conference*, pp. 1524–1529, 2003.

78. M. Gerber, J. A. Ferreira, N. Seliger, and I. W. Hofsajer, Design and evaluation of an automotive integrated system module, *Proceedings of IEEE Industry Applications Conference*, pp. 1144–1151, 2005.

79. X. Ruan, B. Li, Q. Chen, S. C. Tan, and C. K. Tse, Fundamental considerations of three-level DC–DC converters: Topologies, analyses, and control, *Proceedings of IEEE Transactions on Circuits and Systems*, 55(11), 3733–3743, 2008.

80. R. M. Cuzner, A. R. Bendre, P. J. Faill, and B. Semenov, Implementation of a non-isolated three level DC/DC converter suitable for high power systems, *Proceedings of IEEE Industry Applications Conference*, pp. 2001–2008, 2007.

81. Y. Shi, Z. Jin, and X. Cai, Three-level DC-DC converter: Four switches Vsmax=Vin/2, TL voltage waveform before LC filter, *Proceedings of IEEE Power Electronics Specialists Conference*, pp. 1–4, 2006.

82. V. Yousefzadeh, E. Alarcon, and D. Maksimovic, Three-level buck converter for envelope tracking in RF power amplifiers, *IEEE Transactions on Power Electronics*, 21(2), 549–552, 2006.

83. G. Y. Choe, J. S. Kim, H. S. Kang, and B. K. Lee, A optimal design methodology of an interleaved boost converter for fuel cell application, *Journal of Electrical Engineering and Technology*, 5(2), 319–328, 2010.

84. Y. Du, X. Zhou, S. Bai, S. Lukic, and A. Huang, Review of non-isolated Bi-directional DC-DC converters for plug-in hybrid electric vehicle charge station application at municipal parking decks, *Proceedings of IEEE Applied for Electronics Conference and Exposition (APEC)*, pp. 1145–1151, 2010.

85. F. Krismer, J. Biela, and J. W. Kolar, A comparative evaluation of isolated Bi-directional DC/DC converters with wide input and output voltage range, *Proceedings of 40th IEEE IAS Annual Meeting, (IAS)*, pp. 599–606, 2005.

86. L. Zhu, X. Xu, and F. Flett, A 3kW isolated bi-directional DC/DC converter for fuel cell electric vehicle application, *Proceedings of the International Conference, Power Electronics, Intelligent Motion, Power Quality, PCIM 2001*, pp. 77–82, 2001.

87. Y. Srdjan, L. Jacobson, and A. Huang, Review of high power isolated bi-directional DC-DC converters for PHEV/EV DC charging infrastructure, *Proceedings of IEEE Energy Conversion Congress and Exposition (ECCE)*, pp. 553–560, 2011.

88. O. Garcia, L. Flores, A. Oliver, J. Cobos, and J. Pena, Bidirectional DC-DC Converter for Hybrid Vehicles, *Proceedings of IEEE Power Electronics Specialists Conference*, pp. 1881–1886, 2005.

89. L. Zhu, A novel soft-commutating isolated boost full-bridge ZVS-PWM DC-DC converter for bi-directional high power applications, *Proceedings of IEEE Power Electronics Specialists Conference*, pp. 2141–2146, 2004.

90. W. G. Hurley, E. Gath, and J. G. Breslin, Optimizing the AC resistance of multilayer transformer windings with arbitrary current waveforms, *IEEE Transactions on Power Electronics*, 15(2), 369–376, 2000.

91. J. Reinert, A. Brockmeyer, and R. De Doncker, Calculation of losses in ferro- and ferrimagnetic materials based on the modified Steinmetz equation, *IEEE Transactions on Industry Applications*, 37(4), 1055–1061, 2001.

92. S. Inoue and H. Akagi, A bi-directional DC/DC converter for an energy storage system, *Proceedings of IEEE Applied Power Electronics Conference (APEC)*, pp. 761–767, 2007.

93. X. Xu, A. M. Khambadkone, and R. Oruganti, A soft-switched back-to-back bi-directional DC/DC converter with a FPGA based digital control for automotive applications, *Proceedings of Industrial Electronics Society, IECON'2007*, pp. 262–267, 2007.

94. T. Ho, G. Verghese, C. Osawa, B. S. Jacobson, and T. Kato, Dynamic modeling, simulation and control of a series resonant converter with clamped capacitor voltage, *Proceedings of IEEE Power Electronics Specialists Conference (PESC)*, pp. 1289–1296, 1994.

95. Y. Du, X. Bian, S. M. Lukic, B. S. Jacobson, and A. Q. Huang, A novel wide voltage range bi-directional series resonant converter with clamped capacitor voltage, *Proceedings of IEEE Annual Conference of Industrial Electronics, IECON'2009*, pp. 82–87, 2009.

96. A. Esmaili and A. Nasiri, Evaluation of impact of energy storage on effective load carrying capability of wind energy, *Proceedings of IEEE Energy Conversion Congress and Exposition (ECCE)*, pp. 3735–3740, 2012.

97. A. Esmaili and A. Nasiri, A case study on improving ELCC by utilization of energy storage with solar PV, *Proceedings of IEEE Annual Conference of Industrial Electronics (IECON)*, pp. 3957–3962, Portugal, 2009.

98. Q. Fu, L. F. Montoya, A. Solanki, A. Nasiri, V. Bhavaraju, T. Abdallah, and D. C. Yu, Microgrid generation capacity design with renewable and energy storage addressing power quality and surety, *IEEE Transactions on Smart Grids*, 3(4), 2019–2027, 2012.

99. M. Nick, M. Hohmann, R. Cherkaoui, and M. Paolone, On the optimal placement of distributed storage systems for voltage control in active distribution networks, *Proceedings of IEEE PES International Conference and Exhibition on Innovative Smart Grid Technologies*, pp. 1–6, 2012.

100. C. Chen, S. Duan, T. Cai, B. Liu, and G. Hu, Optimal allocation and economic analysis of energy storage system in microgrids, *IEEE Transactions on Power Electronics*, 26(10), 2762–2773, 2011.

101. D. Mukai, K. Kobayashi, T. Kurahashi, N. Matsueda, K. Hashizaki, and M. Kogure, Development of large highperformance lithium-ion batteries for power storage and industrial use, *Mitsubishi Heavy Industries, Technical Review*, 49(1), 6–11, March 2012.

102. B. Misback, *Large Format Li4Ti5O12 Lithium-Ion Batteries Performance and Applications*, Altairnano Inc., New York Hilton, NY, 2010, available at: http://www.alternano.com.

103. Stephentown, New York/Beacon Power, June 2011, available at: http://beaconpower.com/stephentown-new-york/, accessed April 3, 2016.

104. T. Hatta, Recent applications of sodium-sulfur (NaS) batteries system in the United States and in Japan, pp. 43–45, NGK Insulators, Ltd., Nagoya, Japan, December 2011, available at: www.sandia.gov/ess/EESAT/2011_papers/111114_EESAT_Paper_NGK_Hatta.pdf, accessed March 18, 2017.

105. X. Hu and S. Stanton, A total Li-ion battery simulation solution, *SAE World Congress and Exhibition*, Detroit, MI, April 8–10, 2014.

14 Microgrid for High-Surety Power: Architectures, Controls, Protection, and Demonstration

Mark Dehong Xu, Haijin Li, and Keyan Shi

CONTENTS

Abstract 355
14.1 Highly Reliable Microgrid 356
14.2 Architectures 356
    14.2.1 Star Architectures 356
    14.2.2 Multiple-Star Architecture 358
    14.2.3 Ring Architecture 359
    14.2.4 Hybrid Architecture 359
    14.2.5 Evaluation of Architectures 360
14.3 Control 365
    14.3.1 Operation Modes of Super UPS 365
        14.3.1.1 Grid Normal Mode 365
        14.3.1.2 Gas Engine Power Mode 367
        14.3.1.3 Battery Power Mode 367
        14.3.1.4 Fuel Cell Power Mode 367
        14.3.1.5 Bypass Mode 369
    14.3.2 Power Management of Multiple Sources in Super UPS 369
        14.3.2.1 Fuel Cell Unit Power Management 372
        14.3.2.2 Gas Engine Unit Power Management 374
14.4 Protection 376
    14.4.1 Short-Circuit Fault Current Analysis 376
    14.4.2 Solid-State DC Circuit Breaker 380
14.5 Demonstration 382
    14.5.1 Super UPS Platform 382
    14.5.2 Experimental Results 385
14.6 Summary 390
Acknowledgment 390
References 390

ABSTRACT

Microgrid is an emerging and promising field because of the ever-increasing installation of renewable energy and concerns to the environment [1–4]. Microgrid usually comprises energy sources, storage components, and local loads. It can be connected to the outer grid or operates independently. The energy sources include renewable energy sources, such as wind power, photovoltaic power, and fuel cells, as well as traditional power sources such as diesel generation and gas turbine. Since sources such as photovoltaic or wind power in microgrid are fluctuating, the stability and reliability are a big concern for a microgrid system. Reliability is the essential requirement for applications such as data centers, ships, aircrafts, and hospitals [5–9]. The resilience of the microgrid can be achieved through careful design of architectures, controls, protection, etc.

14.1 HIGHLY RELIABLE MICROGRID

A microgrid for high-surety power called super uninterruptible power supply (UPS) is proposed to improve the reliability of the microgrid [10–13]. In traditional UPS, only utility is used as primary power source in grid normal condition. Here besides the grid, natural gas pipeline is introduced as the second energy source, which is complementary to the grid. Actually, the utility and natural gas pipeline are complementary and redundant energy supplies due to their competition. By leveraging these two independent infrastructures, utility and natural gas pipeline, the power system availability is significantly increased. Second, two independent storages are introduced such as battery and fuel cell. Since both sources and storage components have redundancy, power conversion circuits may become the weak point in the complete power system. Therefore, reliable power conversion circuit is introduced into the microgrid. Third, multiple energy sources, multiple storages, and highly reliable power conversion circuit work together to guarantee uninterruptible power supply for the critical load. Figure 14.1 shows the typical structure of Super UPS.

14.2 ARCHITECTURES

There are a variety of architectures for the microgrid, such as star architecture, multiple-star architecture, and ring architecture. In order to get the best architecture for Super UPS, evaluation methods for the microgrid are needed to evaluate the performance of different architectures according to certain criteria. The criteria include system energy efficiency, reliability, and economy.

14.2.1 STAR ARCHITECTURES

The first star architecture is load center architecture. In this architecture, the load is selected as the center and all the other sources are connected with it through power converters. Obviously, the central load has the most accessibility to all energy sources. Therefore, the power to the load is highly guaranteed. Figure 14.2a shows a load-centered architecture, which is very suitable as a highly reliable power supply system.

The second star architecture is DC bus center architecture, shown in Figure 14.2b. All the sources and loads are connected to the central DC bus. Since DC has no frequency and phase, the DC bus

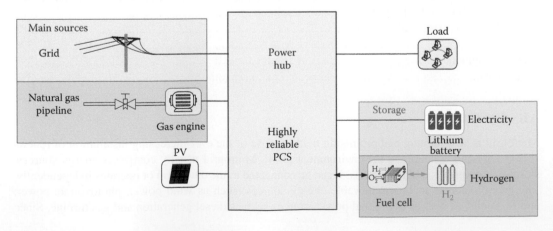

FIGURE 14.1 The concept of Super UPS.

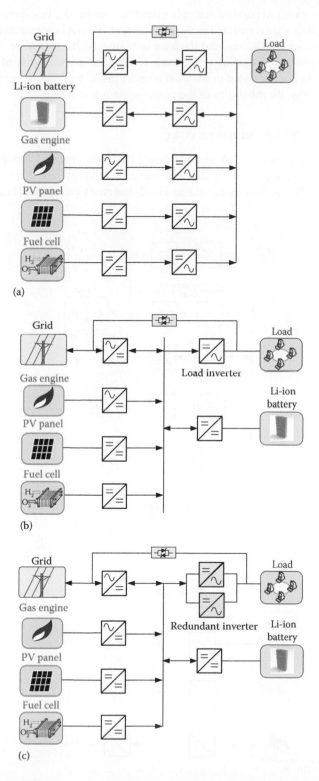

(a)

(b)

(c)

FIGURE 14.2 Star architectures for Super UPS. (a) Star architecture I. (b) Star architecture II. (c) Star architecture III.

center architecture is easier to parallel multiple converters to the DC bus. From the figure, we can see that all energy sources are connected to the DC bus first. Then it is connected to the load through the load inverter between the DC bus and the load in Figure 14.2b. Therefore, the reliability of the inverter is critical to power transferring to the load. To enhance the reliability of bus-centered architecture, a redundant inverter is added to the load inverter in DC bus center architecture as shown in Figure 14.2c. In this way, the reliability of the power system is increased.

14.2.2 MULTIPLE-STAR ARCHITECTURE

The multiple-star architecture is composed of more than one centered topologies. The centered topology may be a source-centered topology or bus-centered topology. A two-star architecture is shown in Figure 14.3. The architecture contains a load-centered topology and bus-centered topology.

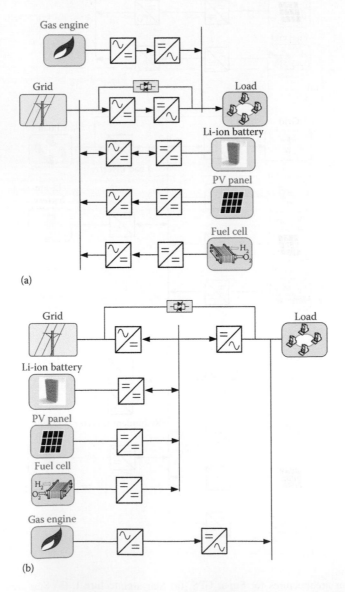

FIGURE 14.3 Multiple-star architectures for Super UPS. (a) Multiple-star architecture I. (b) Multiple-star architecture II.

Energy sources can be integrated into the system more flexible in this architecture. Besides, it introduces redundancy to the architecture. It can be avoided that one centered topology failure results in the failure of the whole system due to the inclusion of multiple centers.

14.2.3 RING ARCHITECTURE

The ring architecture contains one loop. Several sources are connected with each other to form a loop through the converters. The typical ring architecture is shown in Figure 14.4. The advantage of the ring architecture is that it has high reliability. Every source has at least two paths to transfer power to the outside. The energy exchange between the sources is more flexible. However, if we consider the reliability of supplying power to the load, the reliability of the ring architecture is lower than the load center star architecture. All sources must go through either back-to-back converter 1 or back-to-back converter 2 to arrive to the load in Figure 14.4. These two converters are the bottleneck of the power transfer. However, the load center star architecture has five independent paths to the load.

14.2.4 HYBRID ARCHITECTURE

A hybrid architecture is a combination of different architectures including the star ones and ring ones. Figure 14.5 illustrates a hybrid architecture. In this architecture, load, gas engine, fuel cell, and Li-ion battery form a star architecture, while the load, grid, and PV form a ring architecture. The advantage of this architecture is that it combines the merits of different architectures.

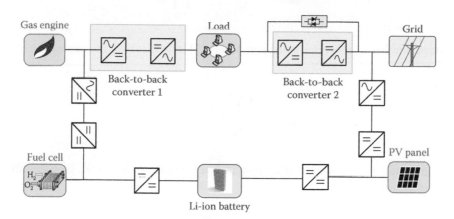

FIGURE 14.4   A ring architecture for Super UPS.

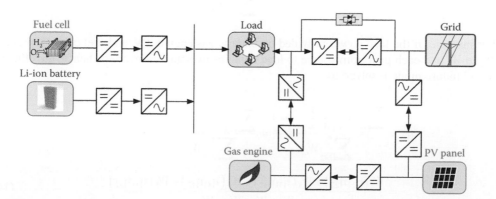

FIGURE 14.5   A hybrid architecture for Super UPS.

14.2.5 EVALUATION OF ARCHITECTURES

Different architectures are compared with regard to efficiency, reliability, and cost. Mathematical models are established to analyze the system efficiency and reliability performance.

The system efficiency includes the efficiency in grid normal mode and the one in grid failure mode. It is assumed that the grid provides the power to load in grid normal mode, and other sources provide the power to load in the grid failure mode according to the power distribution strategy. The system efficiency is an average efficiency of all power transferring paths. Efficiency in grid normal mode is an important indicator for system efficiency because Super UPS operates in this mode for most of the time. In this mode, the efficiency of the path from the grid to the load represents the efficiency in grid normal mode. While in grid failure mode, the system efficiency is the average one of all paths, which then power flows through.

Figure 14.6 is used to explain how to calculate the efficiency and takes star architecture I as an example. W_n represents the power of each source, and W_1 is assumed as the grid. M_n is the number of converters in path n, and η_{ij} is the efficiency of the converter in each path. The parameter i indicates which source it belongs to, and j indicates the exact location in path of source i. The value of i varies from 1 to n, and value of j varies from 1 to M_n. So η_n, the efficiency in grid normal mode, can be expressed as

$$\eta_n = \frac{W_1 \prod_{j=1}^{M_1} \eta_{1j}}{W_1} = \prod_{j=1}^{M_1} \eta_{1j} \tag{14.1}$$

If the stage number of converter in the path from grid to load is two, the expression of η_n can be calculated as

$$\eta_n = \frac{W_1 \prod_{j=1}^{2} \eta_{1j}}{W_1} = \eta_{11}\eta_{12} \tag{14.2}$$

When a grid fails, if the power of other sources is represented by W_2 to W_n, then the efficiency in this mode can be calculated as

$$\eta_f = \frac{\sum_{i=2}^{n} W_i \left(\prod_{j=1}^{M_i} \eta_{ij} \right)}{\sum_{i=2}^{n} W_i} \tag{14.3}$$

For instance, if the load power provided by battery, gas engine, PV, and fuel cell is represented by W_2 to W_5 and each path from source to load contains two-stage converters, hence, the efficiency η_f in grid failure mode is solved as

$$\eta_f = \frac{\sum_{i=2}^{5} W_i \left(\prod_{j=1}^{2} \eta_{ij} \right)}{\sum_{i=2}^{5} W_i} = \frac{\sum_{i=2}^{5} W_i \left(\eta_{i1}\eta_{i2} \right)}{\sum_{i=2}^{5} W_i}$$

$$= \frac{W_2 \left(\eta_{21}\eta_{22} \right) + W_3 \left(\eta_{31}\eta_{32} \right) + W_4 \left(\eta_{41}\eta_{42} \right) + W_5 \left(\eta_{51}\eta_{52} \right)}{W_2 + W_3 + W_4 + W_5} \tag{14.4}$$

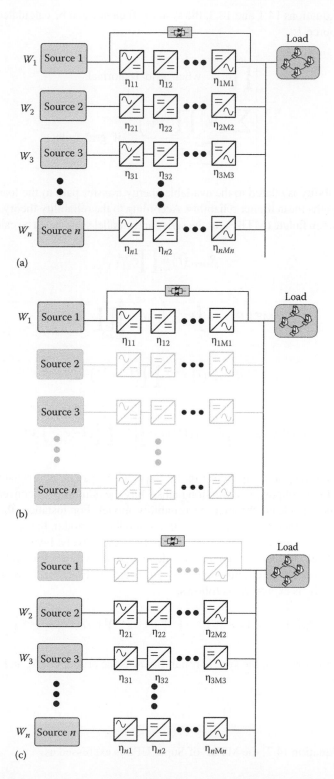

FIGURE 14.6 Efficiency calculation of star architecture I. (a) Efficiency model of star architecture I. (b) Efficiency calculation when grid is normal. (c) Efficiency calculation when grid fails.

By combining Equations 14.1 and 14.3, the system efficiency can be calculated in the following, assuming grid is source 1:

$$
\eta = \begin{cases}
\displaystyle\prod_{j=1}^{M_1}\eta_{1j}, & \text{when grid is normal} \\[2em]
\dfrac{\displaystyle\sum_{i=2}^{n} W_i\left(\prod_{j=1}^{M_i}\eta_{ij}\right)}{\displaystyle\sum_{i=2}^{n} W_i}, & \text{if grid fails}
\end{cases}
\tag{14.5}
$$

The system reliability is related to the available energy transfer path to the load. More redundant power conversion paths mean higher reliability. According to the reliability theory, the reliability $R(t)$ and mean time between failure (MTBF) for the series and parallel system can be calculated as follows:

$$
R_{series}(t) = \prod_{i=1}^{n} e^{-\lambda_i t}
\tag{14.6}
$$

$$
\text{MTBF}_{series} = \int_0^{\infty} R_{series}(t)\,dt = \int_0^{\infty} \prod_{i=1}^{n} e^{-\lambda_i t}\,dt,
\tag{14.7}
$$

$$
R_{parallel}(t) = 1 - \prod_{i=1}^{n}\left(1 - e^{-\lambda_i t}\right)
\tag{14.8}
$$

$$
\text{MTBF}_{parallel} = \int_0^{\infty} R_{parallel}(t)\,dt = \int_0^{\infty}\left[1 - \prod_{i=1}^{n}\left(1 - e^{-\lambda_i t}\right)\right]dt
\tag{14.9}
$$

where λ_i is the failure rate of component i in the Super UPS system, and reliability model can be built based on topologies of power conversion paths. Only the reliability of converters is considered. Then MTBF can be calculated through the reliability model. For instance, the reliability model of star architecture II is shown in Figure 14.7. In the reliability model, the failure rate of grid, gas engine, fuel cell, battery, and PV are λ_1, λ_2, λ_3, λ_4, and λ_5, respectively. Five modules including utility, gas engine, fuel cell, battery, and PV are paralleled first as a whole. Then it has a series connection with the load inverter whose failure rate is λ_6. According to Equation 14.8, the reliability of the five paralleling units R_p is expressed as follows:

$$
R_P(t) = 1 - \left(1 - e^{-\lambda_1 t}\right)\left(1 - e^{-\lambda_2 t}\right)\left(1 - e^{\lambda_3 t}\right)\left(1 - e^{\lambda_4 t}\right)\left(1 - e^{\lambda_5 t}\right)
\tag{14.10}
$$

According to Equation 14.6, the system reliability R_s is the product of R_p and R_{inv} as given in the following equation:

$$
R_s(t) = R_p \cdot R_{inv} = R_p(t)e^{-\lambda_6 t} = \left[1 - \left(1 - e^{-\lambda_1 t}\right)\left(1 - e^{-\lambda_2 t}\right)\left(1 - e^{\lambda_3 t}\right)\left(1 - e^{\lambda_4 t}\right)\left(1 - e^{\lambda_5 t}\right)\right]e^{-\lambda_6 t}
\tag{14.11}
$$

According to Equation 14.7, the MTBF of Super UPS is expressed as

$$
\text{MTBF} = \int_0^{\infty} R_s\,dt = \int_0^{\infty}\left[1 - \left(1 - e^{-\lambda_1 t}\right)\left(1 - e^{-\lambda_2 t}\right)\left(1 - e^{-\lambda_3 t}\right)\left(1 - e^{-\lambda_4 t}\right)\left(1 - e^{-\lambda_5 t}\right)\right]e^{-\lambda_6 t}\,dt
\tag{14.12}
$$

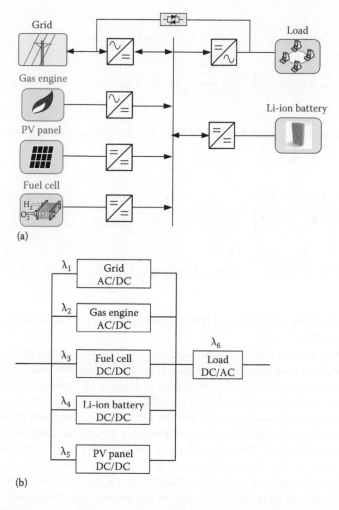

FIGURE 14.7 Reliability model of bus-centered architecture: (a) Bus-centered architecture and (b) the reliability model.

After expanding Equation 14.12, the expression of MTBF becomes

$$
\begin{aligned}
\text{MTBF} = \int_0^\infty \Bigg(& e^{-(\lambda_1+\lambda_6)t} + e^{-(\lambda_2+\lambda_6)t} + e^{-(\lambda_3+\lambda_6)t} + e^{-(\lambda_4+\lambda_6)t} + e^{-(\lambda_5+\lambda_6)t} - \sum_{0<i<j<6} e^{-(\lambda_i+\lambda_j+\lambda_6)t} \\
& + \sum_{0<i<j<k<6} e^{-(\lambda_i+\lambda_j+\lambda_k+\lambda_6)t} - \sum_{0<i<j<k<l<6} e^{-(\lambda_i+\lambda_j+\lambda_k+\lambda_l+\lambda_6)t} + e^{-(\lambda_1+\lambda_2+\lambda_3+\lambda_4+\lambda_5+\lambda_6)t} \Bigg) dt \quad (14.13)
\end{aligned}
$$

Substituting Equation 14.14 into Equation 14.13, the MTBF of the Super UPS can be solved as Equation 14.15:

$$
\int_0^\infty e^{-\lambda t} = \frac{1}{\lambda} \quad (14.14)
$$

$$MTBF = \frac{1}{\lambda_1 + \lambda_6} + \frac{1}{\lambda_2 + \lambda_6} + \frac{1}{\lambda_3 + \lambda_6} + \frac{1}{\lambda_4 + \lambda_6} + \frac{1}{\lambda_5 + \lambda_6} - \sum_{0 < i < j < 6} \frac{1}{\lambda_i + \lambda_j + \lambda_6}$$

$$+ \sum_{0 < i < j < k < 6} \frac{1}{\lambda_i + \lambda_j + \lambda_k + \lambda_6} - \sum_{0 < i < j < k < l < 6} \frac{1}{\lambda_i + \lambda_j + \lambda_k + \lambda_l + \lambda_6} + \frac{1}{\lambda_1 + \lambda_2 + \lambda_3 + \lambda_4 + \lambda_5 + \lambda_6}$$

$$(14.15)$$

The failure rate is the reciprocal of the MTBF. So, the failure rate expression is obtained as follows:

$$\lambda = MTBF^{-1}$$

$$= \left(\frac{1}{\lambda_1 + \lambda_6} + \frac{1}{\lambda_2 + \lambda_6} + \frac{1}{\lambda_3 + \lambda_6} + \frac{1}{\lambda_4 + \lambda_6} + \frac{1}{\lambda_5 + \lambda_6} - \sum_{0 < i < j < 6} \frac{1}{\lambda_i + \lambda_j + \lambda_6} \right.$$

$$\left. + \sum_{0 < i < j < k < 6} \frac{1}{\lambda_i + \lambda_j + \lambda_k + \lambda_6} - \sum_{0 < i < j < k < l < 6} \frac{1}{\lambda_i + \lambda_j + \lambda_k + \lambda_l + \lambda_6} + \frac{1}{\lambda_1 + \lambda_2 + \lambda_3 + \lambda_4 + \lambda_5 + \lambda_6} \right)^{-1}$$

$$(14.16)$$

The economy characteristic of Super UPS is represented by the converter cost which is dominated by the number of converters.

Based on all the evaluation methods mentioned earlier, Table 14.1 shows the system performance of different architectures assuming that each power converter has efficiency η, failure rate λ, and cost c. The architectures discussed earlier are shown in Figures 14.2 through 14.4.

In Table 14.1, star architecture II has the highest efficiency and lowest cost since the minimal converter stages in the path from source to load and the least converter number. However, the different sources share the common inverter; the load inverter became the critical component whose failure can cause the whole system blackout. Thus, the redundant architecture is used in star architecture III (DC bus star architecture with redundant load inverter). It brings the highest reliability, and its other features are relatively good compared with others. Finally, the DC bus centered with redundant inverter architecture star architecture III is selected. Based on this architecture, three-level T-type inverter and three-level buck–boost converter are chosen as bidirectional AC/DC converter and bidirectional DC/DC converter, respectively, according to the loss breakdown. The loss breakdown results are not provided here because of the content limitation. Figure 14.8 shows the power conversion system (PCS) of the overall Super UPS system.

TABLE 14.1
System Performance of Different Architectures

	System Efficiency When Grid is Normal	System Efficiency When Grid Fails	Failure Rate	Cost for Power Converters
Star arch. 1	η^2	η^2	0.88λ	10c
Star arch. 2	η^2	η^2	1.2λ	6c
Star arch. 3	η^2	η^2	0.84λ	7c
Multiple-star arch. 1	η^2	$\left(\eta^2 + 3\eta^4\right)/4$	1.04λ	10c
Multiple-star arch. 2	η^2	η^2	0.97λ	8c
Ring architecture	η^2	$\left(\eta^2 + 2\eta^4 + \eta^5\right)/4$	1.33λ	10c
Hybrid architecture	η^2	$\left(3\eta^2 + \eta^4\right)/4$	1.13λ	12c

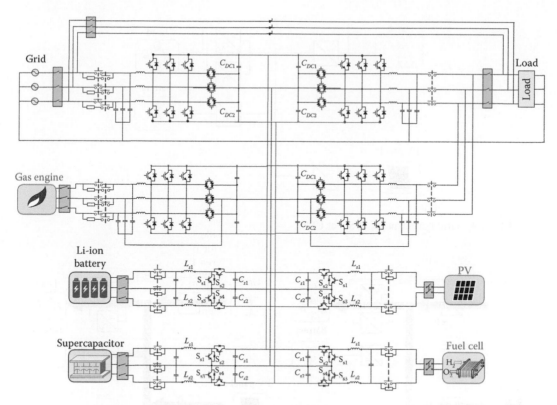

FIGURE 14.8 The power conversion system (PCS) of Super UPS.

14.3 CONTROL

14.3.1 OPERATION MODES OF SUPER UPS

The target of control of Super UPS is to achieve high reliability [15–23]. There are five operation modes in Super UPS, grid normal mode, gas engine power mode, battery power mode, fuel cell power mode, and bypass mode. The mode that the system operates in is determined by the states of sources.

14.3.1.1 Grid Normal Mode

The power distribution in grid normal mode is shown in Figure 14.9. When the utility is normal, the utility provides power for load as the main source. The rectifier connected to the grid regulates the DC bus, while all other converters connected to various sources operate in current control mode or standby mode. The gas engine and fuel cell (FC) are in cold backup mode to extend their lifetime. Photovoltaic (PV) operates in either maximum power point tracking (MPPT) mode or power limiting state according to the system condition. The power flows in this mode are shown in Figure 14.10. When the state of charge (SOC) of the battery is full, the battery operates in float charge mode. If PV power is less than load power, the system is in the state shown in Figure 14.10a, where the PV and utility work together to supply power to load. If PV power is larger than utility, the extra power is injected to grid as given in Figure 14.10c. If there is no sunlight, only the utility provides the power to the load only as Figure 14.10e. When the battery is not full, the system is in states as shown in Figure 14.10b, d, and f according to the PV and the load condition.

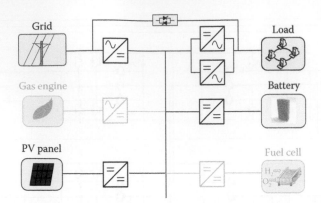

FIGURE 14.9   Grid normal mode of Super UPS.

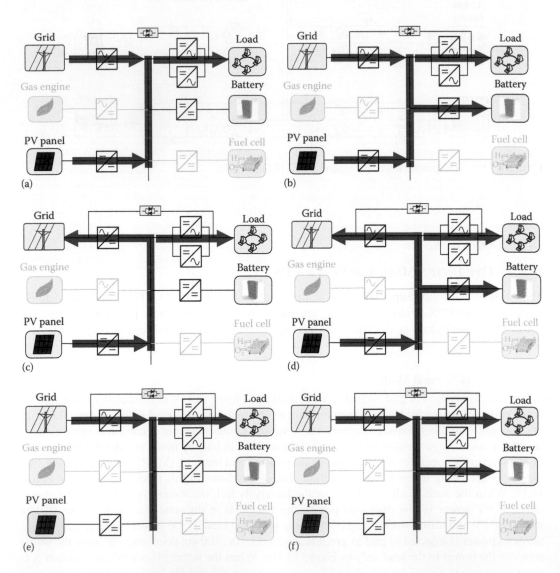

FIGURE 14.10   Power flows of grid normal mode.

14.3.1.2 Gas Engine Power Mode

The gas engine has a lower priority to provide the power to the load than the utility. When utility fails, the gas engine becomes the main source. It is shown in Figure 14.11. During the time when the gas engine is starting, the supercapacitor temporarily regulates the DC bus. After the gas engine finishes starting, the gas engine takes over the right to regulate the DC bus. FC always keeps in cold backup mode. In this mode, only the gas engine regulates the DC bus voltage as a voltage source, while other generation units operate in current source mode to provide power to the load based on the commands from the central controller.

The power flows in this mode are shown in Figure 14.12. When the battery is full, gas engine and PV provide the power to the load as shown in Figure 14.12a. If the PV power is larger than the load power, PV is in limit power point tracking (LPPT) mode; otherwise, it is in MPPT mode. If there is no light, the system is in the state as shown in Figure 14.12c. When SOC of the battery is not full, the system operates in the state as shown in Figure 14.12b.

14.3.1.3 Battery Power Mode

In Figure 14.13, when both utility and gas engine fails, the battery has a higher priority to provide the power for the load than FC. So FC keeps in cold backup mode. And the battery provides the power to the load by the DC/DC converter in this mode. The power flows in this mode are given in Figure 14.14; if the PV power is less than the load power, the system is in the state as shown in Figure 14.14a. However, if the battery is not full, PV provides the power to the load and charges the battery as shown in Figure 14.14b. And if there is no light, only the battery provides the power to the load as shown in Figure 14.14c. It should be noticed that PV operates in LPPT mode if its power is larger than the sum of the load power and the battery charge power.

14.3.1.4 Fuel Cell Power Mode

Because of the limited capacity and life cycle of the battery, when the battery is exhausted or fails, the mode transfers from the battery power mode to the fuel cell power mode as shown in Figure 14.15. If the PV power is larger than the load power, the PV operates in LPPT mode. If its power is less than the load power, FC and PV provide the power to the load together. However, if there is no light, only FC provides the power to the load. The power flows in this case are shown in Figure 14.16.

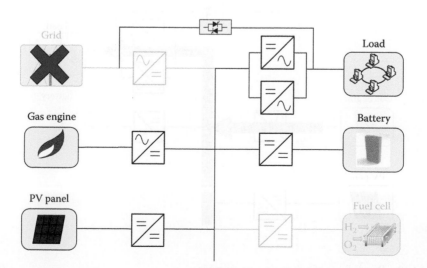

FIGURE 14.11 Gas engine power mode of Super UPS.

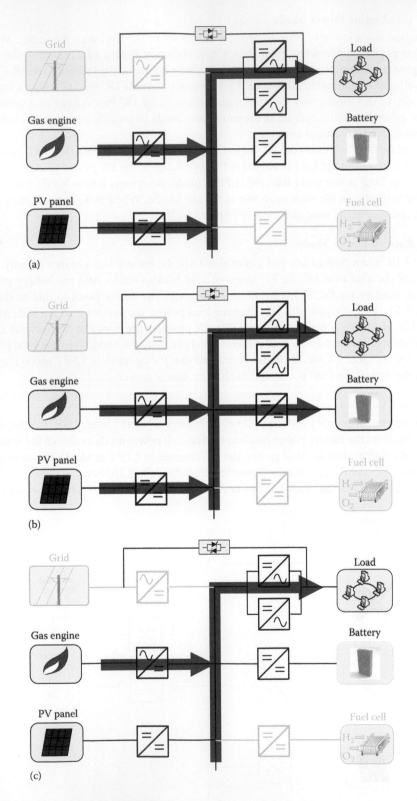

FIGURE 14.12 Power flows of gas engine power mode. (*Continued*)

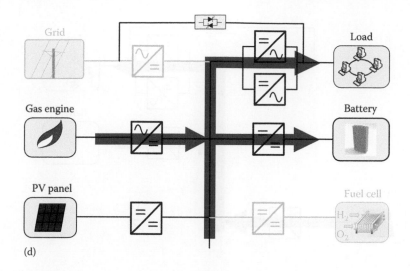

(d)

FIGURE 14.12 (*Continued*)   Power flows of gas engine power mode.

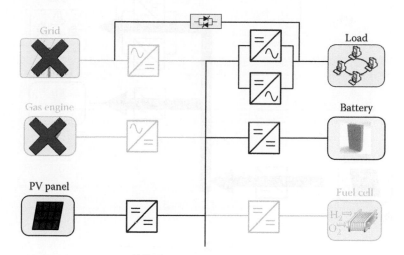

FIGURE 14.13   Battery power mode of Super UPS.

14.3.1.5   Bypass Mode

Two load inverters can be in parallel having a redundant structure. The failure of one inverter cannot lead to the load interruption. But the worst situation is that both load inverters fail. No source can provide power for the load through inverters. In this situation, the system operates in bypass mode, and the utility provides the power to the load through bypass switches as shown in Figure 14.17. This mode is also a unique mode of UPS compared to other power systems. The power flows are shown in Figure 14.18.

14.3.2   POWER MANAGEMENT OF MULTIPLE SOURCES IN SUPER UPS

Power management is important for improving the reliability of multiple sources in Super UPS. In this section, the power management of FC, gas engine unit, is introduced according to the characteristics of the sources. The first priority of power management is to extend the life span of various sources.

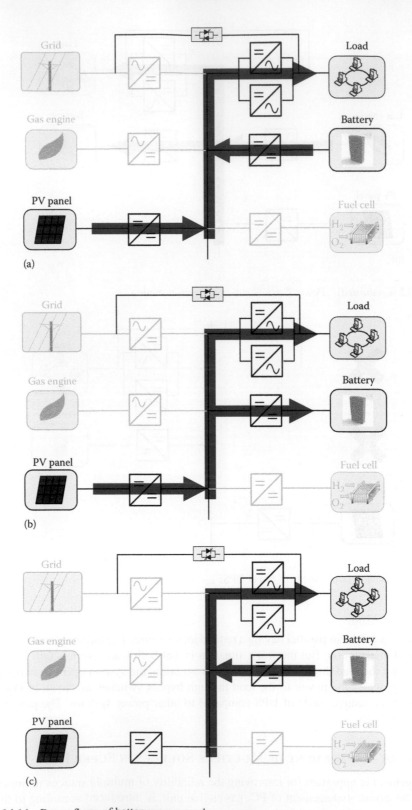

FIGURE 14.14 Power flows of battery power mode.

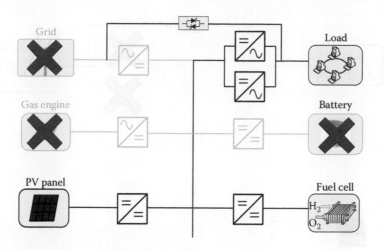

FIGURE 14.15   Fuel cell power mode of Super UPS.

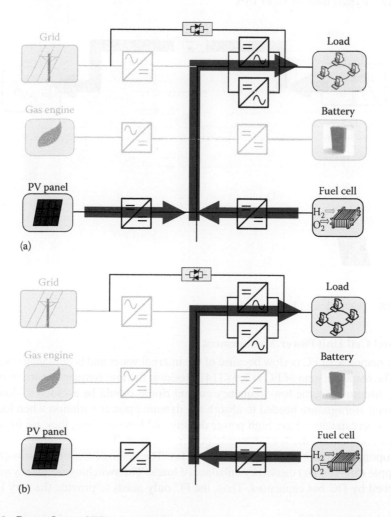

FIGURE 14.16   Power flows of FC power mode.

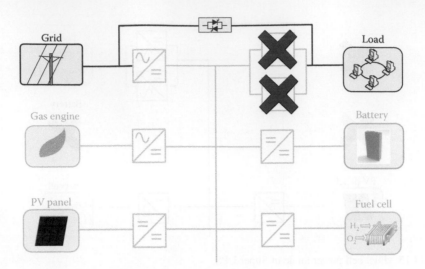

FIGURE 14.17 Bypass mode of Super UPS.

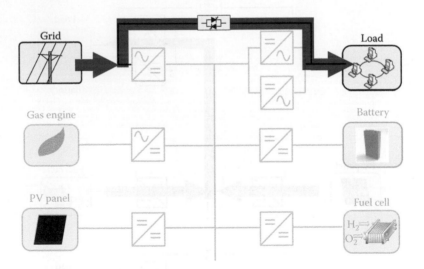

FIGURE 14.18 Power flow of bypass mode.

14.3.2.1 Fuel Cell Unit Power Management

The dynamic response of FC is slow because of the internal water and heat management and airflow regulation. The response time of FC is 1.9 s [14]. Besides, the low-frequency current ripple is harmful to the FC lifetime. So, the low-frequency current ripple should be avoided to flow into the FC. Thus, additional storages are needed to absorb the dynamic power variation when load steps up or down. Since supercapacitors have high power density and long lifetime, they are finally introduced as storages to compensate the load dynamic power variation. The supercapacitor can provide the dynamic component (higher frequency than 0.16 Hz). The range of frequency also covers the low-frequency ripple (50–150 Hz) caused by unbalanced loads. The switching frequency and its harmonics are absorbed by DC bus capacitors. Thus, the FC only needs to provide the very low-frequency component.

The control scheme of FC power management is illustrated in Figure 14.19. It consists of three parts, such as detection loop, inner current loop, and supercapacitor (SC) voltage loop. The detection

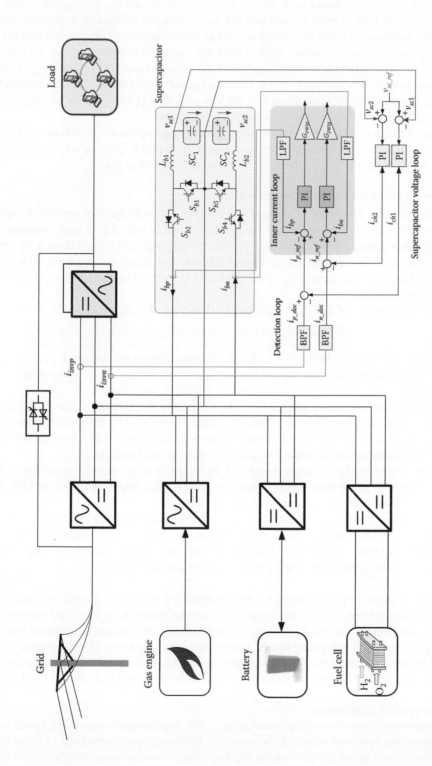

FIGURE 14.19 FC unit power management control diagram.

loop is used to extract the high-frequency components of the load current. The high-frequency components i_{p_dec} and i_{n_dec} are extracted from the inverter DC side current i_{invp} and i_{invn} by a band pass filter. The SC voltage loop is used to maintain the supercapacitor voltage in a constant level. It ensures that the supercapacitor has an enough margin to absorb or supply the power when the next load jumping comes. The inner current loop is used for current reference tracking. Its reference values i_{p_ref} and i_{n_ref} are obtained by the difference of the detection current (i_{p_dec} and i_{n_dec}) and the SC voltage loop output. The feedback output currents i_{bp} and i_{bn} of the DC–DC are sampled and their switching frequency ripple is eliminated by the low-pass filter (LPF). And the proportional and integral (PI) controller is utilized in the inner current loop control.

The lifetime of FC in cold backup is much longer than hot backup mode. Therefore, the FC operates in the cold backup mode if both utility and gas engine are normal. The cold start-up strategy for FC is similar to that for gas engine, which will be discussed in the next part.

14.3.2.2 Gas Engine Unit Power Management

In order to improve the reliability of the gas engine unit, the cold backup strategy is applied. Otherwise, the failure rate of gas engine will obviously increase in hot backup mode because of the mechanical abrasion and aging. The cold start-up time of gas engine is long. Taking a 50 kW gas engine as an example, 8.5 s is needed for gas engine cold start-up. Besides, the gas engine must start without loads during this time. Thus, the supercapacitor is introduced to regulate the DC bus before the gas engine finishes starting.

The detailed procedure of the cold start-up is shown in the right-hand-side waveform in Figure 14.20.

Stage 0: Initial state $(t_0$–$t_1)$—In this stage, the utility is the main power source. It provides the power to the load and regulates the DC bus voltage while the gas engine keeps in the cold backup mode.

Stage 1: Grid failure detection stage $(t_1$–$t_2)$—In this stage, the grid fails. The grid failure detection is based on the DC bus voltage measure. Before the DC bus voltage decreases to the setting threshold, the DC bus capacitors provide the load power. Thus, the DC bus voltage will drop, and the magnitude of drop depends on the threshold.

Stage 2: Gas engine cold start stage $(t_2$–$t_3)$—In this stage, the SC converter turns from the charging mode to the voltage mode to regulate the DC bus. Supercapacitors provide the power to the load while the gas engine begins to start.

Stage 3: Gas engine providing power stage $(t_3$–$t_4)$—In this stage, the gas engine finishes starting and turns to the voltage mode to regulate the DC bus voltage. Meanwhile, the SC converter turns to the current mode. During this time, the output power of supercapacitors decreases gradually; thus, the output power of gas engine increases.

The control scheme of the gas engine power management is shown in the diagram of Figure 14.20. It consists of two outer loops, such as SC voltage loop and DC bus voltage loop, and an inner loop, inner current loop. The SC voltage loop is used to maintain the SC voltage at a certain level. The DC bus voltage loop is used to regulate the DC bus voltage. The references of inner current loop i_{ref}^{+} and i_{ref}^{-} equal to either output of the DC bus voltage loop or output of SC voltage loop depending on the mode control signal. The output currents of the converter are sampled as the feedback of inner current loop. And their switching frequency ripple is eliminated by a low-pass filter (LPF). And PI controllers are applied in all loops.

The most important part of seamless transferring is the mode control signal. The mode control signal determines the operation mode of the converter by changing the reference of inner current loop. The mode control signal depends on the DC bus voltage measurement. Normally, the current reference equals to the output of SC voltage loop, and the SC converter regulates the SC voltage. When the DC bus voltage drops to the threshold V_{th}, the mode control signal reverses immediately,

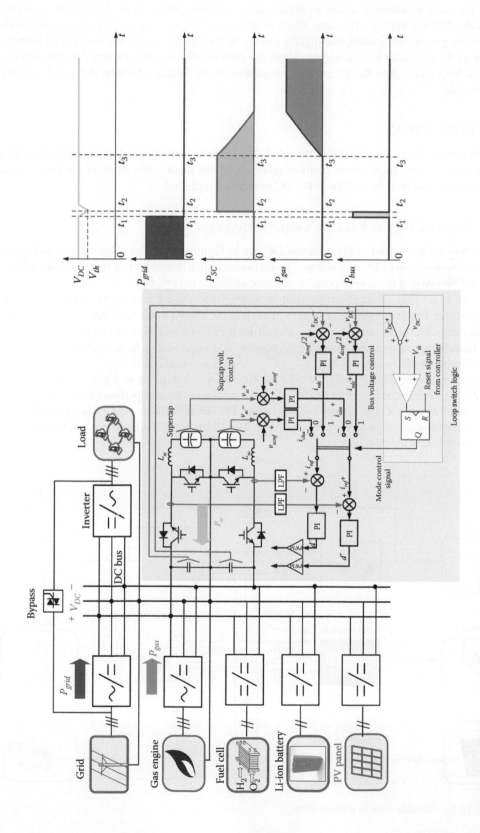

FIGURE 14.20 Cold start control diagram of gas engine.

and then the current reference equals to the output of DC bus voltage loop. And the SC converter regulates the DC bus voltage. The state of the mode control signal will be locked by an RS latch. Only after the gas engine finishes starting, the mode control signal is reset by a reset signal from the controller of the converter for gas engine. Then, the current reference will be equal to the output of SC voltage loop again. And the SC converter regulates the SC voltage. Finally, the seamless transferring finishes.

14.4 PROTECTION

The protection is a key issue regarding the reliability for microgrid. Super UPS is required to have an ability to handle the short-circuit faults of modules. In this section, the short-circuit fault current is analyzed first, and then the design of the DC breaker is explained.

14.4.1 SHORT-CIRCUIT FAULT CURRENT ANALYSIS

All converters are connected to the common DC bus in Figure 14.21. The short-circuit fault in any one can lead to the drop of DC bus voltage. So no matter which converter has short-circuit fault, the fault should be isolated as soon as possible to avoid affecting the DC bus.

There are four types of short-circuit fault in Super UPS: Type I fault is the short-circuit fault between positive bus and negative bus. Type II fault is the short-circuit fault between positive bus and neutral bus. Type III fault is the short-circuit fault between negative bus and neutral bus, and Type IV fault is the short-circuit fault of both positive and negative bus to neutral bus. They are all shown in Figure 14.22.

For instance, the equivalent failure model of Figure 14.22a is given in Figure 14.23. R_s is the short-circuit resistor of the converter module. L_s is the line inductance impedance from bus capacitor to the converter module. C_{DC} is the capacitance of half DC bus capacitors. And $i(t)$ is the short-circuit fault current, and $u(t)$ is the voltage of the DC bus. The short-circuit fault current can be calculated as given in the following equation:

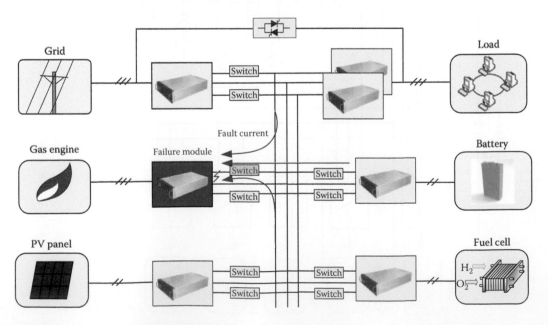

FIGURE 14.21   Module fault in a Super UPS.

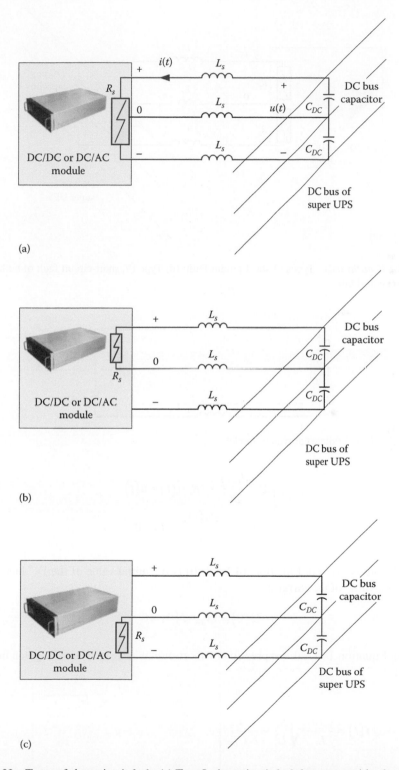

(a)

(b)

(c)

FIGURE 14.22 Types of short-circuit fault: (a) Type I, short-circuit fault between positive bus and negative bus; (b) Type II, short-circuit fault between positive bus and neutral bus; (c) Type III, short-circuit fault between negative bus and neutral bus.   (*Continued*)

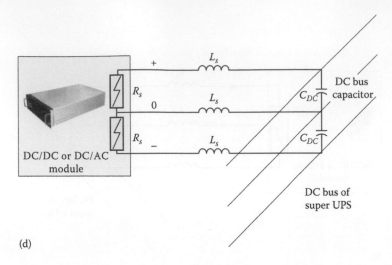

(d)

FIGURE 14.22 (*Continued*) Types of short-circuit fault: (d) Type IV, short-circuit fault of both positive and negative bus to neutral bus.

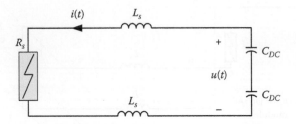

FIGURE 14.23   Type I failure equivalent model.

$$\begin{cases} 2L_s \dfrac{di(t)}{dt} + 2R_s \cdot i(t) = u(t) \\[2mm] 0.5C_{DC} \dfrac{du(t)}{dt} + i(t) = 0 \end{cases} \tag{14.17}$$

The initial state is given in Equation 14.18. $U(0)$ is the initial value of the DC bus voltage, and $I(0)$ is the initial value of the current:

$$U(0) = 2V_{DC}, I(0) = 0 \tag{14.18}$$

By solving Equation 14.17, the expression of short-circuit current $i(t)$ can be given in the following equation:

$$i(t) = \frac{U(0)}{L_s} \cdot \sqrt{\frac{R_s^2}{L_s^2} - \frac{4}{L_s C_{DC}}} \left[e^{\frac{-\frac{R_s}{L_s} + \sqrt{\frac{R_s^2}{L_s^2} - \frac{4}{L_s C_{DC}}}}{2}t} - e^{\frac{-\frac{R_s}{L_s} - \sqrt{\frac{R_s^2}{L_s^2} - \frac{4}{L_s C_{DC}}}}{2}t} \right] \tag{14.19}$$

Based on Equation 14.19, take a 100 kW AC/DC converter module as an example with the specifications of the system shown in Table 14.2. The calculation results are shown in Figure 14.24.

TABLE 14.2

Specifications of Short-Circuit Analysis

Parameter	Value
Positive/negative bus voltage V_{DC}	375 V
DC bus capacitor C_{DC}	10 mF
Line impedance from bus capacitor to submodule L_s	1 µH
Short-circuit resistor of submodule R_s	40 mΩ
Rated power of submodule P_r	100 kW
Rated current of submodule I_r	133 A

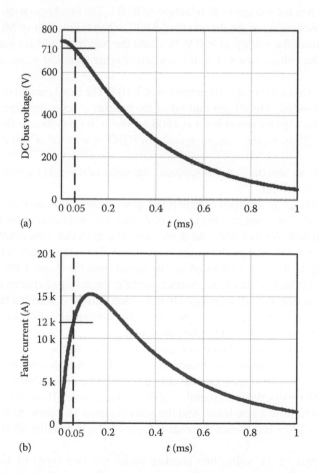

FIGURE 14.24 Short-circuit fault waveform. (a) DC bus voltage waveform. (b) Fault current waveform.

In Figure 14.24, the short-circuit fault occurs in the converter module between positive bus and negative bus. It can be seen that the DC bus voltage drops from 750 to 710 V, and the fault current increases to 12 kA within 50 µs. In order to guarantee the load is uninterruptible, the isolation breaker must be very fast to avoid DC bus voltage drop. However, the trigger time of a mechanical breaker is too slow to fulfill the requirement of this application. Thus, a solid-state DC breaker should be introduced to the system.

14.4.2　Solid-State DC Circuit Breaker

According to the fault current analysis mentioned earlier, a solid-state breaker is designed for Super UPS. A two-series-opposing insulated gate bipolar transistor (IGBT) structure is used in it. Because the large inductor exists in the short-circuit loop, a metal oxide varistor (MOV) is used to protect the IGBT by limiting its voltage stress. The structure of the solid-state DC circuit breaker is shown in Figure 14.26. This structure has been used in [24–28].

The design of the breaker for 100 kW converter module includes three main steps, such as selection of MOV, selection of IGBT voltage level, and selection of IGBT current capacity.

1. The first step is the selection of MOV for the breaker. Because the rated voltage of breaker is 375 V, EPCOS MOV S20K300E2 is selected. Its rated operation voltage is 385 V, and the varistor voltage is 470 V (1 mA).
2. The second step is the voltage level selection of IGBT. The rated operation voltage of IGBT is also 375 V. According to the datasheet of MOV, when the current of MOV reaches 1500 A, MOV can limit the voltage at 950 V. Because the parasitic inductor between IGBT and MOV exists, the voltage stress of IGBT is a little higher than 950 V. So, the 1200 V IGBT is selected.
3. The last step is the current capacity selection of IGBT. The rated current of breaker is 133 A for a 100 kW module. The trigger current of the breaker is set at ten times of rated current normally. So the trigger current is set at 1500 A. The IGBT can handle three times of rated current within 20 μs. So the current capacity of IGBT is selected at 500 A.

A snubber circuit can also be added to suppress the spike of a breaker's voltage stress. The most typical snubber circuit is RCD.

The solid-state breaker is installed between the power conversion module and the DC bus as shown in Figure 14.25. The control scheme of solid-state DC breaker is shown as Figure 14.26. It consists of detection unit, control unit, and drive unit. The detection unit detects the short current within several microseconds and produces the fault signal. The control unit determines the state of the breaker according to the fault signal and enable signal from the Super UPS system.

When a short-circuit fault occurs in a converter module, the voltage between the node C and node E of the breaker in that module will increase. Then the fault current is detected based on this voltage. The voltage is extracted to produce fault signal. Then the control unit disables two breakers in this module based on the fault signal from the detection unit.

The detailed protection logic is shown in Figure 14.26. The fault signals of four IGBTs are connected to an AND gate to generate the protection signal. Then the protect signal and enable signal are connected to an AND gate to generate the driver signal. The driver signal is used to control the solid-state breaker. Normally, the fault signal of IGBT is a high level. If one IGBT has a short-circuit fault, the fault signal turns to a low level. And the protect signal also turns to a low level. Then, the driver signal turns to a low level when the enable signal is high. Finally, all IGBTs in the breaker are disabled.

The test waveforms of the solid-state breaker under the two types of fault are provided in Figure 14.27. The specifications of the short-circuit test are shown in Table 14.2. And the test circuit is shown in Figure 14.22. The waveform under short-circuit fault between positive bus and neutral bus is shown in Figure 14.27a. It can be seen that the voltage stress is 911 V at the fault current 1200 A in this kind of fault. The waveform under short-circuit fault between the positive bus and negative bus is shown in Figure 14.27b. The voltage stress of the positive bus breaker is 940 V, and the voltage stress of the negative bus breaker is 890 V. Furthermore, the test results about the solid-state breaker are given in Figure 14.28. It can be seen that the solid breaker is safe enough based on the design of the solid-state breaker.

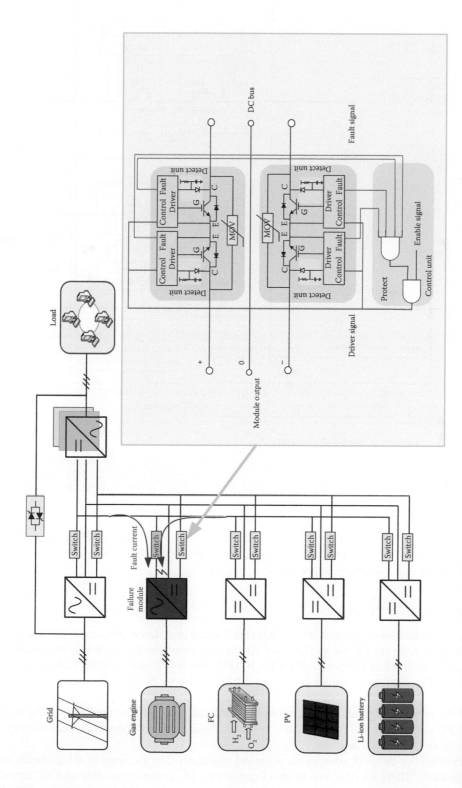

FIGURE 14.25 Installation and location of solid-state breaker.

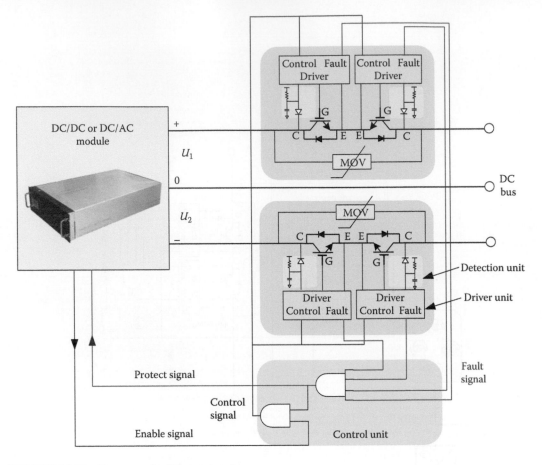

FIGURE 14.26 Structure of solid-state breaker.

14.5 DEMONSTRATION

In order to verify the architecture, the controls, and the protection of Super UPS, a demonstration platform is built in the lab. The demonstration platform of the Super UPS consists of utility, fuel cell, natural gas engine, Li-ion battery, supercapacitor, and PV.

14.5.1 SUPER UPS PLATFORM

In this platform, distributed sources and storages are located in different places, such as the ground floor, the first floor, the third floor, and the roof. All converter modules communicate with each other through the Ethernet. The rated power of load is 100 kW. The specifications of the system are shown in Table 14.3. The platform is shown in Figures 14.29 and 14.30. Besides, the control architecture is given in Figure 14.31. The center controller is an upper computer. The upper computer sends the mode and the power command to all converter modules and receives the feedback of the module state. The FC and gas engine receive the commands through wireless network because they are located far from the center controller.

All converters for different sources are designed modularly, and the sizes of all converter modules are the same. There are only two kinds of converters, DC converter module and AC converter module. The DC converter modules are used for DC sources such as FC, battery, SC, and PV. The

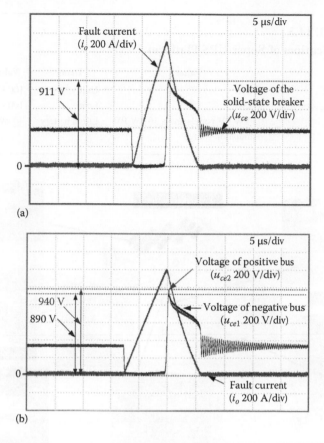

(a)

(b)

FIGURE 14.27 Waveforms of short-circuit protection. (a) Waveform under short-circuit fault between the positive bus and neutral bus. (b) Waveform under short-circuit fault between the positive bus and negative bus.

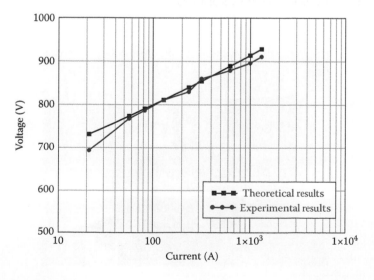

FIGURE 14.28 Voltage stress of solid-state breaker.

TABLE 14.3

Specifications of Super UPS Platform

Items	Value	Items	Value
Fuel cell	30 kW	Utility	100 kW
Natural gas engine	50 kW	Load	100 kW
PV	1.5 kW PV panels and a 30 kW PV simulation source	Li-ion battery	40 kWh

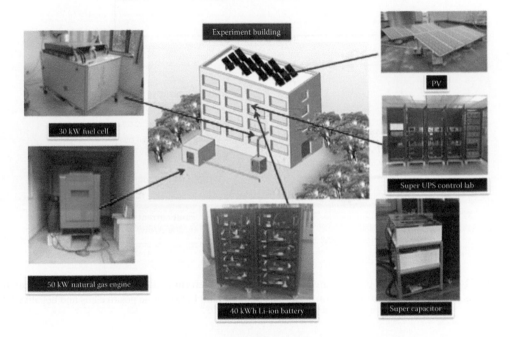

FIGURE 14.29   Demonstration of Super UPS platform.

FIGURE 14.30   The storages and power conversion system (PCS) in the Super UPS.

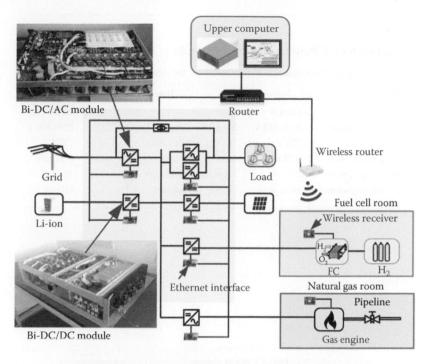

FIGURE 14.31   Control architecture of Super UPS.

FIGURE 14.32   The power conversion system of Super UPS.

AC converter modules are used for AC sources and the AC load. The modular converter can be easily replaced and maintained to achieve high reliability. The prototype of the power conversion system is shown in Figure 14.32. The specifications of the DC and AC modules are listed in Table 14.4.

14.5.2   EXPERIMENTAL RESULTS

The waveforms of the bidirectional DC/AC module are shown in Figure 14.33. The power capacity of DC/AC module is 100 kW. The AC side voltage is three phased 380 VAC, and the DC side voltage

TABLE 14.4

Parameters of Power Converter Module

AC Module Parameters		DC Module Parameters	
Item	**Value**	**Item**	**Value**
Power capacity	100 kW	Power capacity	30 kW
AC input voltage	3-ph 380 V	Input DC voltage	200–400 V
DC voltage	750 V	DC bus output voltage	750 V
Input inductance	0.38 mH	Input low-voltage side inductance	0.32 mH
Input capacitance	200 μF	Input low-voltage side capacitance	800 μF
DC bus capacitance	4400 μF	DC bus capacitance	400 μF

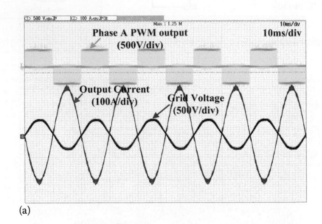

(a)

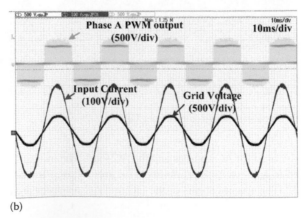

(b)

FIGURE 14.33 Waveforms in the bidirectional DC/AC module. (a) Waveform under rectifier mode. (b) Waveform under inverter mode.

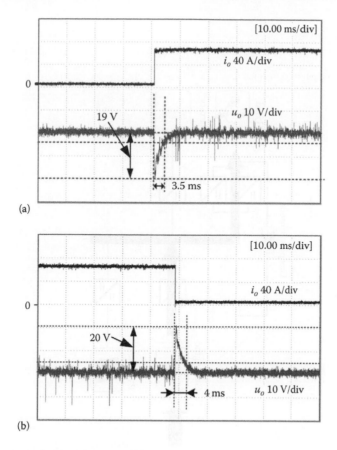

FIGURE 14.34 Waveforms of bidirectional DC/DC module. (a) Load steps up. (b) Load steps down.

is 750 V. The waveform under rectifier mode is shown in Figure 14.33a and the waveform under inverter mode is shown in Figure 14.33b. It can be seen that the power quality of grid-side current in the AC modules is good to fulfill the requirement of the grid.

The waveforms of bidirectional DC/DC are shown in Figure 14.34. The power capacity of DC/DC converter is 30 kW. Input voltage is 400 V and the output voltage is 750 V. The waveform of load step from 0 to 30 kW is shown in Figure 14.34a and that of load step from 30 kW to 0 is shown in Figure 14.34b. The dynamic time of DC module is less than 5 ms when the load steps. The voltage drop or overshoot is less than 5%. The DC module has an excellent dynamic response.

The mode transferring from grid normal mode to gas power mode is shown in Figure 14.35. The waveforms during this period are shown in Figure 14.36. It can be seen that the supercapacitor starts to regulate the DC bus when the grid fails. After the gas engine finishes starting, the gas engine provides the power to the load instead of the supercapacitor. Figure 14.36a is the overall transferring waveform, and Figure 14.36b is the waveform zoomed in. During the mode transferring, the DC bus voltage drop is 40 V. It is small enough to guarantee the load output voltage is uninterruptible. And the seamless transfer is achieved.

The protection waveform of the system is shown in Figure 14.37 with the test condition as shown in Figure 14.38. In this case, a short-circuit fault occurs in the DC converter module between positive bus and neutral bus. It can be seen that when a short-circuit fault occurs in the converter module, solid-state circuit breaker operates within 6 μs to isolate the fault module. The short-circuit fault current reaches 800 A. Finally, both the output voltage and current are continuous and uninterruptible.

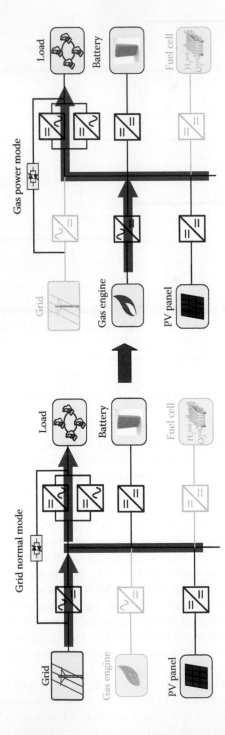

FIGURE 14.35 Mode transferring from grid normal mode to gas power mode.

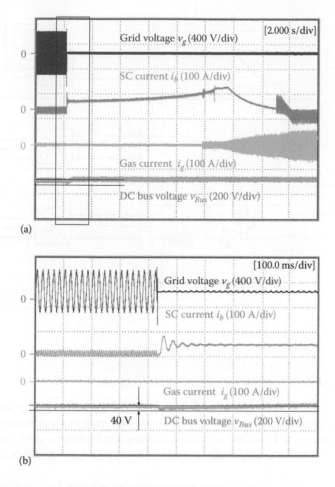

(a)

(b)

FIGURE 14.36 Waveforms of transferring from grid normal mode to gas engine power mode. (a) Transferring waveform. (b) The waveform zoomed in when grid fails.

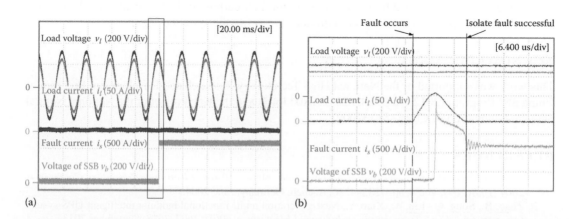

(a)

(b)

FIGURE 14.37 Waveforms of short-circuit fault protection in DC module. (a) Short-circuit protection waveform. (b) The waveform zoomed in.

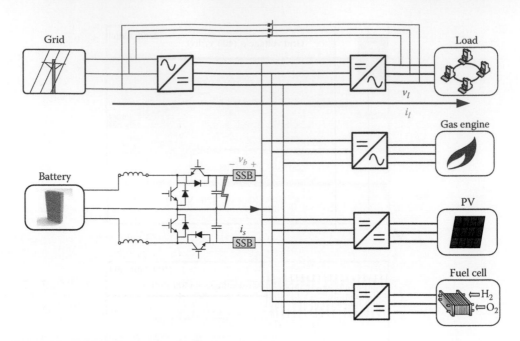

FIGURE 14.38   Test case of short-circuit fault protection.

14.6 SUMMARY

In order to improve the power of microgrid, a microgrid using a high-surety Super UPS concept is proposed. Two independent infrastructures, the utility and gas engine, are introduced in the Super UPS. Renewable energy PV and FC are integrated into the system as well. And the reliability of the microgrid improves obviously with multiple energies, multiple storages, and the highly reliable power conversion system.

According to the architecture deduction and comparison of reliability, cost and efficiency, the DC bus–centered architecture with redundant inverters is selected for the Super UPS as the best choice. The highly reliable power management strategies are presented to extend the lifetime of components and improve the system reliability further. Then a fast isolation solid-state DC circuit breaker with low-voltage stress is designed based on short-circuit fault current analysis. Finally, the demonstration and partial experimental results are introduced to verify the system design.

ACKNOWLEDGMENT

This work was supported by the National High Technology Research and Development Program of China 863 Program (2012AA053602, 2012AA053603) and the Key Project of the National Natural Science Foundation of China (51337009).

REFERENCES

1. Lasseter, R.H., Paigi, P., Microgrid: A conceptual solution, *Power Electronics Specialists Conference, 2004 (PESC 04)*, Vol. 6, pp. 4285–4290, Aachen, Germany, June 2004.
2. Zhao, B., Song, Q., Liu, W., Xiao, Y., Next-generation multi-functional modular intelligent UPS system for smart grid, *IEEE Transactions on Industrial Electronics*, 60(9), 3602–3618, September 2013.
3. Graditi, G., Apicella, A., Augugliaro, A., Dusonchet, L., Favuzza, S., Riva Sanseverino, E., Technical and economical aspects on integrated PV-UPS systems, *19th European Photovoltaic Solar Energy Conference*, pp. 2572–2575, Paris, France, June 2004.

4. Lahyani, A., Venet, P., Guermazi, A., Troudi, A., Battery/supercapacitors combination in uninterruptible power supply (UPS), *IEEE Transactions on Power Electronics*, 28(4), 1509–1522, April 2013.

5. Nguyen, M.Y., Yoon, Y.T., A comparison of microgrid topologies considering both market operations and reliability, *Electric Power Components and Systems*, 42(6), 585–594, 2014.

6. Kwasinski, A., Power electronic interfaces for ultra available DC micro-grids, *Power Electronics for Distributed Generation Systems (PEDG), 2010*, pp. 58–65, Hefei, Anhui, People's Republic of China June 2010.

7. Li, Z., Yuan, Y., Li, F., Evaluating the reliability of islanded microgrid in an emergency mode, *45th International Universities Power Engineering Conference (UPEC) 2010*, pp. 1–5, Cardiff, U.K., August 2010.

8. Burgio, A., Menniti, D., Pinnarelli, A., Sorrentino, N., Reliability studies of a PV-WG hybrid system in presence of multi-micro storage systems, *IEEE Bucharest PowerTech 2009*, pp. 1–5, Bucharest, Romania, June 2009.

9. Salomonsson, D., Soder, L., Sannino, A., Protection of low-voltage DC microgrids, *IEEE Transactions on Power Delivery*, 24(3), 1045–1053, July 2009.

10. Li, H., Zhang, W., Xu, D., High-reliability long-backup-time super UPS with multiple energy sources, *Energy Conversion Congress and Exposition (ECCE), 2013 IEEE*, pp. 4926–4933, Denver, CO, September 15–19, 2013.

11. Zhu, Y., Xu, B., Chen, L., Dong, D., Lin, P., Xu, D., Multi-functional bidirectional AC/DC T-type 3-level converter for super UPS, *IEEE PEAC 2014*, pp. 390–395, Shanghai, China, November 5–8, 2014.

12. Zhou, J., Li, H., Zhou, Z., Xu, D., Control strategy of Li-ion battery module in super UPS, *IEEE PEAC 2014*, pp. 1236–1241, Shanghai, China, November 5–8, 2014.

13. Li, H., Zhou, J., Liu, Z., Xu, D., Solid state DC circuit breaker for super uninterruptible power supply (UPS), *IEEE PEAC 2014*, pp. 1230–1235, Shanghai, China, November 5–8, 2014.

14. Li, X., Zhang, W., Li, H., Xie, R., Chen, M., Shen, G., Xu, D., Power management unit with its control for a three-phase fuel cell power system without large electrolytic capacitors, *IEEE Transactions on Power Electronics*, 26(12), 3766–3777, December 2011.

15. Choi, W., Enjeti, P., Howze, J.W., Fuel cell powered UPS systems: Design considerations, *The 34th IEEE Annual Power Electronics Specialists Conference*, pp. 385–390, Acapulco, Mexico, June 2003.

16. Saghafi, H., Karshenas, H.R., Power sharing improvement in standalone microgrids with decentralized control strategy, *Electric Power Components and Systems*, 42(12), 1278–1288, 2014.

17. Loh, P.C., Li, D., Chai, Y.K., Blaabjerg, F., Hybrid AC-DC microgrids with energy storages and progressive energy flow tuning, *Power Electronics and Motion Control Conference (IPEMC), 2012*, Vol. 1, pp. 120–127, Harbin, Heilongjiang, People's Republic of China, June 2012.

18. Chen, Y.-K., Wu, Y.-C., Song, C.-C., Chen, Y.-S., Design and implementation of energy management system with fuzzy control for DC microgrid systems, *IEEE Transactions on Power Electronics*, 28(4), 1563–1570, April 2013.

19. Valenciaga, F., Puleston, P.F., Supervisor control for a stand-alone hybrid generation system using wind and photovoltaic energy, *IEEE Transactions on Energy Conversion*, 20(2), 398–405, June 2005.

20. Erdinc, O., Uzunoglu, M., Optimum design of hybrid renewable energy systems: Overview of different approaches, *Renewable and Sustainable Energy Reviews*, 16(3), 1412–1425, April 2012.

21. Dagdougui, H., Sacile, R., Decentralized control of the power flows in a network of smart microgrids modeled as a team of cooperative agents, *IEEE Transactions on Control Systems Technology*, 22(2), 510–519, March 2014.

22. Colson, C.M., Nehrir, M.H., A review of challenges to real-time power management of microgrids, *Power & Energy Society General Meeting, 2009 (PES '09)*, pp. 1–8, Alberta, Canada, July 26–30, 2009.

23. Guerrero, J.M., Loh, P.C., Lee, T.-L., Chandorkar, M., Advanced control architectures for intelligent microgrids—Part II: Power quality, energy storage, and AC/DC microgrids, *IEEE Transactions on Industrial Electronics*, 60(4), 1263–1270, April 2013.

24. Schmerda, R., Cuzner, R., Clark, R., Nowak, D., Bunzel, S., Shipboard solid-state protection: Overview and applications, *IEEE Electrification Magazine*, 1(1), 32–39, September 2013.

25. Meyer, C., Kowal, M., De Doncker, R.W., Circuit breaker concepts for future high-power DC-applications, *Fourtieth IAS Annual Meeting. Conference Record of the 2005 Industry Applications Conference 2005*, pp. 860–866, Hong Kong, China, 2005.

26. Krstic, S., Wellner, E.L., Bendre, A.R., Semenov, B., Circuit breaker technologies for advanced ship power systems, *IEEE Electric Ship Technologies Symposium 2007*, pp. 201–208, Arlington, TX, 2007.

27. Xu, D., Li, H., Zhu, Y., Shi, K., Hu, C., High-surety microgrid: Super uninterruptable power supply with multiple renewable energy sources, *Electric Power Components and Systems*, 43(8–10), pp. 839–853, June 15, 2015.
28. Magnusson, J., Saersi, R., Liljestrand, L., Engdahl, G., Separation of the energy absorption and over-voltage protection in solid-state breakers by the use of parallel varistors, *IEEE Transactions on Power Electronics*, 29(6), pp. 2715–2722, June 2015.

Index

A

AC converter modules, 385
AC/DC converters, *see* Bidirectional AC/DC converters
Active zero state PWM (AZSPWM), 73
Alkaline fuel cells (AFCs), 294
Anholt 400 MW wind farm, 9
Annual energy production (AEP), 283–284
ANSYS Maxwell® software, 203
ANSYS RMxprt® module, 203
Aquamarine, 269
Autothermal reformers, 294
Azura device, 269

B

Balance of system (BOS), 147–148
Battery energy storage (BES) system
    compressed air, 324
    electric vehicle charging station
        MATLAB®/Simulink® model, 343–344
        schematics, 343
        simulation results, 343, 345
    flow batteries, 324–325
    flywheels, 324
    lead-acid batteries, 323
    lithium-ion batteries, 323–324
    load leveling/energy time shifting
        simulation results, 343, 347
        Simulink® model, 343, 346
    modeling and test setups, 324
        battery lifetime, 325–326
        electrical models, 325
        equivalent circuit models, 325–326
        mathematical models, 325
        open-circuit voltage *vs.* SOC, 326–327
        physical/electrochemical models, 325
        reduced-order models, 325
        V–I characteristics model, 326
    multiphysics and multidomain simulations, 343, 347–348
    NiMH batteries, 324
    pumped hydro energy storage system, 324
    sodium sulfur (NaS) batteries, 323
Battery management systems (BMS)
    block diagram, 331
    centralized, 330
    distributed, 330–331
    modular/master–slave, 330
Battery power mode, 367, 369–370
Bidirectional AC/DC converters
    boost-type converters, 333–334
    buck–boost converter, 333, 335
    buck-type converters, 333, 335
    multilevel converter, 333, 336
Bidirectional DC/AC module, 385–386
Bidirectional DC/DC converters
    applications, 334

configurations, 334
    isolated DC/DC converters
        configurations, 337
        DAB topology, 339
        SRC topology, 339
        voltage-fed and current-fed FB converter, 338
    non-isolated DC/DC converters
        Cuk and Sepic/Luo converters, 335–336
        HB converter, 335–336
        three-level neutral point clamped bidirectional DC/DC converter, 337
        two-phase interleaved bidirectional converter, 336–337
    requirement, 334
Bidirectional DC/DC module, 387
BMS, *see* Battery management systems
Bypass mode, 369, 372

C

Cage rotor induction generators (CRIGs), 210
    characteristics, 222
    circuit model and performance, 223–225
    control, 225–226
    generic control structure, 222–223
    speed and torque, 222
    stator frequency, 213
    3G transmission drives with 6/8 pole machines, 221
CEC generator, *see* Current energy conversion generator
Cell balancing systems, 332–333
Compressed air energy storage (CAES) systems, 324
Concentrated photovoltaic (CPV) technology
    description, 28
    high-concentration PV technology, 28
    low-concentration PV technology, 28–29
    principle of, 29
    structure, 18
Concentrated solar power (CSP) technology, 6, 8
    advantages, 30
    description, 18, 29
    disadvantages, 30
    grid integration, 34
    large-scale integration, 37–38
    line focusing systems
        linear Fresnel reflector, 31–32
        parabolic trough, 31
    point focusing systems
        parabolic dish reflector, 32–33
        solar towers, 31–32
    stability, 34
    and storage, 33–34
    transient stability analysis
        electrical power, 36
        inertial response, 35–36
        mechanical power, 36–37

single-machine infinite-bus model, 34–35
synchronous machine parameters, 34–35
Constant-voltage and constant-current MPPT
        methods, 97
Converter and electrical control models
    DFIG control, 197
    generator-side converter and control, 197–198
    grid-side converter and control, 198–199
    PMSM control, 197
CPV technology, *see* Concentrated photovoltaic
        technology
CRIGs, *see* Cage rotor induction generators
CSP technology, *see* Concentrated solar power
        technology
Current energy conversion (CEC) generator, 269,
        273–274, 281; *see also* Marine and hydrokinetic
        generation and power plants
    output power and rotational speed, 286–287
    performance coefficient, 285
    TSR, 285–286
    water turbine, hydrokinetic power of, 284

D

DC converter modules, 382
DC-excited synchronous generator (DCE-SG)
    active flux–based sensorless vector control
        cross-coupling saturation, 244
        dq (space phasor) model, 242
        field circuit equations, 244
        magnetic saturation, 244
        stator flux, 245
        unity power factor control, 246
        vector control system, 245–246
        vector diagram, 245
        voltage and current phasor, 242, 244
    active power and torque, 236
    1G/3G transmission drives, 236
    optimal design
        active–reactive power capabilities, 241
        cost function components, 237–239
        efficiency *vs.* load, 241–242
        electric and magnetic loss evolution, 238–239
        excitation current *vs.* load current, 240–241
        FEM key validation, 242–244
        initial cost *vs.* loss cost, 242
        modified Hooke–Jeeves method, 237
        multiobjective optimal design, 241
        parameters and features, 240–241
        rated efficiency, 237–238
        V curves, 241
    phasor diagram, 236
    reactive power, 237
    steady-state phase circuit model, 236
    topology layout, 236
DFIGs, *see* Doubly fed induction generators
Discontinuous PWM (DPWM)-MAX, 72–73
Discontinuous PWM (DPWM)-MIN, 72–73
Distributed MPPT (DMPPT) topology, 113–116
Doubly fed induction generators (DFIGs), 210, 257
    advantages and disadvantages, 213–214
    dq model and control
        FOC, grid-side converter, 219–220
        FOC, rotor-side converter, 219–220

rotor voltages, 219
stator active and reactive powers, 219
in synchronous coordinates, 218
three-phase short circuit, 219, 221
electric active power, 214
with slip rings, 214
small wind energy systems, 153–154
stator frequency, 213–214
subsynchronous mode of operation, 214
supersynchronous mode of operation, 214
technological attempts, 215
topology and circuit model
    equivalent circuit, 215–216
    flux/current relationships, 216–217
    laminated stator and rotor slotting, 215–216
    phase turns ratio, 215
    stator current, 217
Dual active bridge (DAB) topology, 339

E

Electric vehicle (EV) charging station
    MATLAB®/Simulink® model, 343–344
    schematics, 343
    simulation results, 343, 345
Energy storage management systems (ESMS)
    BMSs, 330–331
    cell balancing systems, 332–333
    optimal and safe operation, 330
    SOC
        Coulomb counting, 331
        EKF approach, 332
        OCV *vs.* DOD, 331–332
        rated value of capacity, 330
        table *vs.* OCV value, 331
    SOH, 332
    SOL, 332
    ultracapacitors, 330
Energy storage systems (ESSs)
    advancements, 320
    advantages, 320
    classification
        capacitors, 321–322 (*see also* Ultracapacitors)
        characteristics, 321–322
        electrochemical, 321 (*see also* Battery energy
            storage system)
        mechanical, 321
    electrical utility systems, 320
    ESMS
        BMSs, 330–331
        cell balancing systems, 332–333
        optimal and safe operation, 330
        SOC, 330–332
        SOH, 332
        SOL, 332
        ultracapacitors, 330
    interface systems
        AC/DC converters, 333–336
        isolated DC/DC converters, 337–339
        non-isolated DC/DC converters, 334–337
    renewable energy systems
        long-term support, 320
        short-term support, 320
    total capacity, 321–322

utility-level storage systems
　　ancillary services, 339–340
　　capacity credit, 340
　　energy time shifting, 340
　　field installation, 342
　　microgrids and islanded grids, 341–342
　　renewable energy integration, 340–341
ESMS, *see* Energy storage management systems
ESSs, *see* Energy storage systems
European Marine Energy Center's test site, 269
Evolutionary MPPT algorithms, 110–112
Extended Kalman filter (EKF), 332
Extremum seeking control (ESC) MPPT, 106–107

F

Fatigue, Aerodynamics, Structures, and Turbulence (FAST)
　　software, 202
Finite element analysis (FEA), 246–247
Finite element method (FEM), 231
Fixed speed wind turbines, 257
Flow batteries, 324–325
Flywheels, 324
Frequency-locked loop (FLL), 76–78
Fuel cell–based APU powering, 304–305
FuelCell Energy, 312
Fuel cell power mode, 367, 371
Fuel cell systems, 9–10, 300
　　battery technologies, advancement of, 311
　　definition, 292
　　Doosan Fuel Cell America, 312
　　first fuel cell, 292
　　FuelCell Energy, 312
　　hydrogen, 292
　　modeling
　　　　enthalpy, 299
　　　　entropy, 298
　　　　Gibbs free energy, 299–300
　　　　hydrogen–oxygen fuel cell system, 298
　　　　Nernst equation, 297–298
　　　　open-circuit voltage, 298
　　　　thermal energy, 298–299
　　　　thermodynamic equations, 297
　　MPPT MATLAB®, 313–315
　　Nissan agreement, 311
　　power electronics, 309–311
　　properties and control characteristics
　　　　load transients, 297
　　　　low-frequency ripple current, 297
　　　　MPP, 295–296
　　　　peak-to-peak current ripple, 297
　　　　PEMFC characteristics, 295–296
　　　　polarization losses, 295
　　　　R-1, R-2, and R-3 regions, 295
　　reformer, 294
　　stationary power generation
　　　　AFCs/PEMFCs, 306
　　　　battery, 307
　　　　biological processes, 306
　　　　DC–DC converter, 308
　　　　fuel cell power section, 306
　　　　fuel processor, 306
　　　　high-quality waste heat, 307
　　　　high-temperature fuel cell–based systems, 308–309

　　　　hydrogen, 306
　　　　modular series-connected fuel cell units, 308
　　　　power conversion system, 307
　　　　power inverter, 306–307
　　　　provide buffering and additional power, 306
　　　　SOFC/MCFC, 306
　　　　stand-alone low power fuel cell power generation
　　　　　　system, 307–308
　　　　water management, 307
　　transportation
　　　　APU, 300, 304–305
　　　　FCVs, 300
　　　　high-temperature SOFC, 300
　　　　PEMFCs, 300
　　　　plug-in-fuel cell vehicles, 304–305
　　　　propulsion systems, 301–304
　　types and operation
　　　　AFCs, 294
　　　　MCFC, 293–294
　　　　PAFC, 294
　　　　PEMFC, 292–293
　　　　SOFC, 293
　　water and heat, 292
Fuel cell vehicles (FCVs), *see* Fuel cell systems
Full-bridge zero-voltage rectifier (FB-ZVR) topology, 50

G

Gas engine power mode, 367–369
Generator-side converter and control, 197–198
Gibbs free energy, 299–300
Global MPPT methods, 116–120
Grid normal mode, 365–366
Grid-side converter and control, 198–199
Grid-tied small wind turbine systems
　　angular electrical speed, 168
　　d-q reference voltages, 170
　　electromagnetic torque, 168
　　full-fledged PMSG back-to-back converter, 168–169
　　grid-side converter, 170–171
　　instantaneous power, 168
　　machine-side converter, 170–171
　　power calculation, 170
　　power transfer strategy chart, 170–171
　　proportional–integral control, 170
　　rotating reference frame, 168
　　rotational angular speed, 168
　　speed control, 168
　　TSR and power coefficient, 170
　　wind speed measurement, 170
Ground-mounted PV plants, 68
Gulf Stream, 273

H

Horizontal-axis wind turbines (HAWT), 252
Hybrid architecture, 359
Hydrogen–oxygen fuel cell system, 298
Hydropower, 2–3

I

Incremental-conductance MPPT, 99–100
Induction generator (IG), 152–153

Insulated gate bipolar transistor (IGBT), 380
Isolated DC/DC converters
 configurations, 337
 DAB topology, 339
 SRC topology, 339
 voltage-fed and current-fed FB converter, 338

K

Kuroshio, 273

L

Lead-acid batteries, 323
Levelized cost of energy (LCOE), 70, 269, 278
Lithium-ion batteries, 323–324

M

Marine and hydrokinetic (MHK) power generation and
  power plants
 CEC technologies, 273–274
 cost of energy, 269
 cost reductions, 269
 electrical generation
  control and power conversion, 277–278
  direct drive/gearbox, 277
  generators, 274–276
  power plant, 278–279
  prime mover, 276–277
 energy storage
  CEC generator, 281
  WEC generator, 280–281
 LCOE, 269
 social, economic, regulatory and environmental issues,
  269
 technical challenges, 269
 U.S. electricity demand, 268
 WEC technologies
  OWC and overtopping devices, 270–272
  point absorbers, 270–272
  terminators and attenuators, 270–271
Maximum power point (MPP), 295
Maximum power point tracking (MPPT)
 ANSYS Simplorer® model, 122–124
 block diagram, 92
 controls, 203
 MATLAB®/Simulink® model, 119, 121–122
 power–voltage characteristics, PV array, 92–93
 PV array, nonuniform solar irradiation conditions
  distributed MPPT, 113–116
  evolutionary MPPT algorithms, 110–112
  global MPPT methods, 116–120
  numerical optimization algorithms, 112–113
  overview of, 94
  PV array reconfiguration method, 110
  stochastic and chaos-based MPPT algorithms, 113
 PV array, uniform solar irradiation conditions
  artificial intelligence techniques, 101–102
  constant-voltage and constant-current MPPT
   methods, 97
  control-circuit complexity, 109
  efficiency, 109
  ESC MPPT, 106–107

  incremental-conductance MPPT, 99–100
  low-amplitude switching ripples, 109
  model-based MPPT methods, 100–101
  numerical optimization algorithms, 103–105
  operational characteristics, 107–108
  overview of, 94
  P&O MPPT method, 97–99
  RCC MPPT, 105–106
  single-sensor MPPT, 102–103
  sliding-mode control, 107
 PV module technologies, 95
 single-diode model, 95–96
 solar irradiance and ambient temperature
  conditions, 93–94
 three-phase PV inverters, 82–85
MCFC, *see* Molten carbonate fuel cell
Mean time between failure (MTBF), 362–364
Metal oxide varistor (MOV), 380
MHK generators
 fixed-speed induction generator, 274–275
 variable-slip wound-rotor induction generator, 275
 variable-speed doubly fed induction generator, 275–276
 variable-speed full-power converter, 276
MHK power plant, 278–279
Microgrid, high-surety power, *see* Super uninterruptible
  power supply
Microgrids
 EESs, 341
 fuel cells, 305
 small wind energy system, 151–152
Minicentral inverters (MCI), 69
Model-based MPPT methods, 100–101
Modern power system, 4–5
Module-integrated converters (MICs), 69–70
Molten carbonate fuel cell (MCFC), 293–294,
  306–307, 312
MPPT, *see* Maximum power point tracking
Multiple-star architecture, 358–359

N

National Renewable Energy Laboratory (NREL) TurbSim
  wind simulator, 202
Near-state PWM (NSPWM), 73
Nernst equation, 297–298
Neutral point–clamped (NPC) topology, 50–51
NiMH batteries, 324
Non-isolated DC/DC converters
 Cuk and Sepic/Luo converters, 335–336
 HB converter, 335–336
 three-level neutral point clamped bidirectional DC/DC
  converter, 337
 two-phase interleaved bidirectional converter, 336–337

O

Ocean current, 268, 273, 281
ODAs, *see* Optimal design algorithms
Open-circuit voltage (OCV) value, 331
Optimal design algorithms (ODAs)
 DCE-SG
  active–reactive power capabilities, 241
  cost function components, 237–239
  efficiency *vs.* load, 241–242

electric and magnetic loss evolution, 238–239
excitation current *vs.* load current, 240–241
FEM key validation, 242–244
initial cost *vs.* loss cost, 242
modified Hooke–Jeeves method, 237
multiobjective optimal design, 241
parameters and features, 240–241
rated efficiency, 237–238
V curves, 241
PMSG
on benchmark, 229
comparative results, 231, 234
cost function, 230
energy loss cost, 230
finite element method (FEM), 231
frame cost, 230
loss and efficiency, 231–232
metaheuristic methods, 229
multiple-objective function, 230
penalty component, 231
single-composite objective function, 230
weight and material cost, 231, 233
Oscillating water column (OWC), 270–271

P

Partial oxidation reformers, 294
PCS, *see* Power conversion system
PEMFC, *see* Proton exchange membrane fuel cell
Permanent magnet synchronous generator (PMSG), 210, 257, 276
advantages, 165
bounded NdFeB, 163, 165–166
brushless DC multiphase reluctance machine, 227–229
circuit modeling and control
dq model, 231
generic vector control, 234
practical control system, 234
reactive power, 232
sensorless vector control system, 234–235
underexcited operation mode, 233
voltage-boosting operation, 233
converter losses, 162
copper and iron losses with power level *vs.* wind velocity, 163–164
cost–energy ratio, 165
demagnetization, 166
disadvantages, 166
electromechanical design, 167–168
energy-captured area, 164
ferrite magnet generator, 163, 165
flux reversal tooth-wound coil PMSM, 227–228
full-fledged PMSG back-to-back converter, 168–169
gearbox efficiency, 163
generator losses, 162
high-energy density magnets, 166
mass–energy index, 165, 168
mechanical losses, 162–163
ODAs
on benchmark, 229
comparative results, 231, 234
cost function, 230
energy loss cost, 230
FEM, 231

frame cost, 230
loss and efficiency, 231–232
metaheuristic methods, 229
multiple-objective function, 230
penalty component, 231
single-composite objective function, 230
weight and material cost, 231, 233
power electronic converter topology, 167
sintered NdFeB, 163, 165
stator frequency, 213
transverse flux axial air-gap PMSM, 227
turbine rotation *vs.* power characteristics, 167
Vernier PMSG, 227
Perturb and observe (P&O) MPPT algorithm, 59–60
PFCV, *see* Plug-in hybrid fuel cell vehicle
Phase-locked loop (PLL)-based synchronization system, 53, 57–58
Phosphoric acid fuel cell (PAFC), 294, 307
Photovoltaic (PV) cells
applications, 20
basic structure, 21
categories, 20
electron–hole pair, 20–21
MATLAB®/Simulink® model, 25, 27
modeling parameters, 25–26
mono-crystalline (m-c) cells, 20–21
operating characteristics, 22–25
panel configuration, 26, 28
poly-crystalline (p-c) cells, 20–21
research cell efficiencies, 18–19
roof-mounted residential grid-connected PV system, 20
thin-film solar cells, 18
Plug-in hybrid fuel cell vehicle (PFCV), 304–305, 311
PMSG, *see* Permanent magnet synchronous generator
P&O MPPT method, 97–99
Power conversion system (PCS), 364–365, 382, 384–385
Power electronics, 302
converters, 4–5
fuel cell applications, 309–311
WTS
cascaded H-bridge converter with medium-frequency transformers, 189
evolution of WT size, 180–181
general control structure, 178–179
modular multilevel converter, 189
multilevel converter topologies, 188–189
power semiconductor devices, 185–187
two-level converter topologies, 187–188
for wind farm, 189–193
Power semiconductor devices
module packaging technology, 185–186
press-pack packaging technology, 186–187
silicon power devices, 187
types and characteristics, 185–186
Power take-off system (PTOS), 269–270, 280
Propulsion systems
battery, 301
battery pack and power conditioner, 302
DC–DC converter, 302–303
DC-link voltage, 302–303
design issues, 303–304
fuel cell stack, 301
fuel flow rate, 303
hydrogen fuel input, 301

ICE-based gasoline vehicles, 301
   onboard reformer, 301
   power electronics system, 302
   propulsion motor, 301–302
Proton exchange membrane fuel cell (PEMFC), 306,
      311–312
   characteristics, 295–296
   efficiency, 293
   fuel cell reaction, 292
   fuel cell stack, 301
   hydration, 293
   hydrogen–oxygen reaction, 292
   open-circuit voltage, 298
   propulsion applications, 293, 300
   proton-conducting solid polymer electrolyte
      membrane, 292
Proton exchange membrane (PEM) technology, 9–10
PTOS, *see* Power take-off system
Pulse width modulation–voltage source converter
      with two-level output voltage
      (2L-PWMVSC), 187
Pumped hydro energy storage system, 324
PV cells, *see* Photovoltaic cells
PV energy conversion system, 6–7

R

Rayleigh probability density function, 260–261
Remaining useful life (RUL), 332
Renewable energy technologies
   opportunities and future trends, 12–14
   research challenges, 12–14
Renewables and nonrenewables, annual
      installations, 3–4
Residential photovoltaic systems
   BOS components, 147–148
   cable losses, 145
   design parameters, 132
   financial calculations, 141
   flowchart, design procedure, 141–142
   grid-connected/stand-alone systems, 132–133
   ideal production index, 143
   large utility-scale plants, 132
   load pattern evaluation, 133–134
   monthly energy productions, 145–146
   performance monitoring, 148
   plant specifications, 144
   price, 148–149
   project general information, 143
   PV array sizing, 134–136
   PV inverter
      array yield, 140
      datasheet values, 136–138
      efficiency, 136–137
      final yield, 139
      generator losses, 140
      measured irradiation data, 138–139
      performance ratio, 141
      reference yield, 140
      specifications, 144–145
      string configuration, 137–138
      system losses, 141
      thermal capture losses, 141
      warranty, 148

PV modules
   datasheet, 143–144
   warranty, 148
PV rooftop installations, 132
   SF 150-S module, 147
   solar resource evaluation, 134–135
   string and array configuration, 135
   TF modules, 147
Ring architecture, 359
Ripple correlation control (RCC) MPPT, 105–106
Rooftop-installed PV systems, 6–7

S

Seabased, 269
Second-order generalized integrator (SOGI)-FLL, 76–78
Self-excited induction generators (SEIG)
   capacitor value, 160–161
   design factors, 159–160
   disadvantages of SCIG, 162
   electronic power converters, 161
   IG self-exciting process, 160
   induction machine parameters, 160
   load parameters, 160
   magnetization curve, 160–161
   per-phase equivalent circuit, 161
   per unit (p.u.) frequency, 161–162
   RLC load, 161
   stand-alone operation, 160
   type of primary source, 160
Series resonant converter (SRC) topology, 339
Short-circuit fault current analysis
   module fault, 376
   protection
      test condition, 387, 390
      waveforms, 387, 389
   short-circuit current, 378
   short-circuit fault waveform, 378–379
   specifications, 378–379
   Type I failure equivalent model, 376, 378
   types, 376–378
Sine triangular (ST)-PWM, 72–73
Single-phase grid-connected photovoltaic systems
   central inverters, 44, 46
   control of
      general control structure, 52–56
      MPPT control, 59–60
      objectives, 52
      PLL-based synchronization system, 53, 57–58
      proportional integral (PI) controller, 59
      proportional resonant (PR) controller, 59–60
      simulation models, 59–60
      simulation models and results, 59, 61–62
   demands, 43–44
   double-stage PV technology
      block diagram, 50–51
      conventional double-stage single-phase PV
         topology, 51
      FB inverter, 52
      parallel-input series-output bipolar DC output
         converter, 52
      time-sharing boost converter, 51–52
   evolution of, 42
   high-frequency transformer, 44, 46

low-frequency transformer, 44, 46
multistring inverters, 44–45
string inverter, 44–45
transformerless AC-module inverters
    block diagram, 44, 46
    buck–boost AC-module inverter topologies, 47–48
    with LCL filter, 47
    Z-source inverter, 47–48
transformerless single-stage string inverters
    AC path, 49–50
    bipolar modulation scheme, 48
    FB-ZVR topology, 50
    hardware schematics, 47, 49
    H5 inverter topology, 49
    H6 inverter topology, 49
    NPC topology, 50–51
Single-sensor MPPT, 102–103
Small wind energy systems
    data analysis, 152
    generator selection
        construction criteria, 154, 156
        control criteria, 154, 156
        DFIG, 153–154
        electrical criteria, 154–155
        gearbox, 152
        generating unit cost, 153
        IGs, 152–153
        mechanical criteria, 154–155
        number of poles, 153
        power conversion stages, 152–153
        primary energy source, 153
        SR, 154
        variable-speed features, 152
    grid-tied small wind turbine systems
        angular electrical speed, 168
        d-q reference voltages, 170
        electromagnetic torque, 168
        full-fledged PMSG back-to-back converter, 168–169
        grid-side converter, 170–171
        instantaneous power, 168
        machine-side converter, 170–171
        power calculation, 170
        power transfer strategy chart, 170–171
        proportional–integral control, 170
        rotating reference frame, 168
        rotational angular speed, 168
        speed control, 168
        TSR and power coefficient, 170
        wind speed measurement, 170
    Magnus turbine
        advantages, 173–174
        dimensionless power coefficients, 173–174
        electromechanical power losses, 173
        fuzzy logic–based controller, 175
        lift and drag Magnus effects, 171–172
        mechanical power, 172–173
        with nonsmooth rotating cylinders, 171–172
        parameters, 173–174
        torques, 171, 173
    permanent magnet synchronous generators
        advantages, 165
        bounded NdFeB, 163, 165–166
        converter losses, 162
        copper and iron losses with power level vs. wind velocity, 163–164
        cost–energy ratio, 165
        demagnetization, 166
        disadvantages, 166
        electromechanical design, 163, 167–168
        energy-captured area, 164
        ferrite magnet generator, 163, 165
        gearbox efficiency, 163
        generator losses, 162
        high-energy density magnets, 166
        mass–energy index, 165, 168
        mechanical losses, 162–163
        power electronic converter topology, 167
        sintered NdFeB, 163, 165
        turbine rotation vs. power characteristics, 167
    residential/commercial applications, 152
    SEIG
        capacitor value, 160–161
        design factors, 159–160
        disadvantages of SCIG, 162
        electronic power converters, 161
        IG self-exciting process, 160
        induction machine parameters, 160
        load parameters, 160
        magnetization curve, 160–161
        per-phase equivalent circuit, 161
        per unit (p.u.) frequency, 161–162
        RLC load, 161
        stand-alone operation, 160
        type of primary source, 160
    turbine selection
        annual wind speed distribution, 158
        calm periods, 158
        instantaneous power of turbine, 157–158
        machine frequency and voltage ratings, 159
        power distribution, 158
        Rayleigh probability density function, 158–159
        speed control range, 157
        Weibull probability density function, 158
        wind speed bands, 157
    wind-based grid-connected system, 152
SOC, see State of charge
Societal sustainability, 18
Sodium sulfur (NaS) rechargeable batteries, 323
SOFC, see Solid oxide fuel cell
Solar power generator technologies, 6–7
Solar power sources
    CPV technology
        description, 28
        high-concentration PV technology, 28
        low-concentration PV technology, 28–29
        principle of, 29
        structure, 18
    CSP technology
        advantages, 30
        description, 18, 29
        disadvantages, 30
        grid integration, 34
        large-scale integration, 37–38
        linear Fresnel reflector, 31–32
        parabolic dish reflector, 32–33
        parabolic trough, 31
        solar towers, 31–32

stability, 34
and storage, 33–34
transient stability analysis, 34–37
PV cells
applications, 20
basic structure, 21
categories, 20
electron–hole pair, 20–21
MATLAB®/Simulink® model, 25, 27
modeling parameters, 25–26
mono-crystalline (m-c) cells, 20–21
operating characteristics, 22–25
panel configuration, 26, 28
poly-crystalline (p-c) cells, 20–21
research cell efficiencies, 18–19
roof-mounted residential grid-connected PV system, 20
thin-film solar cells, 18
Solid oxide fuel cell (SOFC), 9, 293, 306–307, 311–312
Solid-state DC circuit breaker
installation and location, 380–381
MOV, 380
short-circuit protection waveforms, 380, 383
short-circuit test specifications, 379–380
structure, 380, 382
two-series-opposing IGBT structure, 380
voltage stress, 380, 383
Space vector modulation (SVM), 72–74
Squirrel cage induction generator (SCIG), 159
Star architecture, 356–358
State of charge (SOC)
Coulomb counting, 331
EKF approach, 332
OCV vs. DOD, 331–332
rated value of capacity, 330
table vs. OCV value, 331
State of health (SOH), 332
State of life (SOL), 332
Stationary power generation
AFCs/PEMFCs, 306
battery, 307
biological processes, 306
DC–DC converter, 308
fuel cell power section, 306
fuel processor, 306
high-quality waste heat, 307
high-temperature fuel cell–based systems, 308–309
hydrogen, 306
modular series-connected fuel cell units, 308
power conversion system, 307
power inverter, 306–307
provide buffering and additional power, 306
SOFC/MCFC, 306
stand-alone low power fuel cell power generation system, 307–308
water management, 307
Steam reformers, 294
Stochastic and chaos-based MPPT algorithms, 113
Super capacitors, see Ultracapacitors
Super uninterruptible power supply (Super UPS)
architectures
hybrid architecture, 359
MTBF, 362–364
multiple-star architecture, 358–359
ring architecture, 359
star architecture, 356–358
system efficiency, 360–362
system performance, 364
system reliability, 362–363
demonstration platform
AC converter modules, 385
bidirectional DC/AC module waveforms, 385–386
bidirectional DC/DC module waveforms, 387
control architecture, 382, 385
DC converter modules, 382
grid normal mode to gas engine power mode transfer, 387, 389
grid normal mode to gas power mode transfer, 387–388
PCS, 382, 384–385
power converter module parameters, 385–386
short-circuit fault protection test condition, 387, 390
short-circuit fault protection waveforms, 387, 389
specifications, 382, 384
storages, 382, 384
operation modes
battery power mode, 367, 369–370
bypass mode, 369, 372
fuel cell power mode, 367, 371
gas engine power mode, 367–369
grid normal mode, 365–366
PCS, 364–365
power management, multiple sources
fuel cell unit, 372–374
gas engine unit, 374–376
life span extension, 369
protection
short-circuit fault current analysis, 376–379
solid-state DC circuit breaker, 380–382
structure, 356
Switched reluctance (SR) generators, 154
Synchronous reference frame (SRF) PLL, 76–77

T

Three-level H-bridge back-to-back (3L-HB BTB) converter, 188
Three-level neutral point clamped back-to-back (3L-NPC BTB) converter, 188
Three-phase PV inverters
active power ramp limitation, 86–87
control scheme, 68–69
current control, 81–82
grid integration functions, 84
grid synchronization
PLL, 76, 78–80
SOGI-FLL, 76–78
SRF PLL, 76–77
internal active power reserve management, 85–86
modulation strategies
AZSPWM, 73
carrier based methods, 74
DPWM-MAX, 72–73
DPWM-MIN, 72–73
implementation of, 74–76
NSPWM, 73
PWM method, 72
ST-PWM, 72–73
SVM, 72–74

MPPT, 82–85
    requirements for, 68
    structures, 69–70
    topologies, 71–72
Tidal currents, 273
Tip-speed ratio (TSR), 163, 194, 285–286
Transportation electrification
    APU, 300
    APU and plug-in-fuel cell vehicles, 304–305
    FCVs, 300
    high-temperature SOFC, 300
    PEMFCs, 300
    propulsion systems
        battery, 301
        battery pack and power conditioner, 302
        DC–DC converter, 302–303
        DC-link voltage, 302–303
        design issues, 303–304
        fuel cell stack, 301
        fuel flow rate, 303
        hydrogen fuel input, 301
        ICE-based gasoline vehicles, 301
        onboard reformer, 301
        power electronics system, 302
        propulsion motor, 301–302
TSR, *see* Tip-speed ratio
Two-level back-to-back structure (2L-BTB) converters, 187

U

Ultracapacitors (UCap)
    behavioral neural network types, 328
    equivalent circuit models, 328–329
    physical/electrochemical models, 328
    Ragone plot, energy and power density, 321–322
    types and characteristics
        banks, 327–328
        electrical double-layer capacitors, 327
        hybrid capacitors, 327
        Li-ion, 327–329
        pseudocapacitors, 327
Uninterruptible power supply (UPS) systems, 9

V

Variable speed wind turbines, 257–259
Verdant Power, 269, 273
Vertical-axis wind turbines (VAWT), 252

W

Wave energy conversion (WEC) generator, 269, 280–281;
        see also Marine and hydrokinetic power
        generation and power plants
    AEP, 283–284
    capture width, 282
    linear wave theory, 281
    mechanical power matrix, 282–283
    OWC and overtopping devices, 270–272
    point absorbers, 270–272
    power density, 281
    sea states, percentage occurrence of, 282
    swell, 281
    terminators and attenuators, 270–271

Wave energy system, 10–11
WEC generator, *see* Wave energy conversion
        generator
Wind energy conversion, 152
Wind farms, power electronic solutions
    grid codes, 192–193
    potential configurations, 189
        DFIG system with AC grid, 190
        full-scale converter system with AC grid, 190
        full-scale converter system with both distribution
            and transmission DC grid, 191
        full-scale converter system with multiple diode
            rectifiers and transmission DC grid, 191
        full-scale converter system with VSC rectifier and
            transmission DC grid, 190–191
Wind turbine (WT)
    blades, 210
    characteristics
        actual mechanical power output, 255
        angular velocity, 255
        assumptions, 254
        control volume of wind, 255
        at different wind speeds, 256
        numerical approximations, 255
        tip-speed ratio, 255
    components
        anemometer, 254
        blades, 252
        brake, 253
        controller, 253
        and cost breakdown, 210, 212
        drivetrain, 254
        gearbox, 253
        generator, 253
        HAWT, 252
        nacelle, 252
        pitch, 252
        rotor, 252
        shaft, 253
        tower, 252
        VAWT, 252
        wind vane, 254
        yaw, 253
    controls
        electrical control, 263–264
        mechanical control, 262–263
        range, 261
        wind turbine torque *vs.* rotational speed, 262
    power rating, 210
    small-scale wind users, 251–252
    storage compensation, 252
    wind energy generator systems
        DC link, 256
        fixed speed wind turbines, 257
        gearbox, 256
        induction machine, 256
        permanent magnet brushless AC generator, 256
        synchronous generator, 256
        variable speed wind turbines, 257–259
    wind speed and generator performance
        air density, 260
        calm periods, 259
        efficiency, 259
        estimate of true parameter, 261

mechanical viscous losses, 259
power distribution, 259
Rayleigh probability density function, 260–261
turbine mechanical power, 259
turbine torque, 260
wind power intensity, 259
wind speed distribution, 259
Wind turbine (WT) generators
with AC–DC–AC bidirectional static power
converters, 210–212
CRIGs, 210
characteristics, 222
circuit model and performance, 223–225
control, 225–226
generic control structure, 222–223
stator frequency, 213
3G transmission drives with 6/8 pole machines, 221
DCE-SG
active flux–based sensorless vector control,
242–246
active power and torque, 236
1G/3G transmission drives, 236
optimal design, 237–242
phasor diagram, 236
reactive power, 237
steady-state phase circuit model, 236
topology layout, 236
DD vs. 3G transmission, 213
DFIGs, 210
advantages and disadvantages, 213–214
dq model and control, 218–221
electric active power, 214
with slip rings, 214
stator frequency, 213–214
subsynchronous mode of operation, 214
supersynchronous mode of operation, 214
technological attempts, 215
topology and circuit model, 215–218
FEA, 246–247
multi-MW wind generators, 213
PMSGs, 210
brushless DC multiphase reluctance machine, 227–229
circuit modeling and control, 231–235
flux reversal tooth-wound coil, 227–228
optimal design, 229–234
stator frequency, 213

transverse flux axial air-gap, 227
Vernier PMSG, 227
types, 210
Wind turbine systems (WTS), 4, 6, 8–9
ANSYS® models, 202–203
components, 178–179
grid codes, 183–184
MATLAB® models, thermal analysis
back-to-back converter data, 199–200
converter and electrical control models, 197–199
generator parameters, 199–200
mechanical model, turbine and generator, 194–196
PMSG system, 199, 201–202
power loss and thermal modeling, 199
wind power generation structure with
simulation files, 194
wind speed profile, 193, 199, 201
wind turbine characteristics, 199–200
mission profiles
balancing control schemes, 182
energy storage, 182
loading conditions, 182
operating conditions, 182
wind speed, 181–182
power electronics
cascaded H-bridge converter with medium-
frequency transformers, 189
evolution of WT size, 180–181
general control structure, 178–179
modular multilevel converter, 189
multilevel converter topologies, 188–189
power semiconductor devices, 185–187
two-level converter topologies, 187–188
for wind farm, 189–193
reliability requirements
junction temperature, 184, 186
one-year thermal profile, 184–185
speed and power curve, 185–186
thermal cycles, 184
top manufacturers, 180–181
types, 181
Worldwide energy demand, 2
Worldwide installations of renewable energy generation
capacity, 2–3
Wound field synchronous generator (WFSG), 257
WTS, see Wind turbine systems

Printed and bound by CPI Group (UK) Ltd, Croydon, CR0 4YY

24/10/2024

01778298-0007